# BASICS OF THERMODYNAMICS

AUTHOR(s)

**Dr. Parmod Kumar**
*Assistant Professor,*
*School of Mechanical and Materials Engineering,*
*IIT Mandi, Mandi*
*Himachal Pradesh*

**Dr. Atul Dhar**
*Associate Professor,*
*School of Mechanical and Materials Engineering,*
*IIT Mandi, Mandi*
*Himachal Pradesh*

REVIEWER

**Prof. Shaligram Tiwari**
*Professor,*
*Department of Mechanical Engineering, IIT Madras*

KHANNA BOOK PUBLISHING CO. (P) Ltd.
4C/4344, Ansari Road, Darya Ganj,
New Delhi, India

All India Council for Technical Education
Nelson Mandela Marg, Vasant Kunj,
New Delhi, India

**ISBN:** 978-93-5538-625-0

**Book Code:** UG407EN

**Basics of Thermodynamics**

*by* Parmod Kumar, Atul Dhar

**First Edition:** 2026

*Originally Published by:*

**All India Council for Technical Education, Ministry of Education, GOI**

*Exclusively Published & Distributed by:*

**Khanna Book Publishing Co. (P) Ltd.**
4C/4344, Ansari Road, Daryaganj,
New Delhi (INDIA)

Visit us at: **www.khannabooks.com**
Write us at: **contact@khannabooks.com**

To view complete list of books, Please scan the QR Code:

KPH

*Printed in Bharat.*

प्रो. टी. जी. सीताराम
अध्यक्ष
Prof. T. G. Sitharam
Chairman

सत्यमेव जयते

आज़ादी का अमृत महोत्सव

अखिल भारतीय तकनीकी शिक्षा परिषद्
(भारत सरकार का एक सांविधिक निकाय)
(शिक्षा मंत्रालय, भारत सरकार)
नेल्सन मंडेला मार्ग, वसंत कुंज, नई दिल्ली–110070
दूरभाष : 011–26131498
ई–मेल : chairman@aicte-india.org

ALL INDIA COUNCIL FOR TECHNICAL EDUCATION
(A STATUTORY BODY OF THE GOVT. OF INDIA)
(Ministry of Education, Govt. of India)
Nelson Mandela Marg, Vasant Kunj, New Delhi-110070
Phone : 011-26131498
E-mail : chairman@aicte-india.org

# FOREWORD

Engineers are the backbone of any modern society. They are the ones responsible for the marvels as well as the improved quality of life across the world. Engineers have driven humanity towards greater heights in a more evolved and unprecedented manner.

The All India Council for Technical Education (AICTE), have spared no efforts towards the strengthening of the technical education in the country. AICTE is always committed towards promoting quality Technical Education to make India a modern developed nation emphasizing on the overall welfare of mankind.

An array of initiatives has been taken by AICTE in last decade which have been accelerated now by the National Education Policy (NEP) 2020. The implementation of NEP under the visionary leadership of Hon'ble Prime Minister of India envisages the provision for education in regional languages to all, thereby ensuring that every graduate becomes competent enough and is in a position to contribute towards the national growth and development through innovation & entrepreneurship.

One of the spheres where AICTE had been relentlessly working since past couple of years is providing high quality original technical contents at Under Graduate & Diploma level prepared and translated by eminent educators in various Indian languages to its aspirants. For students pursuing 2nd year of their Engineering education, AICTE has identified 87 books, which shall be translated into 12 Indian languages - Hindi, Tamil, Gujarati, Odia, Bengali, Kannada, Urdu, Punjabi, Telugu, Marathi, Assamese & Malayalam. In addition to the English medium, books in different Indian Languages are going to support the students to understand the concepts in their respective mother tongue.

On behalf of AICTE, I express sincere gratitude to all distinguished authors, reviewers and translators from the renowned institutions of high repute for their admirable contribution in a record span of time.

AICTE is confident that these outcomes based original contents shall help aspirants to master the subject with comprehension and greater ease.

T. G. Sitharam

(Prof. T. G. Sitharam)

# ACKNOWLEDGEMENT

The authors are grateful to the authorities of AICTE, particularly Prof. T. G. Sitharam, Chairman; Dr. Abhay Jere, Vice-Chairman; Prof. Rajive Kumar, Member-Secretary and Dr Amit Kumar Srivastava, Director, Faculty Development Cell for their planning to publish the books on Basics of Thermodynamics. We sincerely acknowledge the valuable contributions of the reviewer of the book Prof. Shaligram Tiwari, Professor, Department of Mechanical Engineering, IIT Madras for making it students' friendly and giving a better shape in an artistic manner.

Authors also acknowledge the enormous support provided by Mr. Rahul Kumar Mondal and Mr. Saurabh Patel, PhD students at IIT Mandi, in preparation of the figures and formatting of this book. We also acknowledge our families for their unconditional support, continued patience and encouragement throughout the whole process of writing this book.

This book is an outcome of various suggestions of AICTE members, experts and authors who shared their opinion and thought to further develop the engineering education in our country. Acknowledgements are due to the contributors and different workers in this field whose published books, review articles, papers, photographs, footnotes, references and other valuable information enriched us at the time of writing the book.

***Dr. Parmod Kumar***

***Dr. Atul Dhar***

# PREFACE

*The book titled "Basics of Thermodynamics" is an outcome of the well-established role of thermodynamics in the modern engineering practices, our small experience of teaching of Engineering Thermodynamics course, plan of AICTE to develop text books for all affiliated institutions along with constant guidance provided by Prof. Shaligram Tiwari. Keeping in mind the purpose of wide coverage as well as to provide essential supplementary information, we have included the topics recommended by AICTE, in a very systematic and orderly manner throughout the book. Efforts have been made to explain the fundamental concepts of the subject in the simplest possible way by considering the illustrations from the daily life experiences.*

*While preparing the outline of the book, we have strictly adhered to the syllabus prescribed by the AICTE. For comprehensively covering the concepts of thermodynamics starting from the basic introduction to thermodynamic systems and processes, application of these concepts for estimating the performance and efficiency of energy conversion devices such as power plants used for generating electricity, the book has been divided into eight chapters. Chapter one describes the basic concepts of thermodynamics. Chapter two introduces the important concepts of energy and energy conservation. In execution of any energy conversion process, the choice of suitable working fluid is very critical for realising the process. For developing the importance of properties of the working fluids, Chapter three covers the properties of pure substances. Chapter 4 illustrates the applications of the first law of thermodynamics to control volume. Chapter 5 describes the second law of thermodynamics. Concepts of Entropy and Exergy are covered in chapters six and seven, respectively. In the concluding chapter eight, these concepts have been applied for the analysis of various practical power plants and refrigeration systems. We are confident that this sequence of chapters will provide a logical foundation for learning the basic concepts of engineering thermodynamics. The solved examples and numerical problems at the end of each chapter are provided to allow the readers to concentrate on absorbing the theoretical concepts with their practical applications. The content of the book is developed for entry level undergraduate students, however, it can be used by practicing engineers and graduate students for the cohesive learning of the concepts of classical thermodynamics.*

*We sincerely hope that the book will inspire the students to learn the ideas behind various energy systems that they will encounter during their journey of engineering education and practice of engineering knowledge. We are also confident that it will surely contribute to the development of a clear understanding of the concepts of thermodynamics. We are really pleased to place this book in the hands of the teachers and students. We are eagerly waiting for the responses of our readers. We would be really thankful to all critical comments and suggestions for the improvement of the future editions of the book.*

**Dr. Parmod Kumar**

**Dr. Atul Dhar**

# OUTCOME BASED EDUCATION

For the implementation of an outcome based education the first requirement is to develop an outcome based curriculum and incorporate an outcome based assessment in the education system. By going through outcome based assessments evaluators will be able to evaluate whether the students have achieved the outlined standard, specific and measurable outcomes. With the proper incorporation of outcome based education there will be a definite commitment to achieve a minimum standard for all learners without giving up at any level. At the end of the programme running with the aid of outcome based education, a student will be able to arrive at the following outcomes:

**PO-1.** **Engineering knowledge**: Apply the knowledge of mathematics, science, engineering fundamentals, and an engineering specialization to the solution of complex engineering problems.

**PO-2.** **Problem analysis**: Identify, formulate, review research literature, and analyze complex engineering problems reaching substantiated conclusions using first principles of mathematics, natural sciences, and engineering sciences.

**PO-3.** **Design / development of solutions**: Design solutions for complex engineering problems and design system components or processes that meet the specified needs with appropriate consideration for the public health and safety, and the cultural, societal, and environmental considerations.

**PO-4.** **Conduct investigations of complex problems**: Use research-based knowledge and research methods including design of experiments, analysis and interpretation of data, and synthesis of the information to provide valid conclusions.

**PO-5.** **Modern tool usage**: Create, select, and apply appropriate techniques, resources, and modern engineering and IT tools including prediction and modeling to complex engineering activities with an understanding of the limitations.

**PO-6.** **The engineer and society**: Apply reasoning informed by the contextual knowledge to assess societal, health, safety, legal and cultural issues and the consequent responsibilities relevant to the professional engineering practice.

**PO-7.** **Environment and sustainability**: Understand the impact of the professional engineering solutions in societal and environmental contexts, and demonstrate the knowledge of, and need for sustainable development.

**PO-8.** **Ethics**: Apply ethical principles and commit to professional ethics and responsibilities and norms of the engineering practice.

**PO-9. Individual and team work**: Function effectively as an individual, and as a member or leader in diverse teams, and in multidisciplinary settings.

**PO-10. Communication**: Communicate effectively on complex engineering activities with the engineering community and with society at large, such as, being able to comprehend and write effective reports and design documentation, make effective presentations, and give and receive clear instructions.

**PO-11. Project management and finance**: Demonstrate knowledge and understanding of the engineering and management principles and apply these to one's own work, as a member and leader in a team, to manage projects and in multidisciplinary environments.

**PO-12. Life-long learning**: Recognize the need for, and have the preparation and ability to engage in independent and life-long learning in the broadest context of technological change.

# COURSE OUTCOMES

After completion of the course the students will be able to:

**CO-1:** Describe the status of any thermodynamic system in terms of thermodynamic properties

**CO-2:** They shall be able to apply the principle of energy conservation for different systems in terms of first law of thermodynamics

**CO-3:** Apply the second law of thermodynamics for identifying the feasibility and spontaneity of any process

**CO-4:** Understanding the concept of entropy, irreversibility and exergy

**CO-5:** Appreciation of the concepts of energy efficiency, energy availability, exegetic efficiency and limits of energy conservation for best possible designs

**CO-6:** They should be capable to interpret and analyse the energy flow and conversion process in different energy conversion devices through thermodynamic cycles

| Course Outcomes | Expected Mapping with Programme Outcomes *(1- Weak Correlation; 2- Medium correlation; 3- Strong Correlation)* | | | | | | | | | | | |
|---|---|---|---|---|---|---|---|---|---|---|---|---|
| | **PO-1** | **PO-2** | **PO-3** | **PO-4** | **PO-5** | **PO-6** | **PO-7** | **PO-8** | **PO-9** | **PO-10** | **PO-11** | **PO-12** |
| **CO-1** | 3 | 2 | 2 | 2 | 1 | - | - | - | - | - | - | - |
| **CO-2** | 3 | 1 | 2 | 1 | - | - | - | - | - | - | - | - |
| **CO-3** | 3 | 3 | 2 | 1 | - | - | - | - | - | - | - | - |
| **CO-4** | 3 | 3 | 3 | 2 | 1 | - | - | - | - | - | - | - |
| **CO-5** | 3 | 3 | 2 | 1 | 1 | - | - | - | - | - | - | - |
| **CO-6** | 3 | 3 | 3 | 2 | 1 | - | - | - | - | - | - | - |

# GUIDELINES FOR TEACHERS

To implement Outcome Based Education (OBE) knowledge level and skill set of the students should be enhanced. Teachers should take a major responsibility for the proper implementation of OBE. Some of the responsibilities (not limited to) for the teachers in OBE system may be as follows:

- Within reasonable constraint, they should manoeuvre time to the best advantage of all students.
- They should assess the students only upon certain defined criterion without considering any other potential ineligibility to discriminate them.
- They should try to grow the learning abilities of the students to a certain level before they leave the institute.
- They should try to ensure that all the students are equipped with the quality knowledge as well as competence after they finish their education.
- They should always encourage the students to develop their ultimate performance capabilities.
- They should facilitate and encourage group work and team work to consolidate newer approach.
- They should follow Blooms taxonomy in every part of the assessment.

## Bloom's Taxonomy

| Level | Teacher should Check | Student should be able to | Possible Mode of Assessment |
|---|---|---|---|
| Create | Students ability to create | Design or Create | Mini project |
| Evaluate | Students ability to justify | Argue or Defend | Assignment |
| Analyse | Students ability to distinguish | Differentiate or Distinguish | Project/Lab Methodology |
| Apply | Students ability to use information | Operate or Demonstrate | Technical Presentation/ Demonstration |
| Understand | Students ability to explain the ideas | Explain or Classify | Presentation/Seminar |
|  | Students ability to recall (or remember) | Define or Recall | Quiz |

# GUIDELINES FOR STUDENTS

Students should take equal responsibility for implementing the OBE. Some of the responsibilities (not limited to) for the students in OBE system are as follows:

- Students should be well aware of each UO before the start of a unit in each and every course.
- Students should be well aware of each CO before the start of the course.
- Students should be well aware of each PO before the start of the programme.
- Students should think critically and reasonably with proper reflection and action.
- Learning of the students should be connected and integrated with practical and real life consequences.
- Students should be well aware of their competency at every level of OBE.

# ABBREVIATIONS AND SYMBOLS

## List of Abbreviations

| General Terms | | | |
|---|---|---|---|
| **Abbreviations** | **Full form** | **Abbreviations** | **Full form** |
| Conv. | Convective | KE | Kinetic Energy |
| COP | Coefficient of Performance | PE | Potential Energy |
| Heat Exch. | Heat Exchanger | | |

## List of Symbols

| Symbols | Description |
|---|---|
| $A$ | Area normal to the direction of heat transfer |
| $A_c$ | Cross-sectional area |
| $A_s$ | Area of surface exposed to the fluid |
| $C_p$ | Specific heat at constant pressure |
| $C_{p0}$ | Specific heat of an ideal gas at constant pressure |
| $C_v$ | Specific heat at constant volume |
| $C_{v0}$ | Specific heat of an ideal gas at constant volume |
| $E$ | Total energy |
| $F$ | Force |
| $F_{ext}$ | External force |
| $I$ | Electric current |
| $\underline{I}$ | Mass moment of inertia |
| $N$ | Revolution of the shaft per minute |
| $Kn$ | Knudsen number |
| $L$ | Characteristic dimension of the system |
| $P$ | Pressure |
| $P_{MEP}$ | Mean effective pressure |
| $P_{abs}$ | Absolute pressure |
| $P_{atm}$ | Atmospheric pressure |
| $P_{ext}$ | External pressure |
| $P_{gage}$ | Gauge pressure |

| Symbols | Description |
|---|---|
| $Q$ | Heat |
| $\dot{Q}$ | Heat transfer rate |
| $Q_H$ | Amount of heat exchange with high temperature thermal reservoir |
| $Q_L$ | Amount of heat exchange with low temperature thermal reservoir |
| $R$ | Characteristic gas constant |
| $\underline{R}$ | Universal gas constant |
| $\dot{S}_{gen}$ | Rate of entropy generation |
| $T$ | Temperature |
| $T_0$ | Dead state temperature |
| $T_H$ | Temperature of high temperature thermal reservoir |
| $T_L$ | Temperature of low temperature thermal reservoir |
| $T_{condenser}$ | Condenser temperature |
| $T_f$ | Bulk temperature of the fluid medium |
| $T_s$ | Temperature of the solid surface |
| $U$ | Internal energy |
| $V$ | Volume |
| $\dot{V}$ | Volumetric flow rate |
| $\underline{V}_{avg}$ | Average velocity |
| $W$ | Weight |
| $\dot{W}$ | Rate of work transfer |
| $\dot{W}_e$ | Electrical power |
| $\dot{W}_{flow}$ | Flow work |
| $\dot{W}_R$ | Power input to the refrigerator |
| $W_{su}$ | Surroundings work |
| $W_{u,actual}$ | Actual useful work |
| $W_{u,ideal}$ | Maximum useful work |
| $Z$ | Compressibility factor |
| $dT$ | Infinitesimally small temperature difference |
| $d\theta$ | Angular displacement |

| Symbols | Description |
|---|---|
| $e$ | Specific energy |
| $g$ | Gravitational acceleration |
| $\boldsymbol{h}$ | Height |
| $h$ | Specific enthalpy |
| $h_{conv.}$ | Convective heat transfer coefficient |
| $h_f$ | Specific enthalpy of saturated liquid |
| $h_{fg}$ | Enthalpy of vapourization or Latent heat of vapourization |
| $h_g$ | Specific enthalpy of saturated vapour |
| $k$ | Ratio of specific heats |
| K | Thermal conductivity |
| $M$ | mass |
| $\dot{m}_{in}$ | Rate of mass inflow |
| $\dot{m}_{out}$ | Rate of mass outflow |
| $m_p$ | Mass of the piston |
| $n$ | Polytropic index |
| $q$ | Heat transfer per unit mass |
| $\boldsymbol{r}$ | Normal distance from the centre of the shaft |
| $r$ | Compression ratio |
| $r_c$ | Cut-off ratio |
| $r_p$ | Pressure ratio |
| $s$ | Specific entropy |
| $s_f$ | Specific entropy of saturated liquid |
| $s_g$ | Specific entropy of saturated vapour |
| $t$ | time |
| $u$ | Specific internal energy |
| $u_f$ | Specific internal energy of saturated liquid |
| $u_g$ | Specific internal energy of saturated vapour |
| $v$ | Specific volume |
| $v_f$ | Specific volume of saturated liquid |
| $v_g$ | Specific volume of saturated vapour |

| Symbols | Description |
|---|---|
| $w_{pump}$ | Work input to the pump |
| $w_{turbine}$ | Work produced by the turbine |
| $\boldsymbol{x}$ | Displacement |
| $x$ | Quality |
| $z$ | Elevation |
| $\Delta E$ | Change in total energy |
| $\Delta U$ | Change in internal energy |
| $\Delta\chi$ | Change in exergies |
| $\boldsymbol{\kappa}$ | Boltzmann constant |
| $T$ | Torque |
| $\beta$ | Coefficient of performance (COP) |
| $\beta_{Carnot}$ | Carnot COP |
| $\Gamma$ | Specific weight |
| $\delta A'$ | Limiting area |
| $\delta F_n$ | Normal force |
| $\delta m$ | Mass of differential element |
| $\delta V$ | Volume of differential element |
| $\delta V'$ | Limiting volume |
| $\varepsilon$ | Emissivity of the surface |
| $\eta_{Carnot}$ | Efficiency of Carnot cycle |
| $\eta_{Diesel}$ | Efficiency of Diesel cycle |
| $\eta_{Otto}$ | Efficiency of Otto cycle |
| $\eta_{actual}$ | Actual efficiency |
| $\eta_{irrev}$ | Efficiency of irreversible heat engine |
| $\eta_{rev}$ | Efficiency of reversible heat engine |
| $\eta_{II}$ | Second law efficiency |
| $k$ | Spring constant |
| $\lambda$ | Molecular mean free path |
| $\rho$ | Density |

| Symbols | Description |
|---|---|
| $\rho_m$ | Density of the manometric fluid |
| $\sigma$ | Stefan-Boltzmann constant |
| $\phi$ | Exergy per unit mass |
| $\chi$ | Exergy |
| $\chi_{heat}$ | Exergy associated with heat transfer |
| $\chi_{work}$ | Exergy associated with work transfer |
| $\psi$ | Exergy per unit mass of a flowing stream |
| $\omega$ | Angular velocity |

# LIST OF FIGURES

# CONTENTS

# 1 Introduction and Fundamental Concepts

## UNIT SPECIFICS

*Through this unit we have discussed the following aspects:*

- *Introduction to thermodynamics;*
- *Macroscopic and microscopic approaches in the study of thermodynamics;*
- *Thermodynamic systems & control volume;*
- *Thermodynamic states & properties of the system;*
- *Thermodynamic equilibrium;*
- *Process and cycle;*
- *Concepts of specific volume, density, pressure and temperature;*
- *Thermal equilibrium and measurements of temperature.*

*The basic introduction and practical applications of the concepts of thermodynamics are discussed for generating curiosity as well as improving problem solving capacity related with the analysis of energy systems.*

*Besides giving a large number of multiple choice questions as well as questions of short and long answer types, assignments through a number of numerical problems, a list of references and suggested readings are given in the unit so that one can go through these for practice.*

## RATIONALE

*The fundamental science of thermodynamics helps us to define the state of any system and enables us to quantify the feasibility and spontaneity of processes in the nature. For describing the thermodynamic state of any system, both macroscopic and microscopic approaches are used with distinct applications. Both of these approaches have been discussed. For analysis of any system, the identification of the system and classification of the system based on the energy and material flow is very crucial. For describing the thermodynamic states of any system proper system properties need to be identified. Commonly used properties like temperature, pressure, volume has been discussed so that the students can appreciate their importance for the analysis of thermodynamic systems. Temperature measurement methods have been also covered in this chapter to highlight the importance of thermal equilibrium.*

## PRE-REQUISITES

*None*

## UNIT OUTCOMES

*List of outcomes of this unit is as follows:*

*U1-O1: Appreciation of the practical utility of the thermodynamics*

*U1-O2: Understanding the meaning of thermodynamic system*

*U1-O3: Macroscopic and microscopic analysis approaches of thermodynamic systems*

*U1-O4: Thermodynamic states and thermodynamic processes*

*U1-O5: Important system properties such as pressure, volume, temperature, etc.*

*U1-O6: Thermal equilibrium and measurement of temperature*

| Unit-1 Outcomes | EXPECTED MAPPING WITH COURSE OUTCOMES (1- Weak Correlation; 2- Medium correlation; 3- Strong Correlation) | | | | | |
|---|---|---|---|---|---|---|
| | CO-1 | CO-2 | CO-3 | CO-4 | CO-5 | CO-6 |
| U1-O1 | 3 | 2 | 1 | 1 | 1 | 3 |
| U1-O2 | 3 | 1 | 1 | 1 | 1 | 3 |
| U1-O3 | 3 | 1 | 1 | 1 | 1 | 3 |
| U1-O4 | 3 | 1 | 1 | 1 | 2 | 3 |
| U1-O5 | 3 | 1 | 1 | 1 | 3 | 3 |
| U1-O6 | 3 | 1 | 1 | 1 | 1 | 3 |

## 1.1 Introduction

The spontaneous processes in nature involve the transformation of energy from one form to another. For instance, a rock falling from a cliff gains kinetic energy at the expense of its potential energy. There are plenty of examples (Fig. 1.1) in our daily life where we experience the transformation and exchange of energy. A branch of science that elucidates the principles of energy conversion, its storage and utilization in a specific form, is known as Thermodynamics. The laws of thermodynamics help us to determine the feasibility, spontaneity and efficiency of any process occurring in nature. The laws of thermodynamics have a universal applicability in the prescribed circumstances like Newton's laws of motion which are applicable for all kinds of events occurring in nature. In other words, it can be said that no phenomenon can occur in nature without satisfying the laws of thermodynamics. Hence, thermodynamics has formed the basis for the design of different engineering devices to make them behave in a certain way for achieving the desirable tasks. Energy, in one or other form, is the fundamental requirement to operate engineering devices, therefore, it forged the basis in the origination of the subject "Thermodynamics".

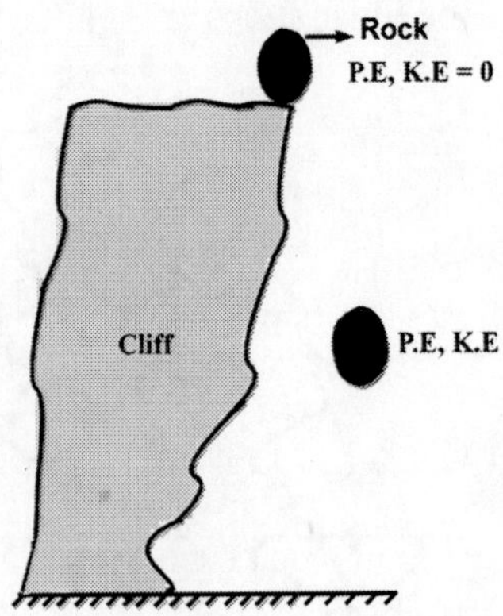


**Fig. 1.1:** A rock falling from a cliff.

The roots of modern thermodynamics are connected to the discovery of the first ideal gas law, which Robert Boyle discovered in 1662. The first book on thermodynamics was published in 1824 by French engineer Sadi Carnot at his own expense which focused on the analogy for the transfer of heat from the hot body to the cold body to a falling entity and hence to mechanical work. He concluded that the motive power increased upon increasing the difference in temperature between the hot and cold bodies. His effort was in a desire to improve the efficiency of the steam engines which were already discovered by Thomas Savery and Thomas Newcomen in 1697 and 1712, respectively. A systematic theoretical elucidation of the steam engine was described in the textbook of thermodynamics written by William Rankine in 1859. The first and second laws of thermodynamics were proposed in the 1850s. The scientists whose names are associated with the discovery of these laws are Julius Mayer, Hermann Helmholtz, Ludvig Colding, James Joule, Rudolf Clausius and Lord Kelvin.

Any engineering device one can think of is based on the principles of thermodynamics. In our kitchen, we use many appliances such as gas stoves, microwave ovens, induction stoves, electric kettles and refrigerators, etc. which operate on the principle of conversion of energy. Electricity has been the main source of input energy for different engineering devices including automobiles (Electric Vehicles). Electricity is generated using coal-based thermal power stations, nuclear power plants, diesel engines, gas turbine engines and solar-powered devices, etc. In general, the layout of such power plants is extremely complex (schematic of typical coal based thermal power plant in Fig. 1.2), however, at the fundamental level, these are developed by the amalgamation of basic components utilizing the laws of thermodynamics. It is almost impossible to list each and every component/system working on the foundation of thermodynamics, whereas to illustrate the importance of the subject, some of the popular applications of thermodynamics are depicted in Fig. 1.3. This book focuses on the fundamental concepts of thermodynamics which are needed for analyzing the thermodynamic systems and wherever possible the illustrations are related to the practical devices.

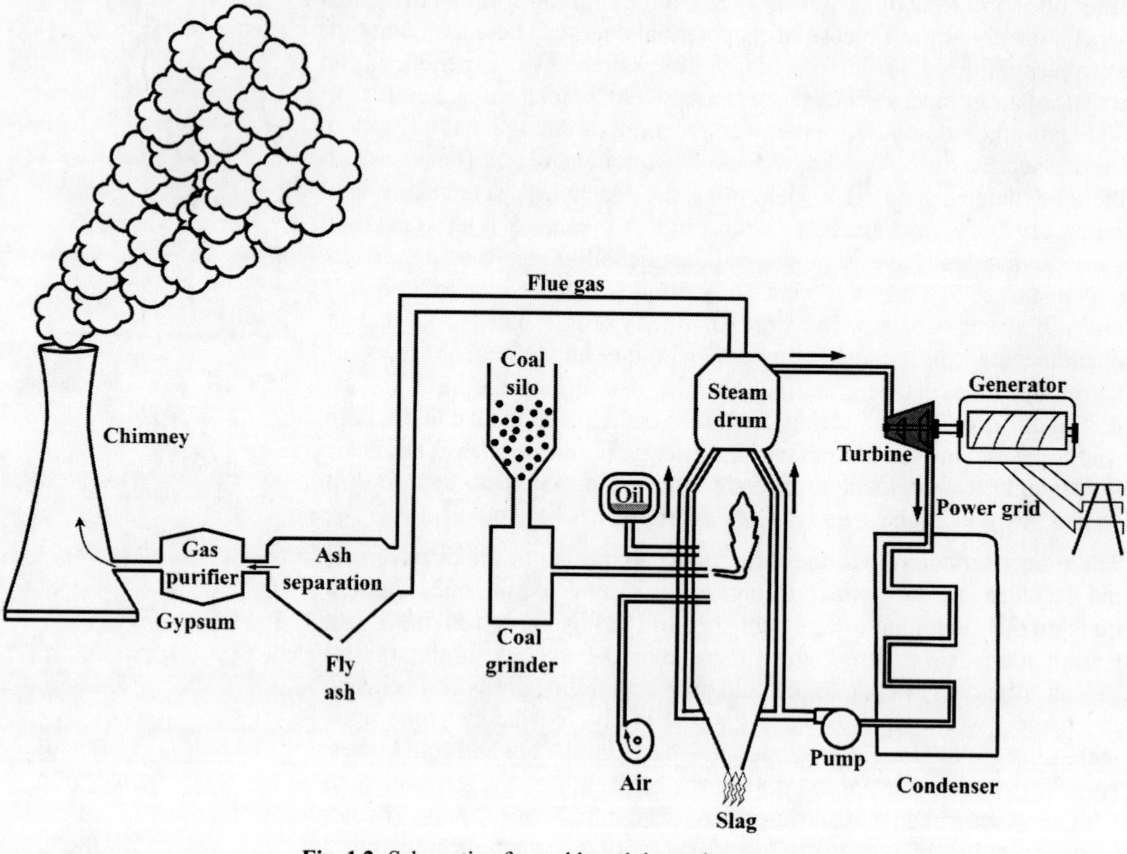


**Fig. 1.2:** Schematic of a coal based thermal power plant.

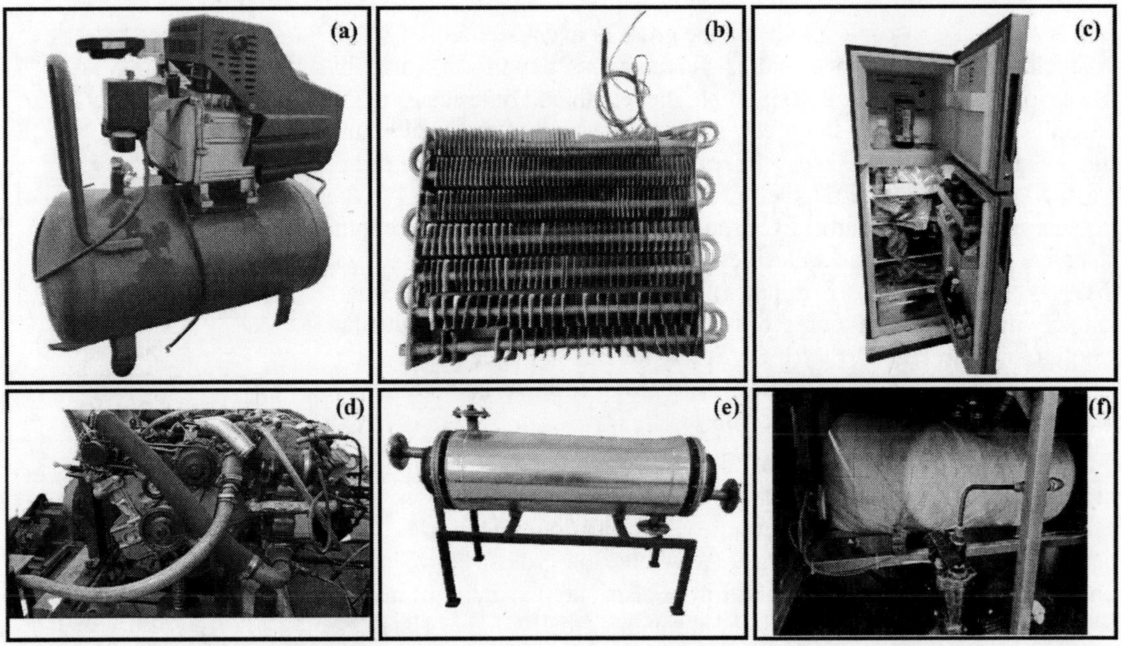


**Fig. 1.3:** Applications of thermodynamics showing: (a) compressor, (b) condenser, (c) refrigerator, (d) internal combustion engine, (e) shell and tube heat exchanger and (e) steam drum of an electric boiler.

## 1.2 Macroscopic versus Microscopic Approach

There exist two fundamental approaches, namely macroscopic and microscopic, to analyze the behaviour of a system. The applicability of the two can be explained with the help of Fig. 1.4. It shows a perfectly sealed room of 25 $m^3$ volume which contains a monoatomic gas such as helium at atmospheric pressure (Fig. 1.4-a) and at a pressure close to absolute zero (Fig. 1.4-b). At standard temperature pressure (STP) condition, the gas contains $\sim 10^{26}$ molecules/atoms (Fig. 1.4-a) whereas in the second case, the number of molecules will be very less (as depicted in Fig. 1.4-b). In case of the microscopic approach, the behaviour of individual molecules is of concern, therefore, to describe the position and motion of each molecule three coordinate positions and three components of velocity are needed at each time instant. This brings in the requirement of $\sim 6 \times 10^{26}$ equations to describe the behaviour of all the molecules. Solving these many equations, even with present day advanced computational facilities, is almost impossible. On the contrary, the less numbers of equations in the second situation make it feasible to determine the behaviour of individual molecules.

In case of the macroscopic approach, the gross or average effect of all the molecules is of importance and is perceived as a uniform value by measuring instruments. In this situation, the length and time scales of the molecules are extremely small compared to the size of the system and time of measurement. Hence the observed quantities are the time-averaged effect of a large number of molecules. In case of Fig. 1.4-a, the pressure acting on the walls of the room is dependent on the momentum exchange between the gas molecules and the walls. At the time scale of observation (in seconds or minutes) enormous numbers of molecules collide with the wall and their gross effect is perceived. Such an outcome is not possible in case of Fig. 1.4-b as the number of molecules are very less and there will be a condition of selective molecular collisions with walls and hence, non uniform pressure.

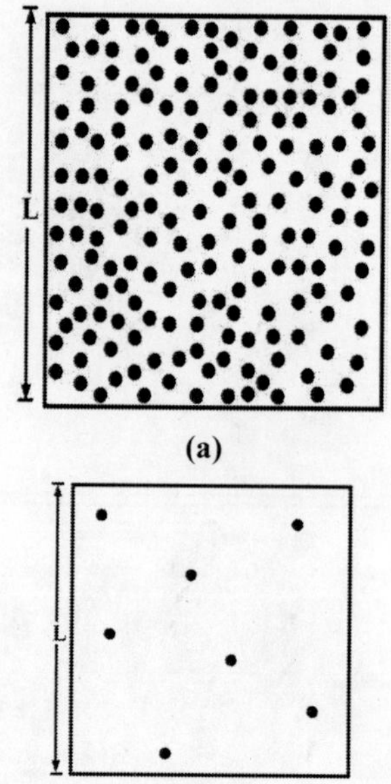


Fig. 1.4: Depiction of gas molecules in a sealed room of 25 $m^3$ volume at (a) 1 atm pressure and (b) near absolute vacuum condition.

If the characteristic dimension of the system ($L$) is significantly larger than the molecular mean free path ($\lambda$) and the time period of observation is significantly larger than intermolecular collision frequencies, then macroscopic approach is applicable where the voids between the molecules can be neglected and uniformly distributed quantity of matter, called continuum, can be considered throughout the volume. Here, the molecular mean free path represents the average distance traversed by the molecules (Fig. 1.4-b) between the successive collisions. For a very small number of molecules in a system, its magnitude is of the order of system size. A dimensionless parameter, the Knudsen number ($Kn$), is defined as the ratio of the molecular mean free path to the characteristic size of the system, holding the assumption of continuum valid for its value less than equal to 0.001. In engineering applications, the assumption of continuum is majory valid, hence the macroscopic approach is convenient. This macroscopic viewpoint of thermodynamics that does not consider the behaviour of individual molecules is termed **classical thermodynamics**. Whereas a probabilistic approach that considers the behaviour of an individual or group of molecules is referred to as **statistical thermodynamics**. Such an approach is beyond the scope of this book.

Scan Me

For more details on macroscopic versus microscopic approach.

## 1.3 Thermodynamic Systems and Control Volumes

In nature and engineering applications, the interactions of energy take place between different entities. Therefore, in order to thermodynamically analyze an entity, it is necessary to separate it from neighboring entities and identify the associated energy interactions and its transformations within the entity.

An entity or region in space or quantity of matter which is of interest is known as a **thermodynamic system** or **system**. The entities or matter outside the system is termed its **surroundings**. The system is separated from its surroundings by a real or an imaginary surface which is called the **boundary**. The system under consideration may be as simple as an air container and it may be as complex as a steam based thermal power plant. Fig. 1.5-a shows a pan over a gas stove containing water at 25°C and 1 atm pressure. If only the behaviour of water is of concern, then the system boundary will include the internal surface of the container and an imaginary surface at the top layer of water within the pan. Such a system is illustrated in Fig. 1.5-b. In this case, the material of the pan is part of the surroundings only and the energy is added to the water through the surface of the pan. Another situation may include where the temperature of the pan, along with the behaviour of water, are of interest. Then pan and water both will become part of the system as depicted in Fig. 1.5-c and energy will enter the system directly from the flame of the gas stove. As the top face of the pan is open in both situations, therefore, up to 100°C (boiling temperature at 1 atm pressure) temperature of water, only energy exchange can take place through the top boundary from water to the surrounding air. Whereas if further heating is done at 100°C, then the mass of water will also pass through the imaginary boundary in the form of water vapours into the surrounding air.

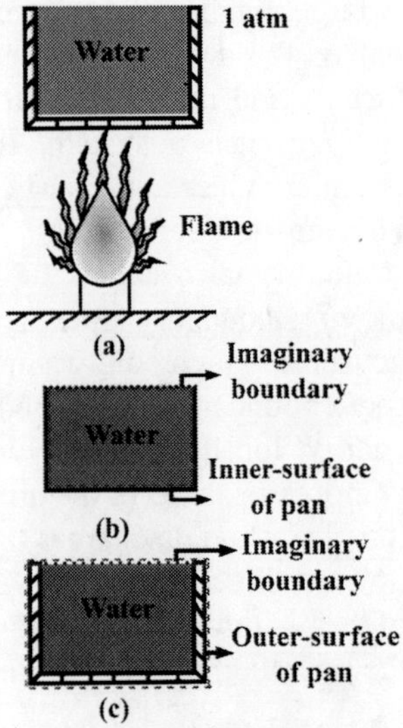


**Fig. 1.5:** (a) A pan of water placed over a gas-stove at 1 atm pressure, (b) a system including only water as its part and (c) a system including both water and pan as its constituents.

In case, the identity of the matter is fixed within the system and no mass exchange takes place through its boundary, then it is called the **fixed mass** or **control mass,** or **closed system**. The heating of water in the pan up to 100°C at 1 atm pressure in Fig. 1.5 is an example of a closed system. In such systems, energy can exchange through the boundary and it does not have to be always fixed. The example of a moving boundary closed system is shown in Fig. 1.6. It includes a piston-cylinder arrangement in which air is filled at atmospheric conditions and occupies an initial volume $V_1$. If only air is part of the system, then the inner surface of the cylinder and pistons act as the boundary of the system and everything outside the boundary, including the cylinder and the piston, is surroundings. When the bottom face of the cylinder is heated, it transfers energy into the air and the piston starts expanding. In this situation, the air occupies more volume and the top boundary of the system has traversed. As no mass exchange has taken place across the boundary, this is an example of a closed system with a moving boundary. A special case of a closed system that does not allow even energy to cross across the system boundary, is termed an **isolated system**. It is almost impossible to construct an isolated system whereas a thermos flask acts as an isolated system for a certain period of time considering the small energy losses are negligible. In a broad sense, the entire universe can be called an isolated system.

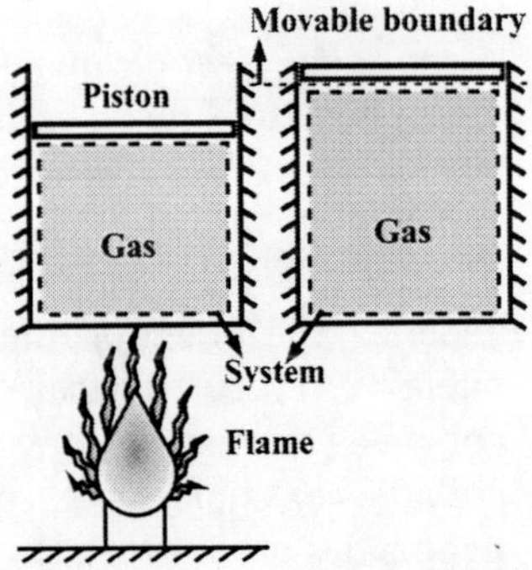


**Fig. 1.6:** A moving boundary closed system.

A large number of engineering devices involve the flow of matter in and out of the system. The centrifugal pump, electric geyser, compressor, turbine and nozzle, etc. are the examples of such devices. In order to analyze such a system, a region is defined in space, across whose boundary mass and energy interaction take place, it is called the **control volume** or **open system**. The boundary of such a control volume is sometimes also called the **control surface**. The examples of fixed and moving boundary open systems are depicted in Fig. 1.7. An electric geyser in which cold water enters and hot water goes out forms a typical fixed boundary open system where electric power supplies the necessary energy for heating the water (Fig. 1.7-a). A reciprocating type air compressor takes in the atmospheric air during the downward motion of the piston and discharges the pressurized air when the piston moves in an upward direction. If behaviour of air is of interest then the inner surface of the piston and cylinder form the control surface which is moving during the operation of the compressor (Fig. 1.7-b).

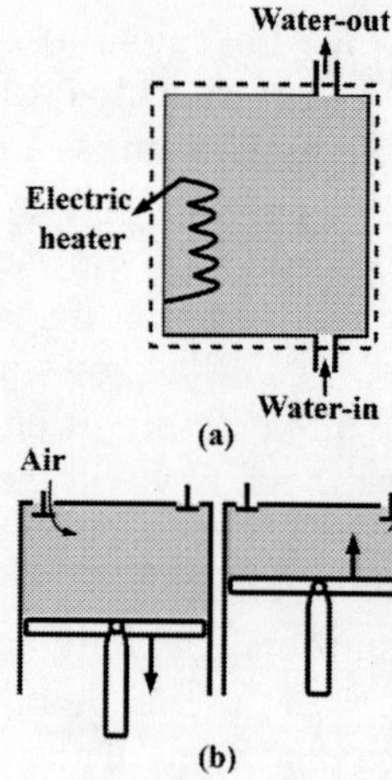


**Fig. 1.7:** (a) An electric geyser as a control volume and (b) an air compressor as a moving boundary control volume.

The system under consideration must be defined carefully while performing the thermodynamics analysis of the engineering devices. In the case of the analysis of simple devices, the system definition appears obvious, whereas the selection of the system is quite tedious for complex devices and may simplify the analysis if specified properly.

## 1.4 Properties and State of a System

A set of macroscopic quantities that defines the unique characteristics of the system are known as its **properties**. The familiar properties of the systems are mass, volume, pressure, temperature, viscosity, thermal conductivity and coefficient of thermal expansion. These are the handful of properties listed here and the list can be extended to many. The two general categories of thermodynamic properties are extensive and intensive. Properties that depend upon the amount of the system are known as **extensive properties**. Volume, mass and energy of the system are a few examples of extensive properties. On the other hand, **intensive properties** are independent of the mass of the system, such as density, pressure, temperature and thermal conductivity. All the extensive properties per unit mass are termed **specific properties** and these are of intensive nature. Specific volume ($v = V/m$) and specific energy ($e = E/m$) are some examples of specific properties. The intensive properties, including the specific properties, are denoted by small letters with temperature T and pressure P being the exceptions. Contrary, the extensive properties are symbolized by uppercase letters with mass m being an exception. An easy method to determine the extensive or intensive nature of properties is demonstrated in Fig. 1.8. Here, a container of volume $V$ has been filled with a gas having mass m (Fig. 1.8-

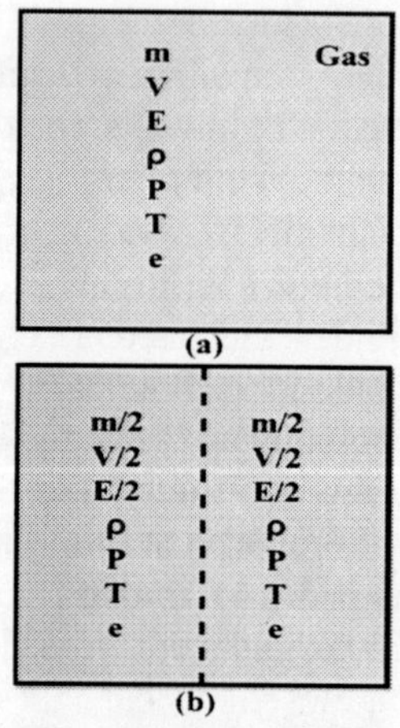


**Fig. 1.8:** Illustration of extensive and intensive properties.

a). The container is now divided into two equal halves by an imaginary partition (Fig. 1.8-b). The extensive properties will reach half of their original value in each partition, however, the magnitude of intensive properties remains unchanged. It is to highlight that the continuum is applicable while defining the properties of the system.

At a **thermodynamic state** or **state** of the system, the values of its properties are the same throughout the entire system. The state of the system represents a unique set of properties that are independent of the fact that how the system has reached a particular state. In other words, the values of properties at a given state are independent of the path which the system has followed to reach a thermodynamic state. The change in any one property of the system results in a change in the state of the system.

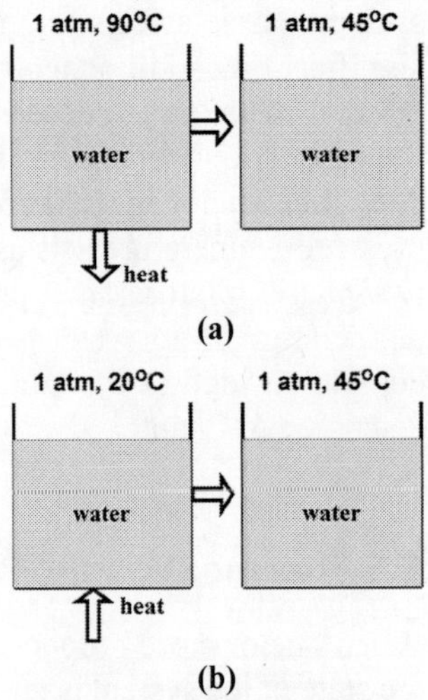


**Fig. 1.9:** (a) A system cooled from 90°C to 45°C at 1 atm pressure and (b) a system heated from 20°C to 45°C at 1 atm pressure.

Water is initially at 90°C and 1 atm pressure in a container and then it is cooled to 45°C (Fig. 1.9-a). Another situation shows the heating of water from 20°C and 1 atm pressure to 45°C (Fig. 1.9-b). In both cases, the final temperature and pressure of the system are the same, therefore, it possesses the same thermodynamic state, irrespective of its history. Moreover, due to the incompressible nature of the water, the change in volume is negligible upon cooling and heating within the specified limits. Therefore, the state of the system has changed solely upon the change in its temperature. It is important to highlight that when the system is heated or cooled, there is no potential (driving force) within the system at its final state to change any of its properties. This signifies that equilibrium within the system is needed to define its state. When a system is in equilibrium it experiences no change if isolated from its surroundings. Typically for thermodynamic systems, there exist three kinds of equilibriums: **thermal equilibrium**, **mechanical equilibrium** and **chemical equilibrium**.

A system is said to be in a complete state of **thermodynamic equilibrium** if each of these is satisfied. If the temperature at all the spatial locations is the same throughout the system, then it is said to be in thermal equilibrium. At this condition, the temperature gradient will not exist within the system which is the driving force for the transfer of heat (a form of energy that is described in later sections). For mechanical equilibrium, the pressure at different locations within the system should not vary with time, however, it may vary with elevation due to gravitational effects within the system. As the pressure at deeper locations within the system supports more quantity of matter, therefore, it is balanced by the weight (detailed in Section 1.6) and results in mechanical equilibrium. Moreover, in the majority of the thermodynamic systems, the variations in pressure due to gravity are comparatively small and often neglected. Finally, the system is said to be in chemical equilibrium, if it does not have any affinity to undergo the chemical reactions. In simpler words, the chemical composition at different spatial locations does not change with time. In the case of systems involving two-phase mixtures, a condition of phase equilibrium should also hold. This is satisfied when the mass of each phase does not change with time.

It is also important to know how many properties are needed to designate the state of a system. As per state postulates, two independent intensive properties are sufficient to completely specify the state of a simple compressible system. The values of all other properties can be obtained using thermodynamic relations with the help of two known independent properties. A system in absence of external effects such as gravitational, electrical, magnetic and surface tension, etc. is called a simply compressible system. These effects are negligible for the majority of engineering problems. In case these become significant, an additional property for each needs to be included. For instance, in the case of gravitational effects, elevation is needed to specify pressure at different locations within the system, in addition to the other two independent intensive properties. The properties can be independent if, upon variation of one, the other can be kept fixed. The combination of pressure-volume, temperature-volume and pressure-temperature acts as independent properties for single-phase systems. Whereas in the case of a two-phase mixture, the pressure and temperature become dependent. For example, the boiling temperature of the water is a function of its pressure. Therefore, the pressure and temperature can not completely define the state of a two-phase mixture. The details of phase-transition processes are discussed in Chapter 3.

## 1.5 Thermodynamic Processes and Cycles

When one or more properties of a system change, it shifts from one state to another. Such a change in the state of the system is termed a **process**. There can be plenty of ways to achieve this change from the same initial to final states. A piston-cylinder containing a gas is shown in Fig. 1.10. A weight $W$ is placed on the piston such that the pressure of the gas is $P$. When this weight is removed from the piston, the gas expands and as a result, its pressure decreases. The system will now occupy a new equilibrium state of decreased pressure and increased volume. The initial and final states of the system are presented on a P-V plot in Fig. 1.10. In this example, as the process of expansion is rapid, therefore, it is not possible to determine the exact path that a system has gone through to change from its initial to the final state.

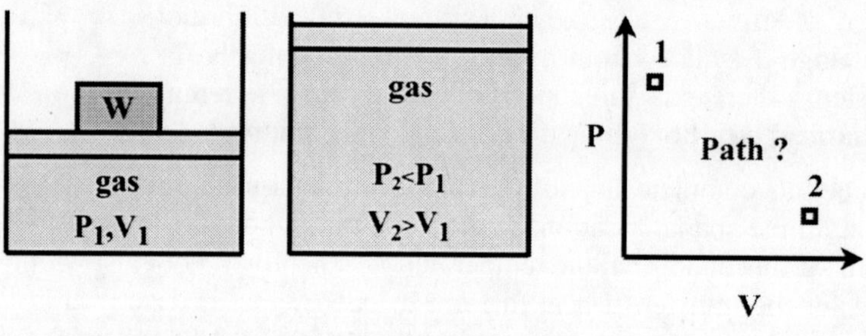


**Fig. 1.10:** Sudden expansion of gas upon release of external pressure.

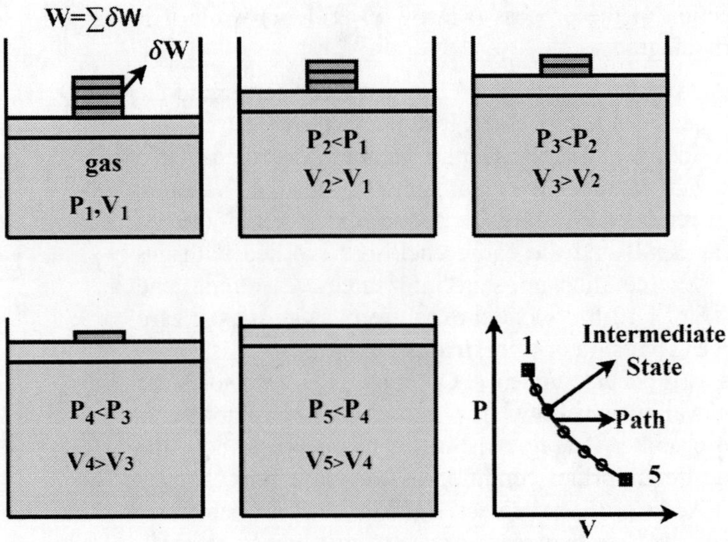


**Fig. 1.11:** Quasi-equilibrium expansion of a gas and process representation on P-V diagram.

Another system that has a change of state between the same initial and final conditions is depicted in Fig. 1.11. In this case, a number of infinitesimally small weights ($\delta W$) are put on the piston such that $W = \sum \delta W$. When one infinitesimal weight ($\delta W$) is removed, the piston will expand slowly to reach a slightly small pressure. If this process of removing the weights is continued, the slow expansion of the system will continue and its pressure will keep decreasing. Once all the weights are removed the system will acquire the final state. As the sum of all the infinitesimal weights is equal to a single weight used in the previous case (Fig. 1.10), therefore, the final state will be the same in both situations. Due to the infinitesimally slow expansion of the system, the properties of the system change uniformly across the entire system, which makes it possible to define the state of the system for all the intermediate states. As the intermediate states are known (depicted by 'o' symbol in Fig. 1.11 on the $P-V$ plot), the exact **path** of the process can be defined. Basically, the sequence of intermediate states, a process has undergone is called the path of the process. The process which takes place at an infinitesimally slow pace such that the system remains at near equilibrium states during the process is referred to **quasi-equilibrium**, or **quasi-static** process. In reality, the processes do not exactly follow the idealized quasi-equilibrium behaviour. However, the majority of the engineering processes can be approximated as quasi-equilibrium with negligible error. It makes the thermodynamic analysis of the process simple and serves as a standard to compare the actual processes. In the case of non-quasi-equilibrium processes, as the exact path of the process is not known, therefore, a dashed line is drawn between initial and final states to indicate such a process. It is also important to put an arrowhead on the process line

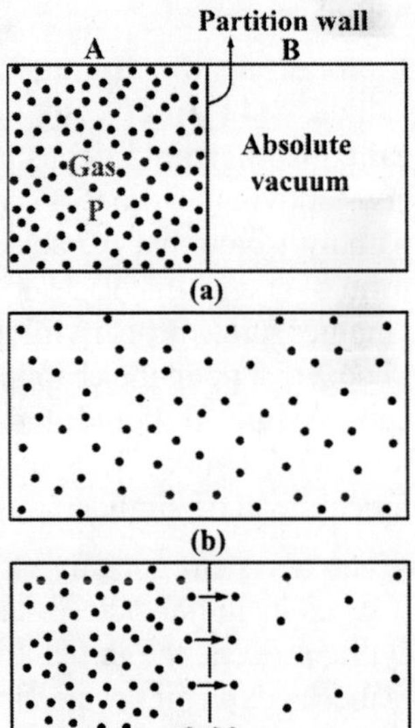


**Fig. 1.12:** (a) Two chambers separated by a partition and maintained at different conditions, (b) depiction of unrestrained expansion and (c) partly restrained expansion.

to indicate the direction of the process (Figs. 1.10 & 1.11) while it is indicated on property diagrams.

Another illustration is presented in Fig. 1.12 to precisely demarcate the quasi-static and non-quasi-static processes. It shows two chambers A and B separated by a partition wall (Fig. 1.12-a). Chamber A is filled with a gas at high pressure and chamber B is maintained at absolute vacuum. When the partition is removed instantly, the sudden expansion of the gas takes place into chamber B and the entire enclosure is filled with gas which occupies a new equilibrium state of increased volume and decreased pressure (Fig. 1.12-b). Such an expansion of gas against zero pressure is called **free expansion** or **unrestrained expansion**. It is a non-quasi-static process as it occurs suddenly. On contrary, a few holes are made in the partition wall which allows slow movement of gas molecules from chamber A to chamber B (Fig. 1.12-c). In this process, both the chambers are in quasi-equilibrium condition as their pressure changes slowly. For chamber A, the pressure decreases and in the case of chamber B, it increases till both the chambers reach a state of equal pressure. Such a process is called the partly restrained expansion and it is of quasi-static nature.

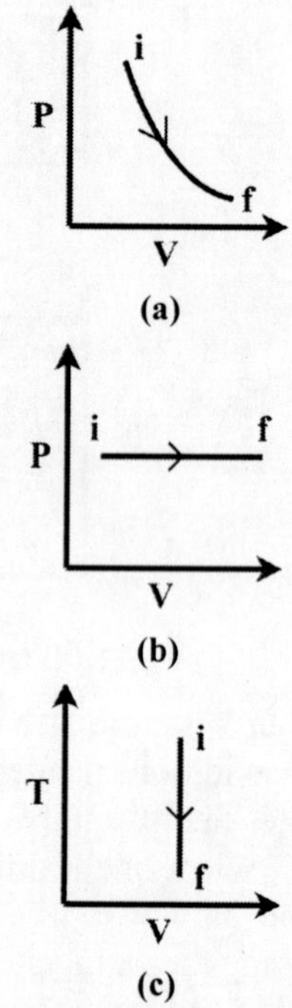


**Fig. 1.13:** Representation of (a) isothermal expansion, (b) isobaric expansion and (c) isochoric heat rejection on property diagrams for an ideal gas.

A process for which any of the thermodynamic properties is kept constant is indicated by prefix **iso**. For instance, a constant temperature process is called **isothermal**, a constant pressure process is termed **isobaric** and a constant volume process is referred to **isochoric** process. Three processes namely isothermal expansion, isobaric expansion and isochoric heat (a form of energy that is described in Chapter 2) rejection are represented on commensurate property plots for an ideal gas in Fig. 1.13. It is important to note that upon the change of working medium (such as real gas, vapour and mixture of liquid and vapour) the specific path of the processes represented in Fig. 1.13 may change whereas the general behaviour of the process will be similar.

When a system goes through a number of processes such that it finally reaches its initial state, is termed a **cyclic process**. This means that for a cyclic process, the initial and final states of the system are identical. For a thermodynamic cycle, the working fluid should be circulated in a closed-loop while it is going through a series of processes.

## 1.6 Primary Measurable Properties

It is described in Section 1.4 that there are a large number of properties that are associated with the thermodynamic state of the system. The direct measurements of some of these properties are possible through standard instruments and for determining a few properties, specific experiments are to be designed along with the use of theoretical relations. Therefore, in thermodynamics, we emphasize using the easy measurable intensive properties to designate the state of the system and utilize thermodynamic property relations to obtain other properties. Specific volume, pressure and temperature are the primary properties that can be measured using standard instruments and are utilized in the thermodynamic analysis. These properties are described in this section.

### 1.6.1 Specific volume and density

The volume of a substance per unit mass is termed **specific volume** and **density** is reciprocal of specific volume. For a differential volume element of volume $\delta V$ and mass $\delta m$, specific volume is expressed as $v = \lim_{\delta V \to \delta V'} \frac{\delta V}{\delta m}$ $(m^3/kg)$. Here, $\delta V'$ is the limiting volume above which a substance can be considered as continuum. For elemental volume less than the limiting value, there exist void spaces between the molecules. Due to this fact, a situation may arise when the elemental volume is so selected that it contains void and hence no mass. In this case, specific volume will approach infinity. In another case, when the elemental volume has completely encapsulated a molecule, the specific volume will reach a small value. Such a variation of specific volume is depicted in Fig. 1.14 for different sizes of elemental volume. The oscillating variation of specific volume has diminished for sizes of control volume above the limiting value as the approximation of continuum holds good.

The specific volume of a substance is also the function of temperature and pressure. In the case of most gases, the specific volume is directly proportional to temperature and is inversely proportional to pressure. The liquids and solids are generally referred to as incompressible substances as the change in their specific volume is very small upon change of pressure and temperature. For example, the specific volume of water has decreased by only 1.91 percent when the pressure is changed from 5 $MPa$ to 50 $MPa$ at 20°C. On the other hand, the increase of specific volume of water is 4.14 percent upon rising temperature from 20°C to 100°C at 5 $MPa$ pressure. This shows that the dependence of the specific volume of liquids on temperature is more pronounced than on pressure. Moreover, for most engineering problems, this variation can be neglected with an acceptable limit of accuracy. The dependence of temperature and pressure on the density of the substance will be the inverse of their dependence on the specific volume. The ratio of the density of a substance to the density of a standard substance at a specified temperature is termed **specific gravity** or **relative density**. Generally, water is used as a standard substance at the temperature of 4°C with density of 1000 $kg/m^3$. Another important parameter linked to density is **specific weight**, which is defined as the weight of unit volume of a substance and is expressed as $\gamma = \rho g$ $(N/m^3)$. Here, $\rho$ is density and $g$ is gravitational acceleration in $m/s^2$.

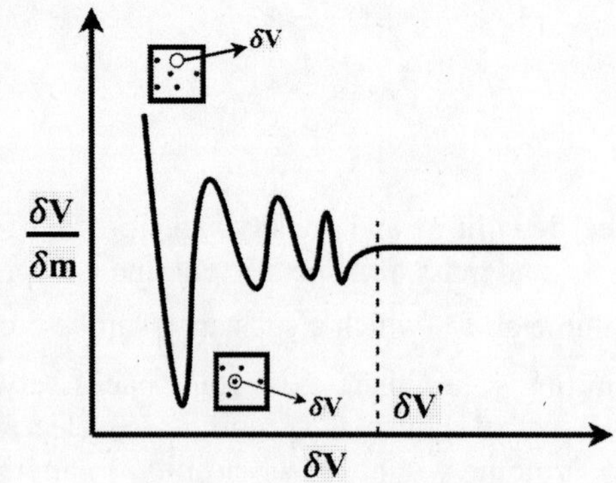


Fig. 1.14: Representation of specific volume to show the limit of the continuum.

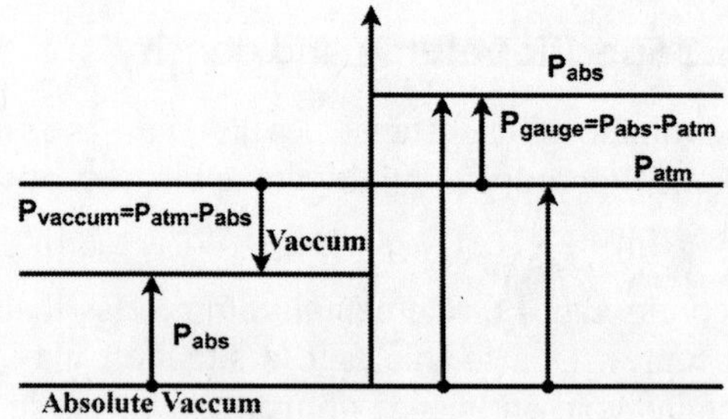


Fig. 1.15: Relation between absolute pressure, gauge pressure and vacuum pressure.

### 1.6.2 Pressure

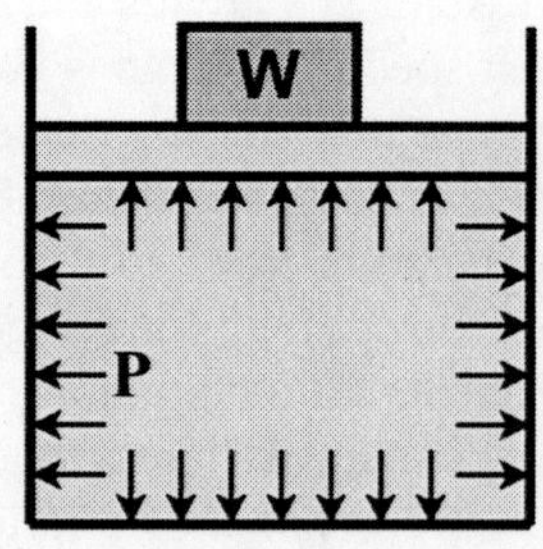


Fig. 1.16: Representation of uniform pressure at all the surfaces of the cylinder and piston.

Pressure is an important property of liquids and gases which is mostly used in thermodynamic analysis. The analogous to pressure is normal stress in solids. Pressure is defined as the normal force exerted by a fluid onto a unit area. For an elemental area $\delta A$ subjected to a normal force $\delta Fn$ by a surrounding fluid, pressure is defined as $P = \lim_{\delta A \to \delta A'} \frac{\delta F_n}{\delta A}$. Here, the limiting area ($\delta A'$) corresponds to the limit of continuum as specified in Fig. 1.14. As per SI system, the unit of pressure is the pascal ($Pa$) and one pascal is equal to a force of one newton acting on an area of one square meter. The value of the pressure at a given point is called the **absolute pressure** and it is measured with reference to absolute vacuum, i.e. the state of absolute zero pressure. As the value of atmospheric pressure is standard, it is also used as a reference in most pressure measurement instruments. The difference of absolute pressure with the local atmospheric pressure is termed the **gage pressure**. Any value of pressure below local atmospheric pressure is termed the **vacuum pressure**. The relations between absolute pressure, gage pressure and vacuum pressure are represented in Fig. 1.15.

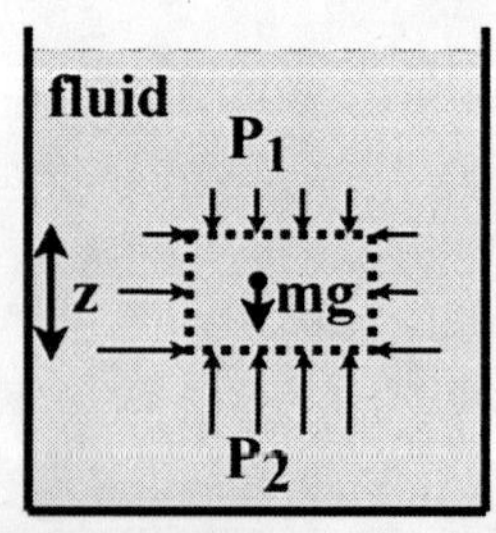


Fig. 1.17: Pressure acting on a fluid element, depicting the effect of elevation.

For a fluid medium at rest, the pressure acting at different points on the surface is the same and always points normal to a surface. An example is shown in Fig. 1.16 which shows a piston cylinder system filled with a gaseous medium and a weight W is placed on the piston such that the system is in equilibrium. The external force acting on the piston is balanced by the gas pressure acting on the inner surface of the piston. For a system in equilibrium, the pressure acting on the walls of the cylinder is also of the same magnitude as indicated by the size of arrows in Fig. 1.16 at different surfaces. Though pressure has both magnitude and direction, still it is a scalar quantity because it always acts in a unique direction normal to the surface. When the gravitational effect is considered into the system, the variation of pressure takes place along the elevation. Fig. 1.17 shows a container filled with fluid at rest. When a fluid element (shown by dashed lines in Fig. 1.17) is subjected to the force balance, in horizontal direction, it results in same pressure on left and right faces of the fluid element, whereas the force balance in vertical direction is expressed as $P_1A + mg = P_2A$. Here mg is the weight of the fluid element which can be given as ρVg where V can be written as $A \times z$. This results in $P_2 - P_1 = \rho gz$. It shows that as we go deeper into the fluid, the pressure increases linearly with depth and it basically balances the weight of the fluid above that depth, therefore, the system remains in equilibrium. In case of fluids for which density is also a function of pressure, the variation of pressure with depth can be presented as $dP/dz = \rho(P)$ and it can be integrated to estimate the pressure at any depth. It is important to mention that the majority of the systems which we study in thermodynamics have almost negligible pressure variation with depth.

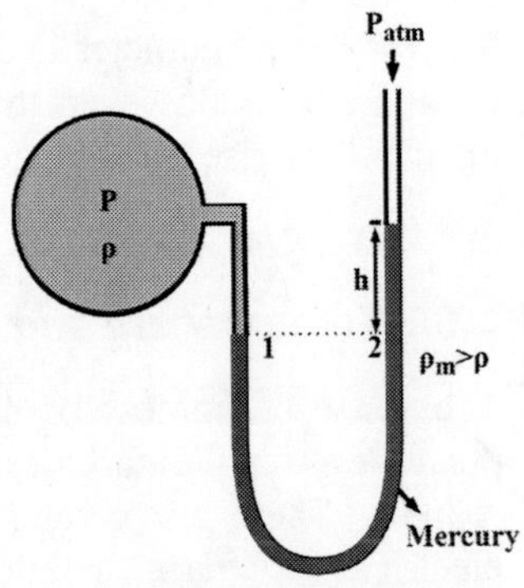


**Fig. 1.18:** A U-tube manometer used to measure the pressure of the system.

A fundamental device that is used to measure the small and moderate pressure differences between two systems or with reference to atmosphere and is based on the idea of pressure variation with elevation, is called the **manometer**. It consists of glass or transparent plastic U-tube which is filled with one or more fluids such as water, mercury, alcohols and oils. The lighter fluids are used for measuring small pressure differences and heavy fluids like mercury is used when large pressure differences are expected. Consider a manometer connected to a system of pressure $P$ and the other end of the manometer is open to the atmosphere as shown in Fig. 1.18. The difference in height of manometric fluid between the two branches of U-tube is h and system contains a gas of density ρ which is significantly small in comparison to the density of the manometric fluid ($\rho_m$). The pressure at point 2 on the right branch is $P_2 = P_{atm} + \rho_m gh$. As the pressure doesn't vary in horizontal direction, hence, $P_1 = P_2$. Assuming the small density of the gas, the elevation effect in the left branch can be neglected, therefore, $P_1 = P$. Thus, $P = P_{atm} + \rho_m gh$ and $P - P_{atm} = \rho_m gh$, which implies $P(gage) = \rho_m gh$. Another illustration of the U-tube manometer is shown in Fig. 1.19 which is used to measure the pressure difference $(P_A - P_B)$ between two systems A and B. The densities of fluids used in systems A and B are $\rho_A$ and $\rho_B$, respectively. The balance of pressure between points 1 and 2 is expressed as $P_A + \rho_A gh_1 = P_B + \rho_B gh_2 + \rho_m gh$. Hence, $P_A - P_B = \rho_m gh + \rho_B gh_2 - \rho_A gh_1$.

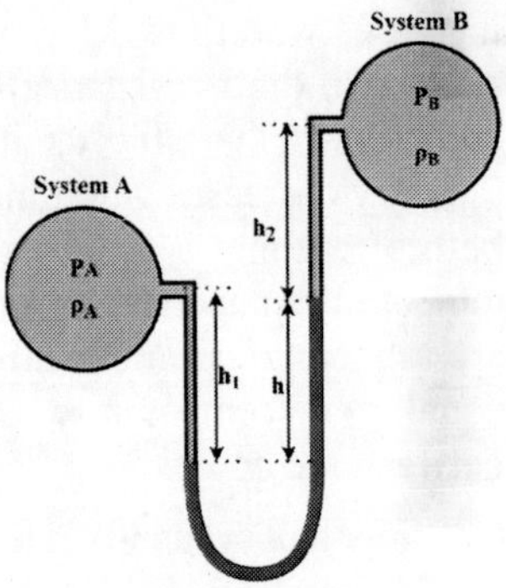


**Fig. 1.19:** A U-tube manometer used to measure the pressure difference between two systems.

A device that is used to measure the atmospheric pressure is known as a **barometer**. A typical barometric configuration is shown in Fig. 1.20. It contains an inverted tube which is dipped into a pool of mercury. The pressure at points A and B is the same. The top of the inverted tube

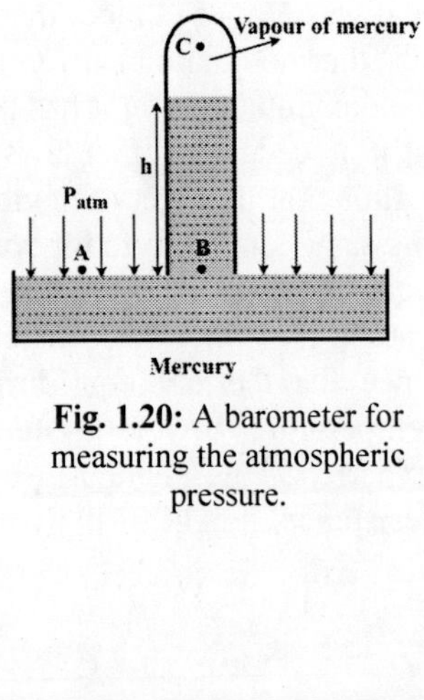


**Fig. 1.20:** A barometer for measuring the atmospheric pressure.

contains the vapor of mercury and vapor pressure of mercury is very small, therefore, pressure at point C can be approximated nearly equal to absolute zero. Thus, $P_B = P_{atm} = \rho_m gh$. The rise in height of fluid in the tube of the manometer and barometer is not the function of its cross-sectional area. However, the size of the tube should not be so small that the capillary rise takes place in it.

### 1.6.3 Temperature and its equality

The sense of hotness or coldness of an object is something that we perceive as temperature. However, our sense of hotness or coldness is not reliable. There are certain physical properties of the matter such as electrical resistance, length, pressure, volume, etc. which change upon change of temperature and form the basis for measurement of temperature. For instance, if we take one hot and one cold aluminum blocks such that these are in thermal contact. We observe a decrease in electrical resistance of the hot block with time and for the cold block, it increases with time. Moreover, after a certain period of time, there is no change in their resistance. Similarly, if we bring a mercury column in contact with these blocks, it is observed that for the hot block, the level of mercury decreases with time and for the cold block, it increases with time before reaching a fixed value. The condition at which the variation of temperature dependent properties diminishes for two bodies when these are in thermal communication is called the equality of temperature. This equality of temperature is actually the state of thermal equilibrium between the bodies.

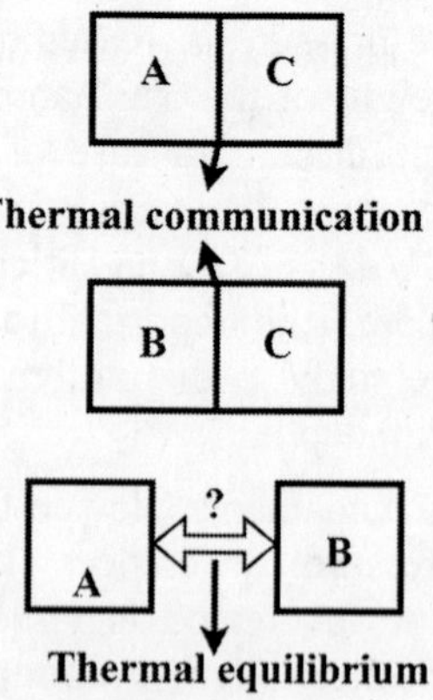

**Fig. 1.21:** Zeroth law of thermodynamics.

Consider that system A is in thermal contact with system C such that these are in thermal equilibrium (Fig. 1.21). Now the same system C is in thermal communication and equilibrium with another system B. Thus, we can say that systems A and B will also be in thermal equilibrium with each other, i.e., there will not be any change in their properties when these are brought into thermal communication with each other. This is called the **zeroth law of thermodynamics**. This statement is very fundamental and was proposed more than half a century after the first and second laws of thermodynamics were formulated. Therefore, due to its very fundamental nature, it has been called the zeroth law of thermodynamics as it precedes the other laws in a logical sequence. If system C is replaced with a thermometer (a device that measures the temperature), it will read the same temperature for both systems A and B. It can be concluded from equality of temperature that for measuring the temperature of a system, it should have thermal contact with the temperature measuring device and once the thermal equilibrium is achieved, the device will reach the temperature of the system. The temperature measurement devices experience the change in properties of their material such as electrical resistance, height of mercury column, pressure, volume, etc., therefore,

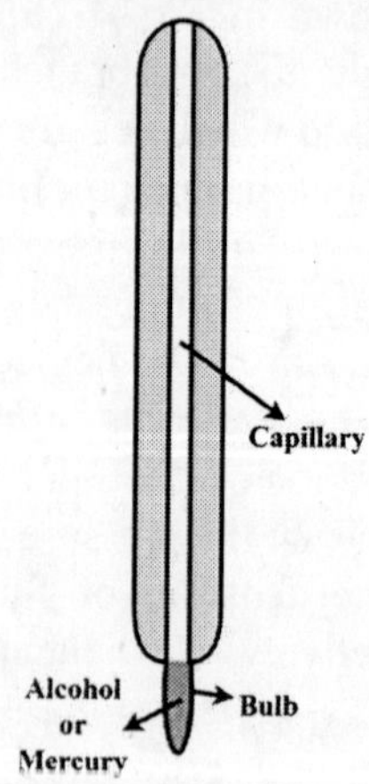

**Fig. 1.22:** A liquid-in-glass thermometer.

the fundamental question is to relate this change to the magnitude of temperature which is described in Section 1.7.

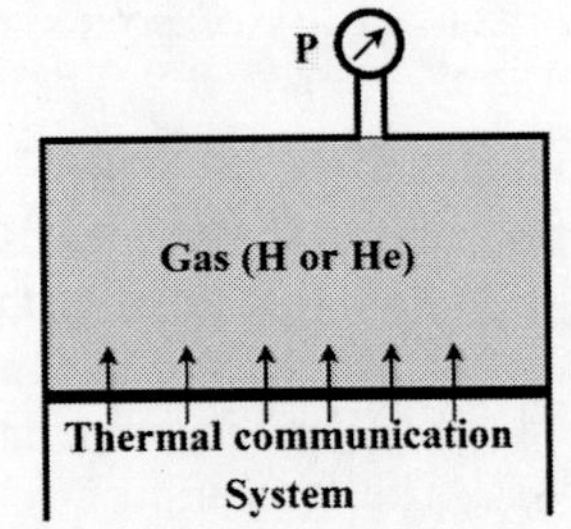


**Fig. 1.23:** Constant volume gas thermometer.

## 1.7 Measurement of Temperature

A number of devices such as liquid-in-glass thermometers, constant volume gas thermometers, constant pressure gas thermometers, electrical resistance thermometers and thermocouples, etc., are used for measuring temperature. Among all, the most popular is the **liquid-in-glass thermometer** (Fig. 1.22) which is used commonly for measuring temperature when extremely high accuracy is not required. It consists of a capillary tube made of glass whose one end is connected to a bulb, often filled with mercury or alcohol and another end is sealed. The portion of the tube above the liquid mostly contains the vapour of the same liquid or in some cases an inert gas. The expansion of the liquid in the tube is the function of temperature. Therefore, the height of liquid in the tube dictates the corresponding temperature of the system when the thermometer is brought to thermal equilibrium with it.

The **constant volume gas thermometer** is another important device that is internationally accepted as the standard for the calibration of other thermometers. The actual setup of this device is elaborate and the procedure of temperature measurement is also cumbersome. Therefore, it is not preferred for routine measurement of temperature. In order to demonstrate the principle of this device, an illustration is depicted in Fig. 1.23. It consists of a constant volume chamber that is filled with a gas (usually hydrogen and helium) at low pressure and a pressure gauge (pressure measuring device) is mounted on the chamber. This chamber is made to contact a system whose temperature is to be measured. For the gaseous medium at low pressure, the temperature is directly proportional to the pressure at constant volume. This results in a linear relationship between the temperature and pressure, therefore, by measuring the pressure of the chamber, the temperature of the system can be obtained. There are also plenty of other devices and sensors for measuring temperature. These are beyond the scope of this book.

After measuring the variation of temperature dependent property, once again we are back to the original question of assigning the corresponding value of temperature using some **temperature scale.** In order to define a temperature scale, a numerical value should be assigned to a standard fixed point with reference to which other temperatures can be related. The standard fixed point should be easily reproducible. For defining all the temperature scales, the ice point and the boiling point of the water are used as the easy reproducible states. The state at which air saturated with vapour is in equilibrium with mixture of ice and water at 1 atm pressure, is called the ice point. The equilibrium between liquid water with its vapour at 1 atm pressure, it termed the boiling point. In SI system, the scale of temperature is **Celsius scale.** As per this scale, the ice point and the boiling point are assigned the values of 0 and 100°C, respectively. The symbol "°" is used in convention of celsius scale because symbol C (Coulomb) is used for unit of electric charge. In English system, the **Fahrenheit scale** is used and it assigns the values of 32F and 212F to the corresponding standard fixed points. Both these scales are defined using a reference fluid (water) and utilize two standard fixed points to define the scale, therefore, these are called the two points scales.

A temperature scale that is independent of the reference fluid is called the **thermodynamic temperature scale**. Such a scale of temperature will be defined using the second law of thermodynamics in Chapter 5. A scale of temperature in SI system that satisfies the thermodynamic temperature scale is **Kelvin scale** and the temperature in kelvin is denoted by K. The smallest temperature for this scale is absolute zero (0K). In English system, a thermodynamic temperature scale is **Rankine scale** and for this scale the temperature is designated by R. The temperature relationship between different scales is depicted in Fig. 1.24 and is expressed as

$$T(K) = T(°C) + 273.15,$$

$$T(R) = T(F) + 459.67,$$

$$T(R) = 1.8T(K),$$

$$T(F) = 1.8T(°C) + 32.$$

Another scale which approximates close to the kelvin scale is the **ideal gas temperature scale**. The temperature for this scale is measured using a constant volume gas thermometer (Fig. 1.23). For such a device, consider the ideal gas behaviour of the medium which gives $T = \alpha + \beta P$. Here, coefficients $\alpha$ and $\beta$ depend on the type of gas and its amount in the chamber. The experiments can be performed for two reference points, ice point and boiling point, to determine the values of coefficients and then the equation can be used to obtain any unknown temperature. When this equation is plotted on the P-T curve (Fig. 1.25) and extrapolated to reach absolute zero pressure, it crosses the temperature axis at -273.15°C or 0K. The slope of the line is different for different gases on the P-T plot, however, upon extrapolation, all lines meet the temperature axis at the same point corresponding to absolute zero pressure. As one point is fixed here, therefore, only a single point temperature is needed to define this temperature scale. A precisely reproducible reference point for defining the temperature scales after 1954 is considered the triple point temperature of water (the temperature at which all three phases, solid, liquid and vapour, coexist in equilibrium) and is assigned the value of 273.16K or 0.01°C. The ideal gas temperature scale cannot be used for extremely low and high temperatures due to issues of gas condensation and ionization, respectively. Moreover, it is also based on the assumption that for all the gas pressures within the range of temperature measurement, the gas behaves as an ideal gas.

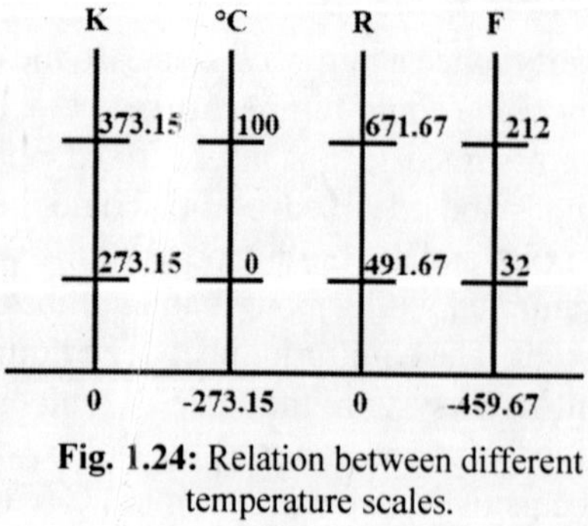


**Fig. 1.24:** Relation between different temperature scales.

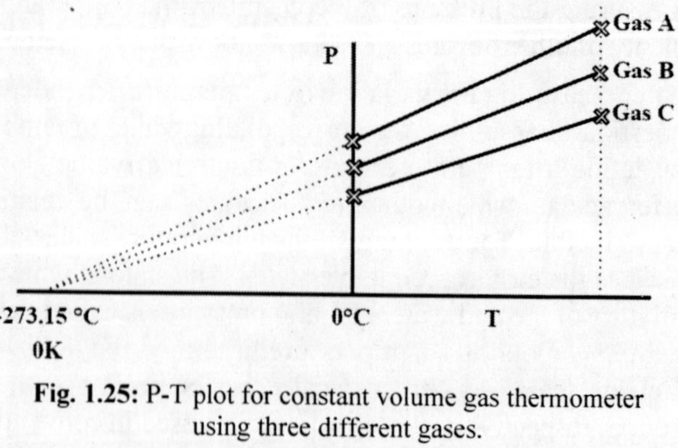


**Fig. 1.25:** P-T plot for constant volume gas thermometer using three different gases.

## UNIT SUMMARY

In this chapter, we introduced the fundamental concepts of the subject thermodynamics. A system of the fixed identity of the matter is called the fixed mass, or control mass, or closed system, whereas a fixed region in space that allows the exchange of mass as well as energy across its boundaries is referred to as the control volume, or open system. A hypothetical system having no exchange of mass and energy across its boundary is called an isolated system. The properties of the system which depend on the mass are called the extensive properties and mass independent properties are termed the intensive properties. The properties of a system are defined only if it is in thermodynamic equilibrium and a unique set of properties of the system, designates its state. When any one of the properties of the system change, it is said to undergo a change of its state. The change of state of the system is called a process and the path of the process is defined by the sequence of intermediate states it crosses before reaching the final state. Such intermediate states are only defined if the process occurs at an infinitesimally slow pace and the process is called the quasi-static, or quasi-equilibrium process. A process for which the initial and final state of the substance becomes the same is called the cyclic process.

The measurable properties which are mostly used to designate the state of the system are specific volume/density, pressure and temperature. The specific is defined as the ratio of system volume per unit mass and its intensive property. A normal force per unit area is referred to as the pressure and it varies with elevation as P = ρgz for an incompressible fluid. For a fluid of variable density with elevation, it is expressed as $dP/dz = \rho(P)$. The pressure of a system close to atmospheric pressure and small pressure differences between systems are measured using a device called the manometer. A device named barometer is used to measure atmospheric pressure. The absolute pressure, gauge pressure and vacuum pressure are related as:

$$P_{gage} = P_{abs} - P_{atm} \text{ (for } P > P_{atm}),$$

$$P_{vac} = P_{atm} - P_{abs} \text{ (for } P < P_{atm}).$$

Temperature is another important property which originates from the zeroth law of thermodynamics. There are a number of devices available to measure temperature and are based on the principle of equality of temperature when two systems are brought in thermal equilibrium with each other. The scales of temperature measurement in different systems are celsius scale, kelvin scale, fahrenheit scale and rankine scale. The relationship between different scales is expressed as

$$T(K) = T(°C) + 273.15,$$

$$T(R) = T(F) + 459.67,$$

$$T(R) = 1.8T(K),$$

$$T(F) = 1.8T(°C) + 32.$$

### Solved Examples

**Example 1.1:** A centrifugal pump is used to elevate water through height $h$ and is powered by an electric motor as shown in Fig. E-1.1. (a) Consider the motor and pump as individual systems and determine the mass and energy interactions. (b) If both the motor and pump are parts of a single system, what will be the type of the system and associated interactions.

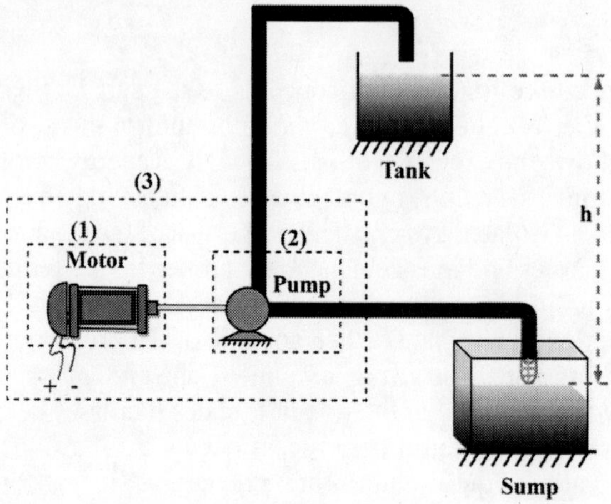


Fig. E-1.1

**Solution:** The schematic of the considered physical situation is given in Fig. E-1.1.

**Assumptions:** In part (a) motor and pump are to be considered individual systems (system 1 and system 2 in Fig. E-1.1).

In part (b), both the motor and pump are part of system 3 as shown in Fig. E-1.1.

**Discussion: (a)** For system 1, which comprises the motor only, there is no mass exchange across its boundary as the flow of electrons is not considered in mass exchange from a macroscopic point of view. Whereas the motor is receiving electrical energy and supplies mechanical energy to the centrifugal pump through a rotating shaft. In addition, there occurs Joule heating in the motor due to the resistance of its windings and hence, there also occurs a transfer of thermal energy to the surroundings. Therefore system 1 is a closed system with no mass interaction and only energy interaction across its boundary.

In the case of system 2, the centrifugal pump continuously takes in the mass of water and supplies it to the discharge line. Therefore, there is mass interaction across its boundary. It also receives energy from the electric motor in the form of mechanical energy, hence there also occurs energy exchange through the boundary of the system. Furthermore, the energy inflow and outflow (which will be detailed in Chapter 2) is also associated with the mass exchange. Thus system 2 is a control volume with mass and energy interactions across its boundary.

**(b)** System 3 includes both motor and pump as its parts. The motor receives electrical energy due to the flow of electrons and there also occurs mass exchange in the pump. Hence, system 3 is also a control volume with mass as well as energy exchange across its boundary.

**Example 1.2:** A pan of water is taken out from an oven and is well insulated such that it doesn't have any energy interactions with its surroundings. The temperature of water within the pan has different values at different locations as depicted in Fig. E-1.2. Can the thermodynamic state be defined if the water inside the container is considered the system?

**Solution:** The thermodynamic state of the given system is to be defined.

**Assumptions:** As the system is perfectly insulated, therefore, there is no mass and energy transfer across its boundary.

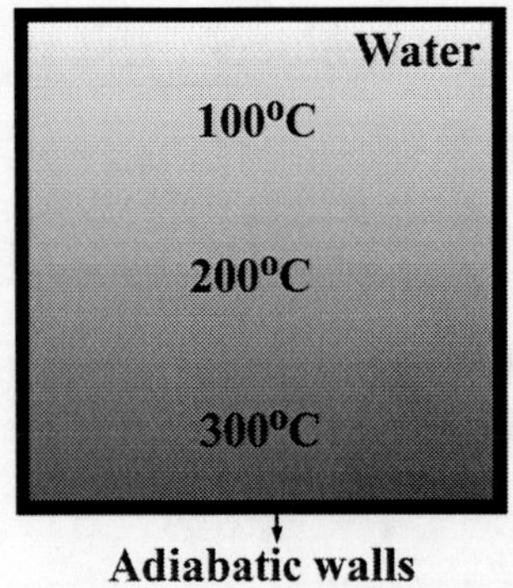

Fig. E-1.2

**Discussion:** The state of the system is only defined when it is in thermodynamic equilibrium. In the given condition, there exist temperature gradients within it, it is not in thermal equilibrium. Therefore, the thermodynamic state of the system cannot be defined.

It is important to note that there will be energy exchange within the system due to temperature differences, till it reaches the condition of thermal equilibrium, i.e. a state of identical temperature throughout the system. After reaching this condition, the state of the system can be defined.

**Example 1.3:** Consider a head of 10 m is maintained in a reservoir by using a dam of 50 m width (Fig. E-1.3-a). Determine the net amount of force exerted by water on the dam.

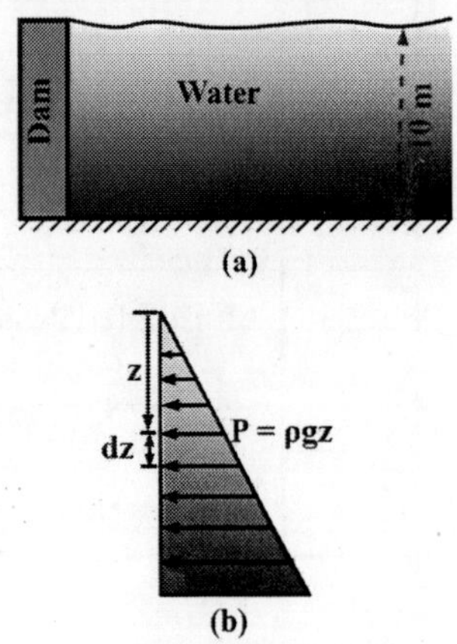

Fig. E-1.3

**Solution:** The net force exerted by water on the dam is to be determined. The dimensions of the dam are 10 m × 50 m.

**Assumptions:** The water inside the reservoir is considered to be static and under this assumption, the pressure exerted by water along the width of the dam is uniform at a given elevation.

Water is considered incompressible, i.e. there is no variation in its density upon change of pressure.

The density of water, $\rho = 1000\ kg/m^3$, $g = 9.81\ m/s^2$.

**Analysis:** The pressure varies in a linear manner with elevation as shown in Fig. 1.3-b. The pressure at the free surface is equal to $P_{atm}$ and at any arbitrary depth $z$ is $P_{atm} + \rho g z$. As the other side of the dam is exposed to atmospheric conditions, therefore, pressure on that side is $P_{atm}$. Thus a differential element of $dz$ thickness at depth $z$ is subjected to a net pressure of $\rho g z$. This pressure will be uniform along the width of the dam at this elevation. Therefore, the net force acting on the differential element is $dF = \rho g z\ dA$ where $dA = b\ dz$. The total force acting on the dam can be obtained by integrating the differential force in the limit of $z = 0$, at the free surface to $z = 10$ m, at the deepest point. Thus $F = \int_{z=0}^{z=10} dF = \int_0^{10} \rho g z \times b\ dz = \rho g b \left[\frac{z^2}{2}\right]_0^{10} = 24525\ kN$.

**Example 1.4:** Two systems A and B having a pressure difference of 25 kPa are connected to a U-tube manometer as shown in Fig. E-1.4. If the density of fluids filled in systems A and B are $\rho_1 = 1000\ kg/m^3$ and $\rho_2 = 700\ kg/m^3$, respectively, determine the density of the manometric fluid. Also, comment which system has more pressure.

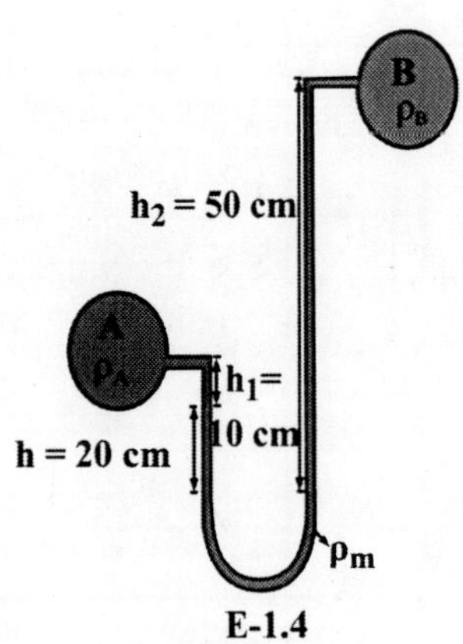

Fig. E-1.4

**Solution:** The density of the manometric fluid is to be determined for the given conditions as shown in Fig. E-1.4.

**Assumptions:** Consider that all the fluids are incompressible. As the manometric fluid has risen in the left branch of the U-tube manometer, therefore, $P_B > P_A$ and $P_B - P_A = 25\ kPa$.

$\rho_1 = 1000\ kg/m^3,\ \rho_2 = 700\ kg/m^3,\ h_1 = 10\ cm,\ h_2 = 50\ cm$ and $h = 20\ cm.$

**Analysis:** The pressure balance can be expressed as

$P_A + \rho_1 g h_1 + \rho_m g h = P_B + \rho\ \ g h_2$. This results in

$\rho_m = \frac{P_B - P_A + \rho_2 g h_2 - \rho_1 g h_1}{gh} = 13992.10\ kg/m^3.$

It is to emphasize that one should be able to determine the system having high pressure by judging the configuration of the manometric fluid.

## EXERCISES

### Multiple Choice Questions

| | | | | |
|---|---|---|---|---|
| 1.1 | Which approach of analyzing a thermodynamic system is called classical thermodynamics? | | | |
| | A. Microscopic | B. Macroscopic | C. Statistical | D. All of the above |
| 1.2 | An isolated system has undergone a process, which of the following statements is correct?<br>1. No heat transfer takes place across the boundary of the system.<br>2. No work transfer takes place across the boundary of the system.<br>3. No mass transfer takes place across the boundary of the system.<br>4. No chemical reaction takes place within the system. | | | |
| | A. 1, 2 & 3 | B. 2, 3 & 4 | C. 1, 2 & 4 | D. 1, 3 & 4 |
| 1.3 | The state of the system is defined only if it is in | | | |
| | A. Thermal Equilibrium | B. Thermodynamic Equilibrium | C. Mechanical Equilibrium | D. Chemical Equilibrium |
| 1.4 | Specific properties of the system are always | | | |
| | A. Extensive | B. Intensive | C. Change upon a change in the amount of the system | D. Depend on the mass of the system |
| 1.5 | A system is divided into four equal halves, the intensive properties of the system will | | | |

| | | | | |
|---|---|---|---|---|
| | A. become half | B. become quarter | C. remain unchanged | D. depend on which property (P or T or e, etc.) is being studied |
| 1.6 | The zeroth law of thermodynamics specifies | | | |
| | A. Thermodynamic equilibrium | B. Thermal Equilibrium | C. Mechanical equilibrium | D. Chemical equilibrium |
| 1.7 | A single point temperature scale is defined is based on | | | |
| | A. Triple point of water | B. Boiling point of water | C. Critical temperature of water | D. Ice point of water |
| 1.8 | Which of the following dimensionless number defines the limit of the continuum? | | | |
| | A. Reynolds number | B. Knudsen number | C. Relative density | D. Specific gravity |
| 1.9 | Which instrument is used for measuring atmospheric pressure? | | | |
| | A. Manometer | B. Thermometer | C. Barometer | D. Rheometer |
| 1.10 | Consider the following properties:<br>1. Temperature, 2. Viscosity, 3. Specific entropy, 4. Thermal conductivity, 5. Energy, 6. Volume<br>Which of the above properties of a system are intensive? | | | |
| | A. 1, 2, 3, 4 | B. 2, 3, 4, 5 | C. 3, 4, 5, 6 | D. 1, 2, 5, 6 |
| 1.11 | A control volume/open system is | | | |
| | A. a specified amount of mass in space | B. an isolated system | C. a closed system that allows energy interaction | D. a fixed region in space across whose boundary mass and energy can cross |
| 1.12 | Hot water contained in a thermos flask is an example of | | | |

| | | | | |
|---|---|---|---|---|
| | A. Non-flow diabatic system | B. Isolated system | C. Open system | D. Closed system |
| 1.13 | For a thermodynamic cyclic process, the difference of properties between final and initial states should be | | | |
| | A. positive | B. zero | C. negative | D. non-zero |
| 1.14 | A submarine of 12 m height travels at a depth of 200 m. Determine the pressure difference between its bottom and top surfaces in kPa. | | | |
| | A. 177.72 | B. 1962 | C. 2079.72 | D. 1844.28 |
| 1.15 | The temperature of an object has decreased by 20°C during its cooling. The equivalent decrease in temperature of the object is | | | |
| | A. 293.15K | B. 253.15K | C. 20K | D. 38K |

**Answers of Multiple Choice Question**

**1.1 B, 1.2 A, 1.3 B, 1.4 B, 1.5 C, 1.6 B, 1.7 A, 1.8 B, 1.9 C, 1.10 A, 1.11 D, 1.12 B, 1.13 B, 1.14 A, 1.15 C**

**Short and Long Answer Type Questions**

1.1 Explain the difference between macroscopic and microscopic approaches for analyzing a thermodynamic system.

1.2 Explain different kinds of thermodynamic systems with real-life examples.

1.3 List down three situations in which conversion of potential energy takes place into kinetic energy and vice versa. Also, illustrate the detailed energy conversion mechanism.

1.4 Explain precisely the relationship between state and property of the system.

1.5 State the clear distinction between a process, a quasi-static process, a cyclic process, an intermediate equilibrium state and the path of the process.

1.6 Define continuum and explain how it assists in the thermodynamic analysis of a system.

1.7 Explain the variation in the specific volume of liquids and gases upon change of pressure and temperature.

1.8 Consider a tank of height $h$ is filled with water. Draw the pressure profile acting on the sidewall of the tank considering the free surface of the water is exposed to atmospheric pressure.

1.9 Explain how a static column of water stays in equilibrium under the varying state of pressure with elevation.

1.10 Can a manometric fluid of lighter density than the system fluid be used for the measurement of the pressure difference between two systems? If yes, explain how, using a schematic of the arrangement.

1.11 For the same range of temperature measurements, the length of the alcohol thermometer is more than the mercury thermometer. Explain why?

1.12 Which among the alcohol and mercury thermometers is more sensitive and why?

**Numerical Problems**

1.1 Determine the number of $O_2$ molecules in a container of 10 Liters volume at standard temperature and pressure condition. Also, comment if this system can be analyzed using macroscopic or microscopic approach.

1.2 A container of 1 $m^3$ volume is filled with 0.3 $m^3$ solid of density 1500 $kg/m^3$. The remaining portion of the container is filled with water having 1000 $kg/m^3$ density. Determine the average specific volume of the container.

1.3 In an automobile workshop, a hydraulic lift of 0.25 $m$ cylinder diameter and piston arm of 150 $kg$ is used to lift a car of 1000 $kg$. Determine the required fluid pressure to lift the car.

1.4 Consider a container filled with different liquids as shown in Fig. P-1.4. Determine the pressure at points A and B and indicate its direction. The density of water is 1000 $kg/m^3$ and specific gravity of mercury is 13.6.

1.5 The density of a liquid varies with elevation as $\rho = \rho_o(1+z)^{0.25}$ where $\rho_o = 950\ kg/m^3$ at the free surface of the liquid. Determine the pressure difference between a point at 10 $m$ deep in the liquid column and the free surface.

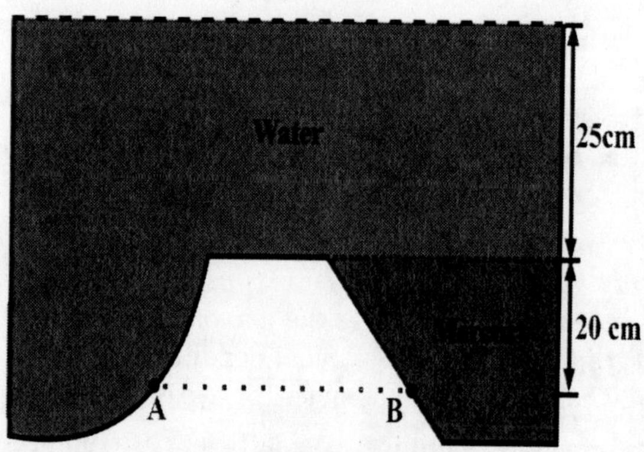


Fig. P-1.5

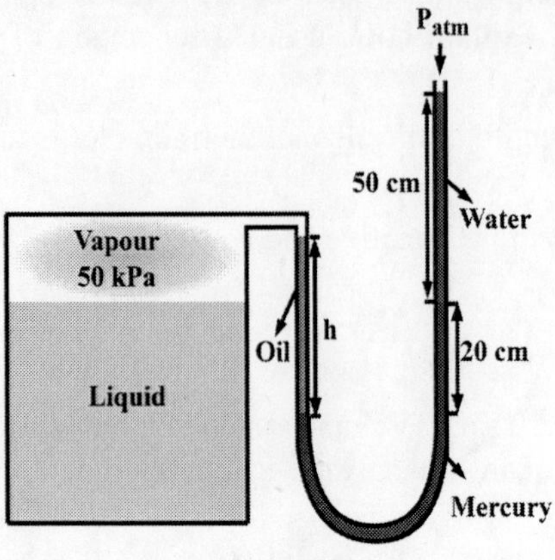


Fig. P-1.6

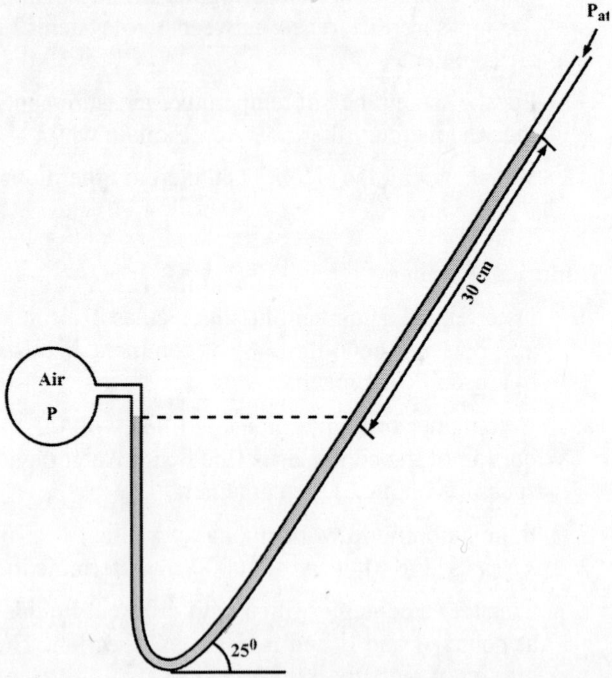


Fig. P-1.7

1.6 A rigid tank contains a mixture of liquid and vapor of a substance at a pressure of 50 $kPa$. The tank is connected to a manometer as shown in Fig. P-1.6. Determine the height $h$ of oil of specific gravity 0.75 in the manometer tube.

1.7 A system containing air is connected to an inclined U-tube manometer as shown in Fig. P-1.7. If the manometric fluid is water, determine the absolute pressure of the system.

1.8 A dam is constructed to store water in a 10 $m$ deep reservoir. A gate of 5 $m$ width and 5 $m$ height is installed at the bottom of the dam to control the water level within the reservoir. Determine the net horizontal force required to support the gate if the other side of the gate is exposed to atmospheric pressure.

1.9 The density of mercury varies with temperature as $\rho = 13600 - 2.75T\ (kg/m^3)$. Here, T is in °C. Consider a manometer with mercury as the manometric fluid is used to measure a pressure difference of 150 $kPa$ in two different ambient conditions of 40°C and 4°C. What will be the difference of mercurey height in the manometer for two different conditions?

1.10 An experiment is performed to determine the coefficient of volume expansion of mercury using a liquid-in-glass thermometer. In this experiment, initially, the thermometer is dipped into a pool of 30°C and rise in height of the thermometer is recorded. The temperature of the water pool is raised at a constant step of 10°C till 70°C and the corresponding height of the mercury level is recorded (Table P-1.10). Determine the sensitivity of the thermometer.

Table P-1.10

| Temperature (°C) | Height of mercury (cm) |
|---|---|
| 30 | 11.2 |
| 40 | 12.6 |
| 50 | 14.1 |
| 60 | 15.7 |
| 70 | 17.1 |

## PRACTICAL

**Activity 1:** List down all the engineering devices being used in your home and categorize each device into different kinds of thermodynamic systems. Also, list all the mass and energy interactions across the boundary of the selected system.

**Activity 2:** Find the different methods used to measure the specific volume/density of the liquids and gases. Determine experimentally the density of water using household stuff.

**Activity 3:** Consolidate a list of different pressure measurement devices and also comment on their accuracy of measurement. Determine the length of the tubes of a U-tube manometer that can be used to measure the pressure difference between the suction and discharge sides of a centrifugal pump used for pumping water at home. Consider the manometric fluid being used in mercury.

**Activity 4:** List down the different temperature measurement devices and determine the associated thermometric property. Use this information to develop a crude instrument for measuring the temperature difference between hot and cold glasses of water.

## REFERENCES AND SUGGESTED READINGS

1. Girolami, G.S., 2019. A brief history of thermodynamics, as illustrated by books and people. *Journal of Chemical & Engineering Data, 65*(2), pp.298-311.
2. Battino, R., Wood, S.E. and Strong, L.E., 1997. A brief history of thermodynamics notation. *Journal of chemical education, 74*(3), p.304.
3. Bejan, A., 2016. *Advanced engineering thermodynamics*. John Wiley & Sons.
4. Saslow, W.M., 2020. A history of thermodynamics: the missing manual. *Entropy, 22*(1), p.77.

# 2 Energy and the First Law of Thermodynamics

## UNIT SPECIFICS

*Through this unit we have discussed the following aspects:*

- *Energy and its different forms;*
- *Constituents of internal energy;*
- *Heat and work;*
- *Different forms of work transfer;*
- *First law of thermodynamics;*
- *Enthalpy;*
- *Specific heats;*
- *Internal energy, enthalpy and specific heats of solids, liquids and gases.*

*The basic introduction to energy and its interaction between system and surroundings in different forms are discussed for generating the basis for analyzing the fundamental thermodynamic systems and related practical applications.*

*Besides giving a large number of multiple choice questions as well as questions of short and long answer types, assignments through a number of numerical problems, a list of references and suggested readings are given in the unit so that one can go through these for practice.*

## RATIONALE

*The fundamental understanding of energy at the macroscopic and microscopic level is crucial to analyse the thermodynamioc systems and processes. The processes involve energy interaction between system and surroundings in the form of heat, work and mass flow are discussed. The different mechanisms of heat and work exchange are described to precisely understand the difference between heat and work transfer. In addition to internal energy, enthalpy is recognized as an important property of the system which is relevant for analyzing thermodynamic processes often in power generation and refrigeration applications. Specific heat of a substance is another important property which tells about the heat storage capacity of materials. Specific heat in its two different forms has been elucidated for solids, liquids and gases and its relevance is established here for obtaing the changes in internal energy and enthalpy of the substance while analyzing the thermodynamic processes.*

## PRE-REQUISITES

*Understanding of thermodynamic systems, thermodynamic state and properties.*

## UNIT OUTCOMES

*List of outcomes of this unit is as follows:*

*U2-O1: Appreciation of the energy at macroscopic and microscopic level*

*U2-O2: Understanding the precise distinction between heat and work*

*U2-O3: Learn the modes of heat transfer and different forms of work*

*U2-O4: First law of thermodynamics and its applications*

*U2-O5: Study the applicability of internal energy, enthalpy and specific heat for different substances*

| Unit-2 Outcomes | EXPECTED MAPPING WITH COURSE OUTCOMES (1- Weak Correlation; 2- Medium correlation; 3- Strong Correlation) | | | | | |
|---|---|---|---|---|---|---|
| | CO-1 | CO-2 | CO-3 | CO-4 | CO-5 | CO-6 |
| U2-O1 | 3 | 3 | 1 | 1 | 1 | 2 |
| U2-O2 | 2 | 3 | 2 | 1 | 1 | 2 |
| U2-O3 | 2 | 3 | 2 | 1 | 2 | 3 |
| U2-O4 | 3 | 3 | 2 | 2 | 3 | 3 |
| U2-O5 | 3 | 3 | 2 | 2 | 3 | 3 |

## 2.1 Energy and Its Different Forms

Energy is the fundamental requirement for all the spontaneous as well as engineering processes to occur in a certain way. In fact, all the devices which we use on a daily basis utilize energy in one form or another. A quantity of matter always carries a certain amount of energy and in the subject of thermodynamics, we are more interested in the change of energy when a system undergoes a certain process, rather than the absolute value of energy. The energy associated with a substance is categorized as macroscopic and microscopic energy. The organized form of energy that a system carries relative to some reference is called **macroscopic energy**. For instance, the kinetic energy, a system possesses due to its motion and potential energy due to its elevation in the presence of the gravitational field, are examples of macroscopic energies. The system also possesses some energy that is inherent to its molecular structure and activity, which is called **microscopic energy**. The sum of all microscopic forms of energy is known as the **internal energy** of the system and is directly associated with its thermodynamic state because molecular structure and activities vary upon changes in the thermodynamic properties of the system. The internal energy is denoted by the symbol $U$ and its specific form is designated as $u$. Therefore, the total energy of the system and its specific form can be expressed as

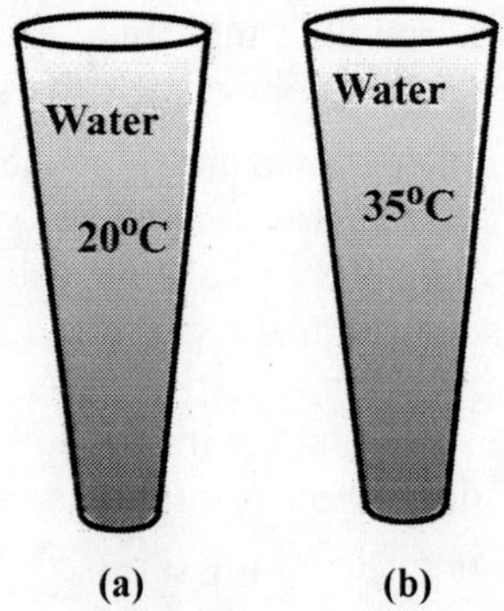


**Fig. 2.1:** A glass of water at (a) 20°C and (b) 35°C held by a person walking at 3 kmph.

$$E = (KE + PE) + U \qquad (J)$$

$$e = E/m = \frac{1}{2}\underline{V}^2 + gz + u \qquad (J/kg)$$

Here, $\underline{V}$ refers to the velocity of the system. If the system is subjected to angular motion, then the rotational kinetic energy of the system is expressed as $\frac{1}{2}\underline{I}\omega^2$ where $\underline{I}$ is the mass moment of inertia and $\omega$ is the angular velocity. The energy equation described here considers only gravity as the external field. If electric, magnetic and surface tension effects are to be considered then more terms are needed in the total energy equation to account for their contributions. However, for the majority of the systems that we study in this subject, such effects are negligible and are not described here. In order to understand the precise distinction between macroscopic and microscopic forms of energy, consider a person walking at 3 kmph on the top floor of his home, holding a glass of water which is at 20°C. Let's consider water within the glass as our system (Fig. 2.1-a). The whole mass of water is moving at a speed of 3 kmph, therefore, it possesses kinetic energy and with reference to the ground, the water has

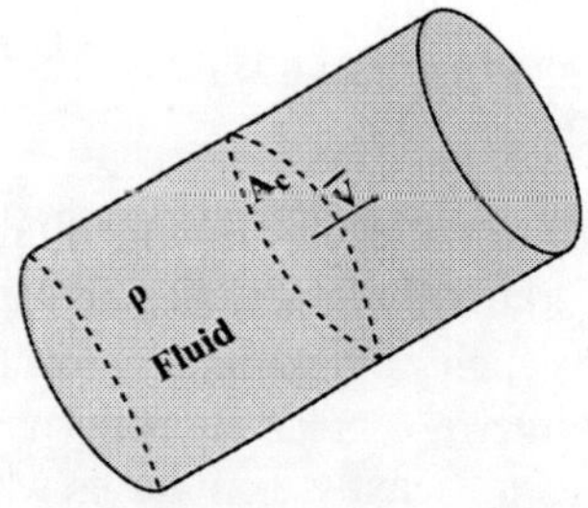


**Fig. 2.2:** Depiction of flow of a fluid of density $\rho$ through a pipe of cross-sectional area $A_c$ moving with velocity $\underline{V}$.

also acquired potential energy. The temperature of 20°C of water corresponds to a certain set of molecular activities within the water and hence, internal energy. If under similar conditions, the temperature of the water is changed to 35°C (Fig. 2.1-b), the molecular activities within the water will change and hence the internal energy of the system will also change, for the same values of its kinetic and potential energies.

In thermodynamics, the majority of the systems remain stationary while a process occurs, therefore, there is no change in their kinetic and potential energies. For such systems, the change in their total energy ($\Delta E$) is equal to the change in internal energy ($\Delta U$). While analyzing the open systems, there exists mass flow across the boundary of the system, it also possesses some inflow and outflow of energy. The mass flow rate of any fluid of density $\rho$ through a pipe of cross-sectional area $A_c$ at an average velocity of $\underline{V}$ is expressed as $\dot{m} = \rho A_c \underline{V}\ (kg/s)$ (Fig. 2.2). Here $\dot{m} = \frac{dm}{dt}$ is the mass flowing through the cross-section of the pipe per unit time. The symbol dot over any quantity is used to represent the time rate of change of that quantity throughout the book. The rate at which energy carried by this flowing mass will become $\dot{E} = \dot{m}e\ (J/s\ or\ W)$ where $e\ (J/kg)$ is the specific total energy possessed by the fluid at the given physical and thermodynamic conditions. The physical conditions such as velocity and elevation of the flowing stream, control the macroscopic energies and the thermodynamic conditions govern the internal energy.

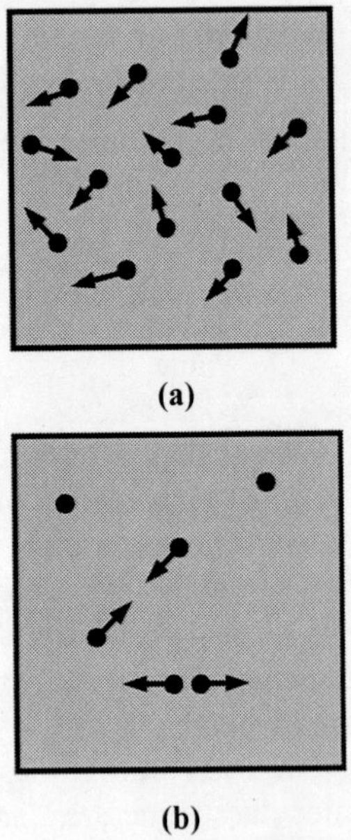

**Fig. 2.3:** (a) Representation of translational motion of molecules and (b) depiction of intermolecular attraction and repulsion forces.

### 2.1.1 Constituents of internal energy

At a macroscopic level any medium contains many molecules. The energy associated with a single molecule and between molecular interactions is the fundamental step in constituting the total internal energy of a substance. Therefore, total internal energy can be expressed as

$$u = u_{trans,molecule} + u_{inter-molecule} + u_{intra-molecule}.$$

The molecules in a medium move with some velocity of their center of mass (Fig. 2.3-a) and possess translational kinetic energy ($u_{trans,molecule}$). There also exits interaction forces between the molecules which contribute to potential energy between them($u_{inter-molecule}$). Such forces depend upon the intermolecular distances as depicted in Fig. 2.3-b. For instance, these forces for solids and liquids differ significantly from gases where molecules are far apart and for a thin gas, inter-molecular potential can be neglected except for the molecular collisions. The term $u_{intra-molecule}$ includes all the energies associated with atoms, nucleus, electrons, etc. and is given as

$$u_{intra-molecule} = u_{potential} + u_{rotation} + u_{vibration} + u_{atom}.$$

The binding forces between the atoms of a polyatomic molecule results in potential energy within a molecule (Fig. 2.4-a). The atoms within a polyatomic molecule also rotate about an axis and contribute to rotational kinetic energy (Fig. 2.4-b). Moreover, the atoms may also vibrate about the center of mass of the molecule and this to and fro motion results in vibrational kinetic energy (Fig. 2.4-c). This is mainly observed in solids and liquids and for gases, it becomes considerable only at high temperatures. Apart from these, there occurs a lot of activities within the atom and are accounted for together in $u_{atom}$ (Fig. 2.5). These include the orbiting motion of electrons about the nucleus that causes rotational kinetic energy and the spin motion of electrons about their axis, which contributes spin energy. Spin energy is also possessed by the other particles in the nucleus of an atom. Moreover, the neutrons in the nucleus are bound with positively charged protons by strong nuclear forces.

The sum of all forms of kinetic energies (described above) that are associated with molecules, is called the **sensible energy**. The average velocity of gas molecules is proportional to their temperature. Therefore, upon the increase of temperature, the sensible energy, hence, the internal energy of the gas increases. The energy required to break the binding force between the molecules of a substance is called **latent energy**. This energy breaks the bond between molecules of solids and liquids, and molecules of such substances behave like gas with increased internal energy. On the other hand, the energy associated with the breakage of bonds between atoms of a molecule and to form new molecules is called **chemical energy**.

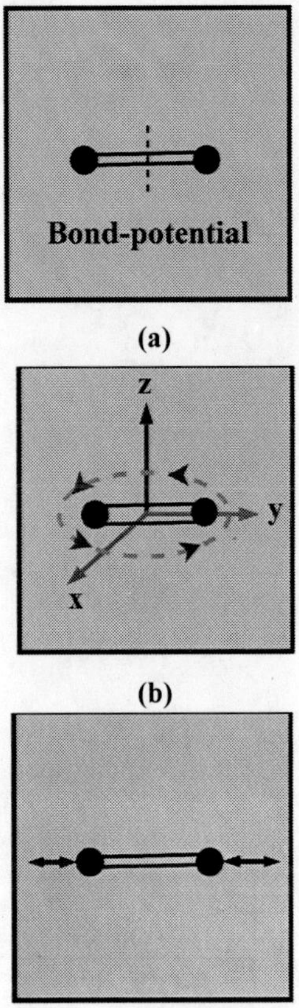


**Fig. 2.4:** Depiction of (a) bond potential, (b) rotational motion and (c) vibrational motion for a polyatomic molecule.

Such a process involves chemical reactions which govern the breakage of reactant molecules and formation of new molecules as product and hence, changes in internal energy of the system. Finally, the enormous amount of energy associated with the strong bonding between protons and neutrons within a nucleus of the atom is called **nuclear energy**. The changes in internal energy due to changes in these components mainly occur during the nuclear fission and fusion reactions. However, these are beyond the scope of this subject. Unlike chemical reactions, the atom loses its identity during a nuclear reaction. The atoms also change their activity upon the inclusion of electric and magnetic fields, however, for thermodynamic analysis, these effects will not be considered.

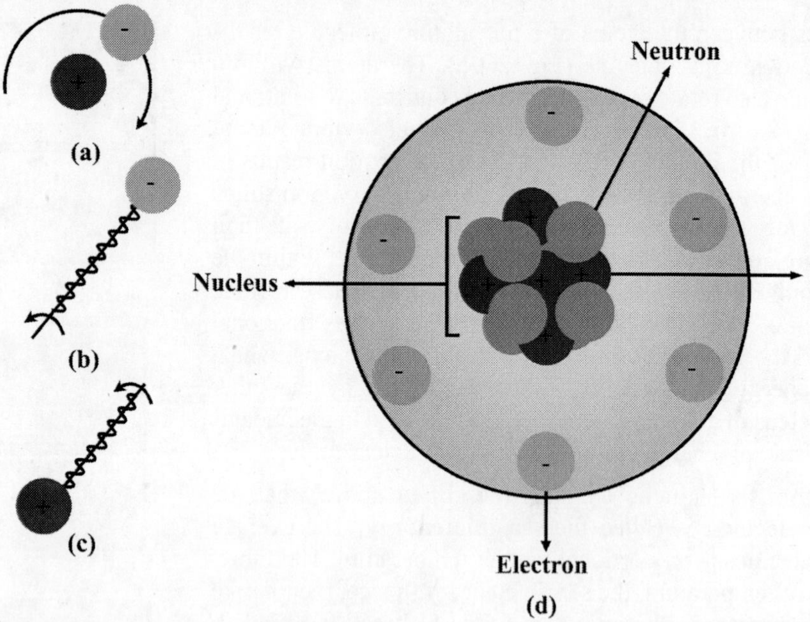


**Fig. 2.5:** (a) Orbiting motion of electron about the nucleus, (b) spin motion of electron, (c) spin motion of proton and (d) the binding of protons and neutrons in the nucleus of an atom.

### 2.1.2 Heat and work

The different constituents of the internal energy discussed in the previous section, represent the energy that is stored in a system. When a system undergoes energy interactions across its boundary, the internal energy of the system changes. The energy interactions across the boundary of the system, always occur during a process and result in an increase or decrease in the internal energy content of the system, depending upon the direction of net energy exchange, in and out of the system. There may exist situations in which the net inflow of energy is equal to the net outflow across the system boundary during a process, hence its internal energy content remains unchanged. It is important to mention that the internal energy stored in a system corresponds to its thermodynamic state, whereas the energy interactions across the boundary of the system, come into action only when a system is undergoing a process, therefore, such energy exchanges cannot be associated to a unique thermodynamic state of the system. There are two kinds of energy interactions, **heat** and **work**, that take place across the boundary of the closed systems. A transfer of energy through a system boundary that is derived by the temperature difference is called heat transfer. Energy exchange other than temperature difference is referred to as the work transfer. In the case of control volumes, there occurs mass exchange across the system boundary and it also causes energy inflow and outflow in addition to heat and work transfer.

Scan Me

For more details on internal energy.

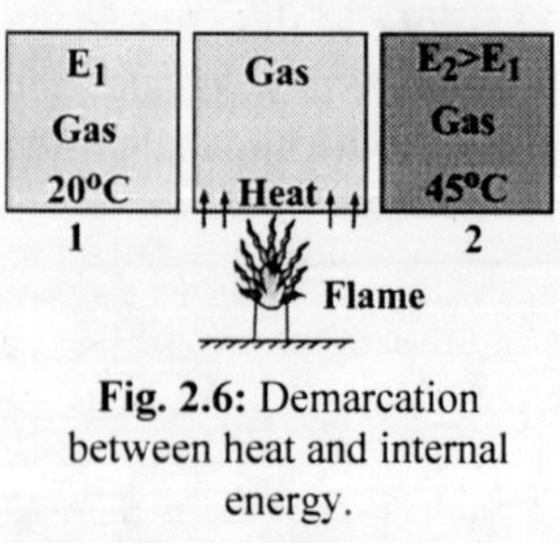


**Fig. 2.6:** Demarcation between heat and internal energy.

An illustration is presented in Fig. 2.6 to demarcate the difference between internal energy and heat transfer. It shows a fixed boundary closed system filled with a gas at a temperature of 20°C and the total energy of the system corresponds to this state is $E_1$. Now the bottom surface of the system is made to have thermal contact with another object which is at a higher temperature than the system temperature. This temperature difference results in the transfer of energy from high-temperature object to the system. Such energy exchange is heat transfer. During the heat transfer process, the energy content of the system increases continuously. The thermal contact is broken when the system has reached 45°C. Corresponding to this point, the energy of the system is $E_2 > E_1$. Therefore, the change in energy of the system between state points 1 and 2 can be expressed as $\Delta E = E_2 - E_1 = \Delta U$ because there is no change in the physical state of the system i.e. its motion and elevation, therefore, the changes in kinetic and potential energies are negligible. If a thermometer is connected to the system, it will read fixed values of temperatures at the initial and final states of the system when thermal contact is not there to facilitate the heat transfer. However, its reading will keep increasing continuously during the heat addition process. During the heat addition process, the system does not stay at unique properties, therefore, its state cannot be defined. Moreover, the amount of heat transfer cannot be linked to the initial and final points of the system as it did not exist corresponding to these points. Therefore it has an association with the process that occurred between the initial and final points. Due to this reason, the heat is termed **dynamic energy** which comes into consideration when the system is in the transient phase of changing its thermodynamic states. Whereas the energy stored in the system, in the form of its internal energy, is called **static energy** and it can be defined at the initial and final state points of the system. The work transfer is also a form of dynamic energy and it is discussed in the later section.

## 2.2 Heat versus Work

It is well understood from the previous section that heat and work are transitional energies that are recognized only when passing through the boundary of the system, otherwise, the energy within the system is stored in the form of its internal energy. First, we will emphasize a better understanding of the energy exchange in the form of heat transfer. For instance, if we place a heated pan on water and a cold container of water taken out from the refrigerator in the surrounding air (Fig. 2.7), both reach the temperature of the surrounding air after some time. In the case of the hot water pan, it loses its energy content to reach the state of thermal equilibrium with its surroundings, whereas the cold water container gains the energy from the surrounding air till it reaches the temperature of its surroundings. This exchange of energy between the system and surroundings is caused due to differences in their temperature, therefore, called heat transfer. This also explains that heat always transfers from high temperature to low temperature and vanishes when the thermal equilibrium is reached between the participating media. It means that after the heat has been exchanged through the boundary, it becomes part of the internal energy of the system and surroundings. In other words, if 5 $kJ$ of heat has been transferred from a system to its surroundings, then after the process of heat transfer is completed, the system suffers a loss of 5 $kJ$ in its internal energy, whereas the internal energy of the surroundings increases by the same amount. It once again emphasizes that the heat is recognized only while it crosses the boundary of the system and hence, is commonly referred to as heat transfer instead of heat.

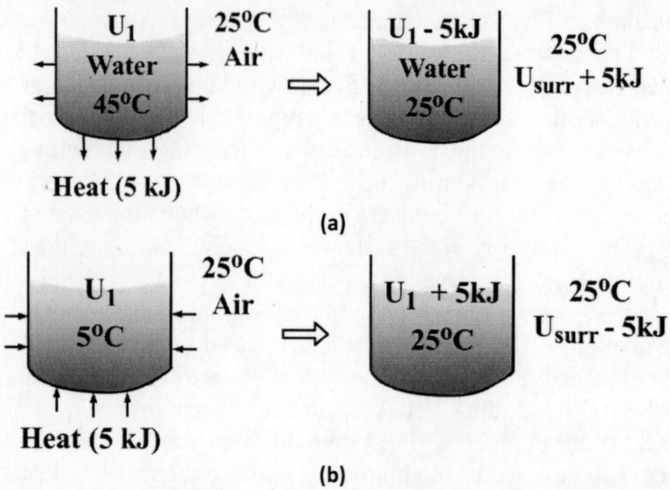


**Fig. 2.7:** Energy exchange in the form of heat transfer when it transfers from (a) system to surroundings and (b) surroundings to system.

Heat is a directional quantity as its direction is also important along with magnitude. Heat added to the system is considered positive and leaving from the system is considered negative. A process in which there is no heat transfer is called the **adiabatic process**. Such processes are possible if there is insulation around the system boundary so that no heat can exchange between the system and surroundings. The same temperature of the system and surroundings during a process also makes it adiabatic as the driving force for heat transfer is zero in this situation. In the case of adiabatic processes, the internal energy content of the system changes due to other means than heat transfer such as work transfer, etc. Heat is denoted by $Q$ and its unit is joule ($J$) as it is a form of energy only. The heat transfer per unit mass is denoted by $q$ and is expressed as $q = Q/m\ (J/kg)$. During the analysis of open systems, it is often required to express heat transfer per unit time, called the heat transfer rate and is denoted by $\dot{Q}\ (J/s\ or\ W)$. For such situations, if the state of the system changes from 1 to 2 in time interval $t_1$ to $t_2$, then the amount of heat transfer is estimated as $Q = \int_{t_1}^{t_2} \dot{Q}\ dt$.

In the thermodynamic analysis of the systems, we are only concerned with the amount and rate of heat transfer, however, the mechanism of heat transfer is not considered. Moreover, for the sake of completeness, the fundamental modes of heat transfer are briefly elucidated here. The heat transfer within a substance due to the molecular interactions from high temperature region to low temperature region is called **conduction**. The rate of heat transfer in conduction is given by Fourier's law and is expressed as $\dot{Q} = -ƙA\frac{dT}{dx}$. Here, ƙ is the thermal conductivity of the substance, $A$ is area normal to the direction of heat transfer and $\frac{dT}{dx}$ is the temperature gradient which is a driving force for heat transfer. The negative sign is used to indicate that heat transfer takes place in a direction of decreasing temperature. The heat exchange between a solid surface and adjacent moving fluid is called **convection** or **convective heat transfer**. In this mode, the heat transfer takes place due to the combined action of conduction and bulk motion of the fluid. The rate of heat transfer in convection is given by Newton's law of cooling and is expressed as $\dot{Q} = h_{conv.}A_s(T_s - T_f)$. The parameter $h_{conv.}$ is convective heat transfer coefficient, $A_s$ is the area of surface exposed to the fluid, $T_s$ is the temperature of the solid surface and $T_f$ is the bulk temperature of the fluid medium. Another mode of heat transfer where the electromagnetic waves emitted by the surface result in the exchange of energy is called **radiation**. It does not require any material medium for the exchange of heat between surfaces. The radiative emissive power of a surface is calculated using Stefan-Boltzmann law and is given as $\dot{Q} = \varepsilon\sigma A T_s^{\,4}$. Here, $\varepsilon$ is the emissivity of the surface, $\sigma$ is Stefan-Boltzmann constant and is equal to $5.67 \times 10^{-8}\ W/m^2 - K^4$, $A$ is area of the surface from where radiations are emitted and $T_s$ is the temperature of the surface. A representation of all three modes of heat transfer is presented in Fig. 2.8.

Work is another form of energy exchange between the system and surroundings which is also dynamic energy like heat. There are a number of mechanisms for work transfer and some of these are discussed in the next section. It is easy to identify the heat transfer as it always occurs due to temperature differences. Therefore, energy exchange, which is caused due to any mechanism other than temperature difference for closed systems, comes under the category of work transfer. Moreover, in the case of an open system, energy exchange also takes place due to mass interaction between the system and its surroundings. Hence, the interaction of energy across the boundary of control volume caused due to any mode other than temperature difference and mass exchange is called the work transfer. Fundamentally the energy associated with a force acting through a certain distance is called the work transfer. Work transfer is denoted by $W$ and its unit is joule $(J)$. The work transfer per unit mass is expressed as $w = \frac{W}{m}\ (J/kg)$. The rate of work transfer is called the power, denoted by $\dot{W}$ and its unit is $J/s$ or $W$. Like heat, work is also a directional quantity, work done by the system is considered positive and work done on the system is considered negative.

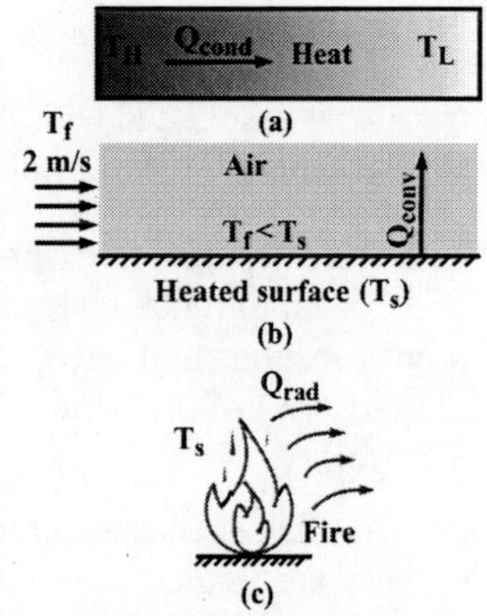


**Fig. 2.8:** Depiction of different modes of heat transfer: (a) conduction, (b) convection and (c) radiation.

It is important to precisely distinguish between work and heat transfer, therefore, an illustration is presented in Fig. 2.9 to elucidate this distinction. It shows a closed container of water in which an electric heater is immersed to heat the water. In the first situation, consider that only water is the system and the heater is part of the surroundings (Fig. 2.9-a). In this situation, the temperature of the heater surface rises when the electric current is passed through it. The difference in temperature between the heater surface and the water results in the transfer of energy from the heater to the water through the system boundary. Therefore, such an exchange of energy is called heat transfer. On the other hand, if both heater and water are considered the system (Fig. 2.9-b), then upon passing the electric current, the flow of electrons takes place through the system boundary into the heater. This generates thermal energy in the heater which itself is part of the system. Such an energy interaction has not occurred due to temperature difference, therefore, termed the work transfer.

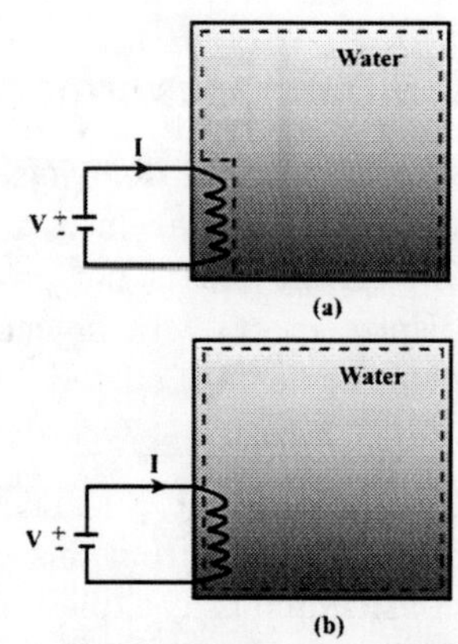


**Fig. 2.9:** Energy exchange through the system boundary in the form of (a) heat and (b) work.

Though the mechanism of energy transfer is different during heat and work transfer, still both occur at the boundaries of the system during a process. These cannot be associated with a unique state of the system, therefore, are not the properties of the system. It signifies that the system can store energy, but not work or heat and there is no meaning of heat and work corresponding to a state of the system. It is also to be mentioned that the magnitude of heat and work depends upon the path that a system undergoes between its initial and final states. In other words, if a system moves through different processes between the same initial and final states, then the amount of heat and work is different for each process (Fig. 2.10). Due to this reason, the work done during different processes A, B and C for a simple compressible substance in Fig. 2.10 can be expressed as $W_{12,A}$, $W_{12,B}$ and $W_{12,C}$, respectively. The letters in subscript are used after the state points to represent the associated process. If any heat transfer is associated with these processes, it is also to be designated in a similar manner, for example $Q_{12,A}$ for process A.

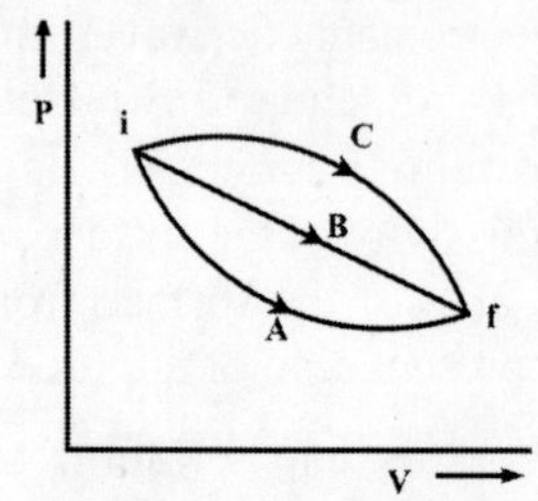

**Fig. 2.10:** Representation of three different processes for a simple compressible substance between the same states.

The system parameters that depend only on the state of the system and not on the path that it has undertaken to reach that state are called **state** or **point functions**. All the properties of the system are point functions and have **exact differentials** represented by the symbol "$d$". The small change in any property ($X$) of the system is denoted by $dX$ and the finite change between the given state points 1 and 2 is expressed as $\Delta X = X_2 - X_1 = \int_1^2 \; dX$ which is independent of the nature of the process that occurred between the same states. On the other hand, the parameters that depend on the path of the process are termed **path functions** and follow **inexact differentials** represented by the symbol "$\delta$". Hence, a small amount of work or heat is denoted as $\delta W$ or $\delta Q$, respectively. The work done between states 1 and 2 during a process is $\int_1^2 \; \delta W = W_{12} \neq W_2 - W_1$ as work corresponding to state points has no meaning. Therefore, the total work done is to be obtained by adding the differential works, $\delta W$ along the path of the process. Hence, the integration of differential amounts of work is path specific and will result in different amounts of total work done for different processes shown in Fig. 2.10.

## 2.3 Different Forms of Work Transfer

The work can be done in several ways, however, as per the classical definition, mechanical work is the product of displacement $x$ of a body with the constant force $F$ responsible for this displacement, i.e., $W = Fx$ ($J$). It is not possible, majority of the time, to have a constant force through the entire displacement of the object. Therefore, the differential amount of work ($\delta W$) associated with a small displacement $dx$ caused due to a force $F$ is given as $\delta W = F\,dx$. In such a situation, the work done is obtained by integrating the differential work, hence, $W = \int_1^2 \; F\,dx$. In order to perform this integration, force $F$ should be known as the function of displacement $x$. Along with the magnitude of work obtained from this equation, it is also important to know its direction based on the physical considerations as explained in Section 2.2. Work done by the system against an external force in the direction opposite to that of displacement is considered positive and work done on the system by an external force acting in the direction of displacement is given the negative sign. It is important to mention that irrespective of the nature of the system, the mechanical work interaction will be zero in situations of displacement of system boundary against no force (such as free expansion or unrestrained expansion of gas) or no displacement of the system boundary. The occurrence of mechanical work in the form of motion of the system boundary or the motion of the entire system is often in thermodynamic systems. Some of the commonly observed work interactions in thermodynamics are elucidated next.

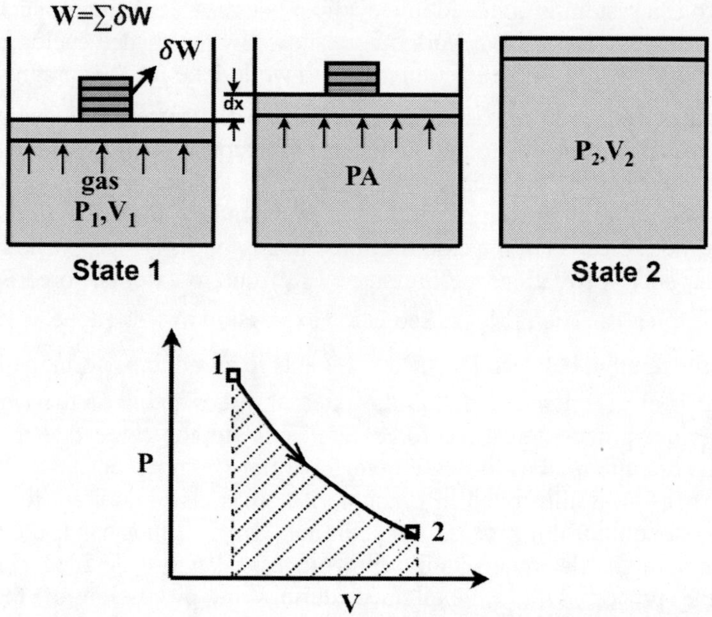


**Fig. 2.11:** Depiction of boundary work during the quasi-equilibrium expansion of a gas.

### 2.3.1 Displacement work

The mechanical work associated with the moving boundary of a system is often called **displacement work** or **moving boundary work** or **boundary work**. It is detailed in this section with the help of a piston cylinder system shown in Fig. 2.11. The gas contained in the cylinder is initially at pressure $P_1$, volume $V_1$ and a number of infinitesimally small weights $\delta W$ are placed on the piston. When one such weight is removed, the gas expands over a small distance $dx$. Considering the quasi-equilibrium expansion of the gas, the external force acting on the piston must be equal to the force exerted due to the gas pressure ($P$) and it is given as $PA$ where $A$ is the area of the piston. The work $\delta W$ is expressed as $\delta W = PA\,dx$. Since $dV = Adx$, therefore, $\delta W = PdV$. When all thc weights are removed one by one, the system reaches state 2 through path 1-2 represented on $P - V$ diagram in Fig. 2.11. Hence the total amount of displacement work can be obtained by integration of differential amounts of work $\delta W$ along the process path between states 1 and 2 as

$$W_{12} = \int_1^2 \delta W = \int_1^2 PdV \tag{2.1}$$

In order to perform this integration, the relationship between $P$ and $V$ should be known for a particular process. Mathematically the area under $P - V$ diagram represents work done as shown by the shaded region in Fig. 2.11. The shaded area in the present case shows the work done by the system as the volume of the system has increased which signifies that the expansion is caused due to the pressure of the gas within the system. If the process is reversed along the same path, then the same area will represent the work done on the system. Another illustration in the earlier Fig. 2.10 shows three different expansion processes. The area under the $P - V$ diagram is different for each process and as the magnitude of work. The path dependence of work is the key characteristic which makes it possible for cyclic devices (such as power plants and engines, etc.) to produce work, otherwise the net work output would have been zero if a system is operated in a cyclic process. A depiction of such a cycle is shown in Fig. 2.12 which produces a net positive work output shown by the shaded enclosed region. Here, the work done during the expansion is more in comparison to work done for the compression process.

The automotive engines and compressors primarily involve the piston cylinder system, hence the moving boundary work. However, due to the high speed of piston in such systems, it is not possible to calculate the boundary work using the thermodynamic analysis as the process is not quasi-equilibrium. During the non-quasi-equilibrium movement of the system boundary, the forces due to gas pressure $PA$ inside the piston will not be equal to the external force acting on it. Hence, work cannot be obtained using Eqn. 2.1. Under such a situation, the pressure ($P_{ext}$) due to external force $F_{ext}$ is obtained by dividing it with the piston area. Then $P_{ext}$ is used in the expression of work done as $W_{12} = \int_1^2 \delta W = \int_1^2 P_{ext} dV$. Here, the relation between $P_{ext}$ and volume is required to perform the integration. As the amount of work done is always associated with the external forces acting on the system boundary and for the quasi-equilibrium process, external forces are equal to the force due to gas pressure $PA$. Therefore, Eqn. 2.1 is often termed as **indicated work**. A more comprehensive understanding of non-quasi-equilibrium expansion is illustrated in Example 2.1. It is also important to note that the work output during the quasi-equilibrium process is maximum. On the other hand, the work required for compression is minimum for the quasi-equilibrium process. A more detailed elucidation for such characteristics of the process will be explained during the discussion of the second law of thermodynamics in Chapter 5.

In general, if we measure the pressure and volume during a process, it can be either numerically integrated to calculate the boundary work or a relation can be obtained between pressure and volume using curve-fitting, which can be further substituted for integration in expression of work. A more useful way of mathematically obtaining the boundary work is to estimate the path of the process with the help of an analytical expression. One such simple mathematical relation to represent expansion and compression of gases is expressed as $PV^n = C$ and is called a **polytropic process**. It involves two constants $n$ and $C$. For a polytropic process $P = CV^{-n}$, the boundary work can be expressed as

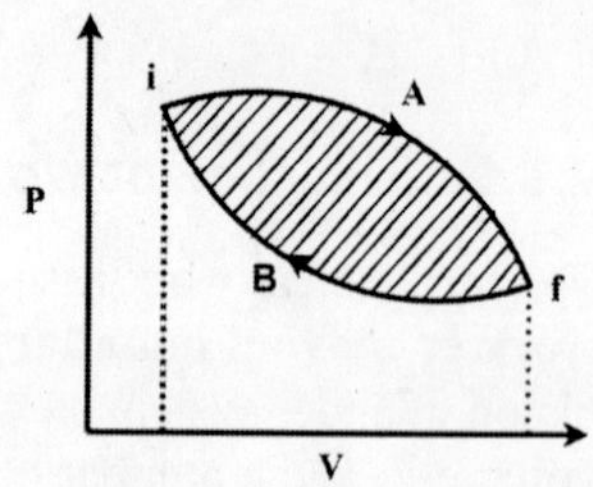

**Fig. 2.12:** Representation of net work done in a thermodynamic cycle.

$$W = \int_1^2 PdV = \int_1^2 CV^{-n}dV = C\left[\frac{V^{-n+1}}{-n+1}\right]_1^2 = C\left[\frac{V_2^{-n+1} - V_1^{-n+1}}{1-n}\right].$$

Since $P_1V_1^n = P_2V_2^n = C$, therefore,

$$W = \frac{P_2V_2 - P_1V_1}{1-n} \tag{2.2}$$

Equation (2.2) holds good for $n \neq 1$ and for $n = 1$, $P = CV^{-1}$, therefore,

$$W = \int_1^2 CV^{-1}dV = C\, ln\frac{V_2}{V_1} = P_1V_1\, ln\frac{V_2}{V_1} \tag{2.3}$$

The polytropic process is usually followed by a simple compressible substance, however, if the working fluid is an ideal gas, then it follows equation of state $PV = mRT$, the boundary work can be expressed as

$$W = \frac{mR(T_2-T_1)}{1-n}; \qquad n \neq 1 \qquad (2.4)$$

$$W = mRT\, ln\frac{V_2}{V_1}; \qquad n = 1 \qquad (2.5)$$

In the expression of the ideal gas, $m$ is the mass of the gas, $R$ is characteristic gas constant and $T$ is temperature in kelvin (K). The characteristic gas constant is given as $R = \frac{\underline{R}}{M}$, where $\underline{R} = 8.3145\ kJ/kmol - K$ is the universal gas constant and $M$ is the molecular mass of the gas. Equations (2.3 & 2.5) are equivalent to the expression of work in an isothermal process, for which the ideal gas equation results in $PV = C$. Therefore, for an isothermal process, polytropic index $n = 1$. The expression of work and associated polytropic index for isobaric, isochoric and adiabatic processes are elucidated in Example 2.2.

## 2.3.2 Shaft work

The transmission of energy by rotating shaft is widespread in engineering devices such as automotive engines, power plants, compressors, etc. The torque ($T$) responsible for the rotation of the shaft is often constant for the majority of the applications and is associated with a force $F$ acting at a normal distance $r$ from the center of the shaft as shown in Fig. 2.13. As per the fundamental definition of torque $T = Fr$. Consider that shaft has gone through an angular displacement $d\theta$ and corresponding linear displacement at the rotor surface is $r\ d\theta$. Thus the differential amount of work associated with this displacement is $\delta W = F\ rd\theta = Td\theta$. If the angular displacement $d\theta$ has taken place in time $dt$, then the rate of work transfer or power is expressed as

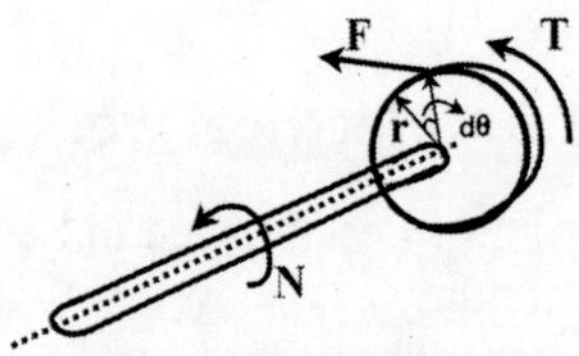


**Fig. 2.13:** A rotor subjected to torque and produces shaft work.

$$\dot{W} = \frac{\delta W}{dt} = T\frac{d\theta}{dt} = T\omega \qquad (2.6)$$

Here $\omega$ is the angular velocity of the shaft and is given as $\omega = \frac{2\pi N}{60}\ rad/s$ where $N$ is the revolution of the shaft per minute.

### 2.3.3 Spring work

When spring is subjected to a force, it changes its length. The displacement $dx$ in the spring corresponds to a force $F$ results $\delta W = F\,dx$. The spring stays at its rest position when it is not subjected to any force as shown in Fig. 2.14. The displacement of spring corresponding to any force is always measured with respect to its neutral position. In the case of linear elastic spring, the applied force is proportional to its displacement, i.e. $F = kx$ where $k$ is the spring constant and its unit is $N/m$. Consider a spring is subjected to displacements $x_1$ and $x_2$ and corresponding forces are $F_1$ and $F_2$, respectively (Fig. 2.14). The spring work for this displacement can be expressed as

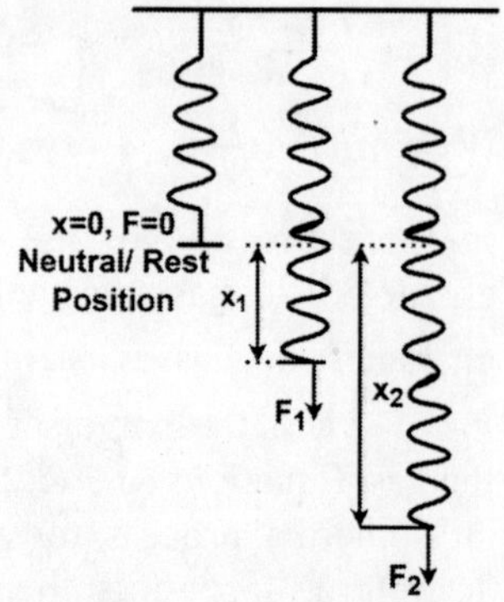


Fig. 2.14: A linear elastic spring at its neutral position corresponding to zero force and show different displacements when subjected to different forces.

$$W = \int_{x_1}^{x_2} F\,dx = \int_{x_1}^{x_2} kx\,dx = \frac{1}{2}k(x_2^2 - x_1^2) \quad (2.7)$$

Next we describe electrical work which is a non-mechanical form of work.

### 2.3.4 Electrical work

It is well established in Section 2.2 that any energy exchange across the boundary of the system that is not due to temperature difference is called work. Therefore, the energy exchange associated with the flow of electrons across the boundary of the system is termed electrical work (Fig. 2.15). In the presence of the electric field, electrons move in a wire under the influence of electromotive forces and do work. The electrical work done by $N$ coulombs electric charge moving under the potential difference $V$ is expressed as $W_e = VN$. The rate of electrical work is given as

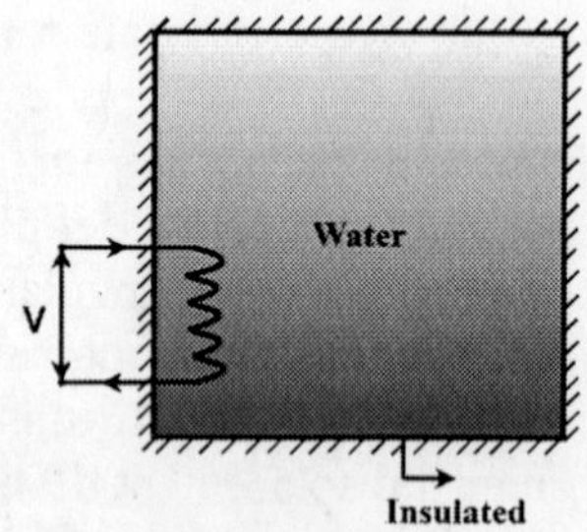


Fig. 2.15: Work done by flow of electric charge through the boundary of the system.

$$\dot{W}_e = VI \quad (2.8)$$

Here, $\dot{W}_e$ is called the electrical power, the number of electrical charges moving per unit time is electric current ($I$) and its unit is ampere ($A$).

There are other forms of non-mechanical work such as magnetic work and electrical polarization work, etc. which are beyond the scope of this book.

## 2.4 First Law of Thermodynamics

We described the various forms of energy in the previous sections, however, the relationship between these and their interactions for different processes is formed by the first law of thermodynamics. In the nineteenth century, James Prescott Joule demonstrated that the different forms of energy are fundamentally the same and can be converted from one form to another. This discovery formed the basis for the first law of thermodynamics which is nothing but the law of conservation of energy. The famous experimental demonstration by Joule involved a paddle wheel within an insulated container of water and a set of pulleys and strings to which weights are attached, connected to the paddle wheel. A simplified schematic representation of Joule's experiment is shown in Fig. 2.16. In his experiment, Joule raised the weights to a suitable height and allowed them to fall slowly. In this process, the paddle wheel rotated and stirred the water. The frictional heat generated within the system due to the stirring of the water caused a rise in its temperature. The total energy change associated with the change in the thermodynamic state of water is found equivalent to the mechanical work done on the system in terms of lowering the weights. The change in the total energy of the water is only dependent on the magnitude of the work and is independent of the mechanism by which work is done. This experiment established an important finding that **the net amount of work done for a closed system is the same for all adiabatic processes occurring between two specified states, irrespective of the nature of the system and the kind of the processes.**

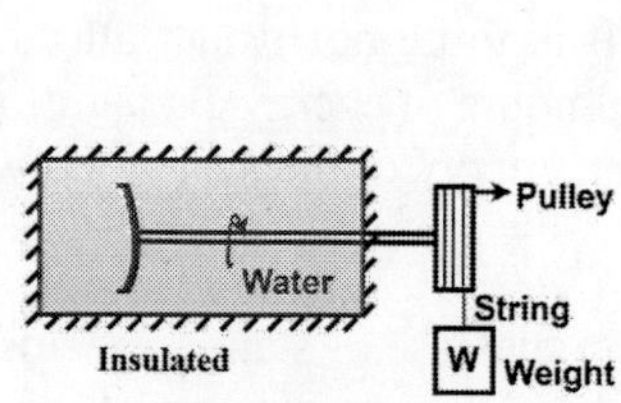


**Fig. 2.16:** Schematic representation of Joule's experiment.

This experimentally driven statement cannot be obtained from any known physical principle, therefore, recognized as the fundamental principle, called the **first law of thermodynamics**. As the work done for all adiabatic processes between two specified states is the same for a closed system, hence it must only depend on the end states of the system. Thus, it is associated with a change in the property of the system, recognizing the total energy as the consequence of the first law of thermodynamics. It is important to mention that as we referred to change in the property of the system between two state points, therefore, defining the absolute value of total energy at a thermodynamic state is not of relevance here. However, any arbitrary value assigned for total energy at a state acts as a reference for defining total energy at other state points.

In addition to defining the total energy, the first law of thermodynamics also holds the conservation of energy principle. It states that the difference between the total energy entering and total energy leaving the system during a process is equal to the net change in total energy of the system. Thus $E_{in} - E_{out} = \Delta E_{system}$ where $E_{in}$and $E_{out}$ represent the total energy entering and leaving the system, respectively across its boundary by means of heat and work transfer, etc. The term $\Delta E_{system}$ represents the change in energy of the system due to the net exchange of energy between system and surroundings. For instance, consider a rigid container of water at 25°C contains 5 $J$ of energy. Now a heat of 15 $J$ is added to raise its temperature and a heat loss of 3 $J$ takes place from system to surroundings during the heating process. In this situation, the net heat transfer to the system is 12 $J$. Therefore, the final energy of the system will be 17 $J$. Further, in the rate form, the energy conservation principle can be expressed as $\dot{E}_{in} - \dot{E}_{out} = dE_{system}/dt$. This principle holds good for any type of the process irrespective of the nature of the system. The energy can enter and leave the system through its boundary in the form of heat transfer, work transfer and energy carried by flowing mass (Fig. 2.17). Thus energy conservation can be written as

$$Q_{in} - Q_{out} + W_{in} - W_{out} + E_{mass,in} - E_{mass,out} = \Delta E_{system} \tag{2.9}$$

It is to be noted that all the quantities on the left hand side of Eqn. (2.9) are positive, represent the amount of energy exchange and their direction is accounted for in subscripts "in" and "out". The change in energy of the system between thermodynamic states 1 and 2 can be given as

$$\Delta E_{system} = E_2 - E_1 = \Delta U + \Delta KE + \Delta PE \quad (2.10)$$

In the case of stationary closed systems, the change in kinetic energy and potential energy will be zero. Also, the energy exchange associated with mass flow will be zero as mass cannot pass across the boundary of the closed systems. Thus energy balance for such system can be expressed as

$$Q_{in} - Q_{out} + W_{in} - W_{out} = \Delta U \quad (2.11)$$

The heat transfer for an adiabatic process becomes zero. As the heat and work transfer are inexact differentials, therefore, Eqn. (2.11) in differential form can be given as

$$\delta Q - \delta W = dU \quad (2.12)$$

Consider a substance in a closed system undergoing a cyclic process. For such a substance the change in energy of the system is zero, therefore, the energy balance can be written as

$$\oint \delta Q - \oint \delta W = 0 \quad (2.13)$$

Here, $\oint \delta Q$ represents the net heat transfer during the cyclic process and is called the **cyclic integral of heat transfer** and $\oint \delta W$ denotes the net work output of the cycle and is termed the **cyclic integral of the work**. Equation (2.13) gives $\oint \delta Q = \oint \delta W$ and is called the first law of thermodynamics for a cyclic process. This equation is a consequence of the energy balance, however, historically it was proposed first for heat engines, where the work output is obtained upon adding heat from a source and the energy equation was derived from it. Refrigerator and heat pump are the other cyclic devices where work is input and heat transfer is the output. Therefore, it can be generalized that all the cyclic processes represent the energy conversion devices where the energy is conserved, however, it changes from one form to another.

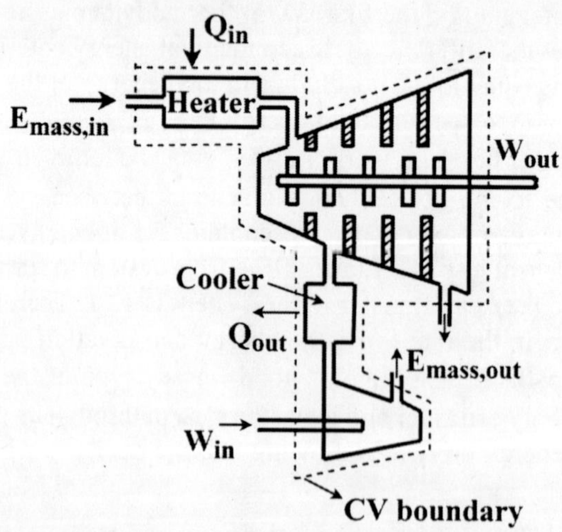


**Fig. 2.17:** Different possible modes of energy exchange through the boundary of the system.

## 2.5 Enthalpy: A Thermodynamic Property

While analyzing the certain type of processes, often in power generation and refrigeration applications, we come across the combination of the properties, which can be regarded as the property of the substance undergoing the change of state. One such property is illustrated here with the help of a quasi-equilibrium process as shown in Fig. 2.18. It shows a piston cylinder system filled with gas which is subjected to simultaneous heat and work transfer. The work is associated in terms of the expansion of the gas under the constant pressure condition. If the changes in the kinetic and potential energy are neglected, then the first law of thermodynamics for this process can be expressed as $Q_{12} - W_{12} = \Delta U = U_2 - U_1$. For a constant pressure process, $W_{12} = \int_1^2 P\,dV = P(V_2 - V_1)$.

This results in $Q_{12} = U_2 - U_1 + PV_2 - PV_1 = (U_2 + P_2V_2) - (U_1 + P_1V_1)$ since $P_1 = P_2 = P$. Therefore, it can be said that heat transfer for this process is equal to the change in parameter $U + PV$ between the states 1 and 2. As the parameter $U + PV$ involves all the state functions, therefore, it should also exhibit the same characteristics and can be designated as a property of the substance. Thus, this combination of the properties is defined as a new extensive property called the **enthalpy**, $H = U + PV$. In specific form, enthalpy is expressed as $h = u + Pv$. It is important to note that the heat transfer equals to change in enthalpy between the initial and final states in not a general statement and is applicable only for the constant pressure process described here. However, the enthalpy, being a state function, is a property of the system which is applicable for any process irrespective of its nature. The details about the physical understanding of the property enthalpy will be elucidated in Chapter 4 during the energy analysis of the control volume.

## 2.6 Specific Heats

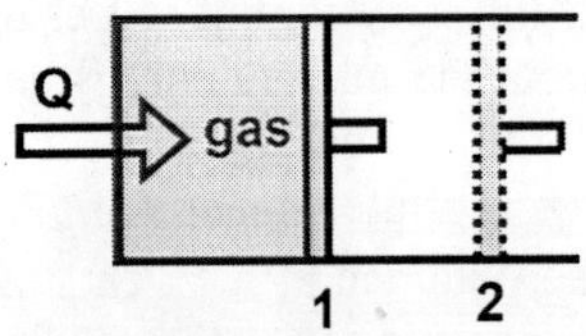


**Fig. 2.18:** Constant pressure expansion of gas with simultaneous heat and work interaction.

If we consider a homogeneous substance of constant composition without any phase change, it requires a certain amount of heat to raise its temperature by 1°C. However, it is known from our experience that different substances require different amounts of heat to change their temperature between the same limits of temperature. For example, 1 $kg$ of water requires 4.18 $kJ$ of energy to raise its temperature by 1°C whereas the energy required by 1 $kg$ of kerosene oil for the same temperature rise is 2.01 $kJ$. This ascertains that the energy storage capabilities are different for distinct substances. A thermodynamic property that defines this characteristic of the substance is called specific heat. The amount of heat required to raise the temperature of the unit mass of a substance by one degree is defined as the **specific heat**. It is important to note that the amount of specific heat also depends upon the nature of the heat addition process. Based on the practical experience, in thermodynamics, mainly our interest is in two kinds of specific heats, i.e. **specific heat at constant volume** ($C_v$) and **specific heat at constant pressure** ($C_p$). The amount of heat needed to raise the temperature of 1 $kg$ of a substance by one degree when its volume is maintained constant is called the specific heat at constant volume. The amount of heat required to do the same at constant pressure is termed the specific heat at constant pressure. An illustration depicting constant volume and constant pressure heating of 1 $kg$ of air is shown in Fig. 2.19. As the specific heat at constant pressure also includes the additional energy needed to expand the substance, therefore, it is always greater than specific heat at constant volume.

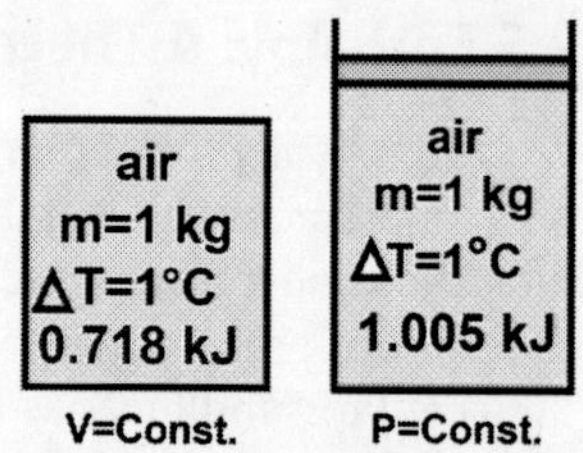


**Fig. 2.19:** Constant volume and constant pressure heating of air.

The specific heat at constant volume is defined for the special case of a simple compressible system when other forms of work including displacement work are absent during the heating process, therefore, $\delta Q = dU$ and $C_v$ is expressed as

$$C_v = \frac{1}{m}\left(\frac{\delta Q}{\delta T}\right)_v = \frac{1}{m}\left(\frac{\partial U}{\partial T}\right)_v = \left(\frac{\partial u}{\partial T}\right)_v \quad (2.14)$$

For the specific heat at constant pressure, the boundary work comes into consideration and heat transfer results in $\delta Q = dU + PdV = dH$ as explained in Section 2.5 for a constant pressure heating process. Thus, specific heat at constant pressure is

$$C_p = \frac{1}{m}\left(\frac{\delta Q}{\delta T}\right)_p = \frac{1}{m}\left(\frac{\partial H}{\partial T}\right)_p = \left(\frac{\partial h}{\partial T}\right)_p \quad (2.15)$$

It is observed that final terms in Eqns. (2.14) and (2.15) contains only the thermodynamic properties, therefore, $C_v$ and $C_p$ should also be thermodynamic properties. Though specific processes are used to derive these parameters, these are applicable for any substance irrespective of the nature of the process. An example is shown in Fig. 2.20 which shows two identical gaseous masses. A heat of 150 $kJ$ is supplied to one mass whereas a work of equivalent amount is done on the other. In both the cases, the change in internal energy of the substance is the same and both occupy the same final state. Therefore, specific heat at constant volume is the same for both cases irrespective of the fact how energy is added. As both the internal energy and enthalpy can be changed by transfer of any form of energy, therefore, the term specific energy seems more appropriate than specific heat. Like other properties, $C_v$ and $C_p$ can also be obtained at a given state by knowing the two independent intensive properties. Therefore, their values depend upon temperature and pressure. For instance, the specific heat at constant volume of air is 0.718 $kJ/kg-K$ at 300K and it is 0.855 $kJ/kg-K$ at 1000K. In SI system, the unit of specific heat is either $J/kg-K$ or $J/kg-°C$ as change in temperature is equal in both the scales of temperature measurement.

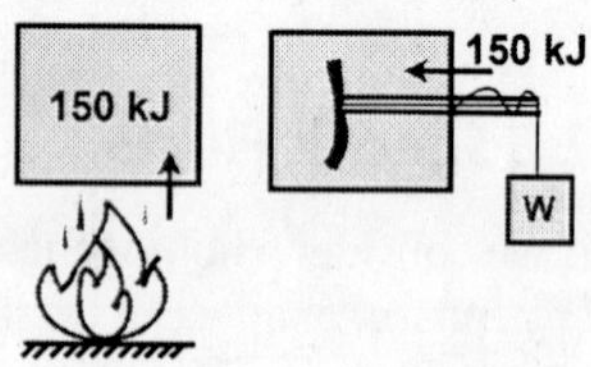


**Fig. 2.20:** Same amount of energy added to identical mass of a gas by heat and work transfer.

## 2.7 Internal Energy, Enthalpy and Specific Heats of Solids and Liquids

A substance of constant density (or specific volume) is referred to as an incompressible substance. The specific volume of solids and liquids remains almost constant during a process, therefore, these can be considered incompressible with sufficient accuracy. Fundamentally, solids and liquids expands upon change of temperature and this property is mainly utilized in liquid-in-glass thermometers. Therefore, the main reason behind calling these incompressible is that the change in energy associated with volume change is negligible when compared to other types of energy. Moreover, the specific heats of incompressible substances are the only function of temperature. Therefore, partial differentials in Eqns. (2.14) and (2.15) can be replaced with ordinary differential. This expresses Eqns. (2.14) and (2.15) as

$$du = C_v(T)dT \text{ and } dh = C_p(T)dT \quad (2.16)$$

As $h = u + Pv$, therefore, its differential form is given as$dh = du + Pdv + vdP$. As $v$ is constant, therefore, $dv = 0$. Also, the specific volume $v$ of solids and liquid is very small, hence, the term $vdP$ can be neglected. This yields $dh \approx du$, therefore, using Eqn. (2.16), it can be inferred that $C_p \approx C_v = C$. The change in internal energy and enthalpy between states 1 and 2 can be obtained as

$$h_2 - h_1 \simeq u_2 - u_1 \simeq \int_1^2 \quad C(T)dT \quad (2.17)$$

Here $C$ should be known as the function of temperature to perform integration of Eqn. (2.17). In the case of small temperature interval between states 1 and 2, $C_{avg}$ can be obtained at average temperature and can be considered constant which yields $h_2 - h_1 \simeq u_2 - u_1 \simeq C_{avg}(T_2 - T_1)$.

## 2.8 Internal Energy, Enthalpy and Specific Heats of Ideal Gases

Like any other property of the system, internal energy also depends upon two independent intensive properties defining the state of the system. However, it has been established experimentally and theoretically that internal energy for an ideal gas solely depends upon temperature. The famous experiment of Joule includes two vessels connected by a pipe and a valve, submerged in a water bath as depicted in Fig. 2.21. Initially, one vessel was evacuated and another was filled with air at high pressure. Then the vessels were allowed to attain thermal equilibrium with the water bath and finally the valve was opened to allow the air flow in the evacuated vessel till the pressure equalized in both vessels. Joule recorded the temperature of the water bath and found that there is no change in temperature of water after the thermal equilibrium was achieved. This established that there is no heat transfer between water and air during its flow into the evacuated vessel. Also, there is no work transfer between air and water. Thus the internal energy of the air did not change when its pressure and volume changed. Therefore, Joule concluded that internal energy of air is a function of temperature only and does not depend upon pressure and volume. Usually the ideal gas behaviour is observed at low pressures and high temperatures, therefore, the low density gases can be treated as ideal gases with sufficient accuracy. The relationship between specific volume, pressure and temperature for an ideal gas is given as $Pv = RT$ where $R$ is characteristic gas constant. The functional dependence of internal energy for an ideal gas can be represented as $u = f(T)$. This functional relationship is not valid for gases which deviate from ideal gas behaviour. The equations of state for such gases will be elucidated in Chapter 3.

Enthalpy $h = u + Pv$ using ideal gas equation $Pv = RT$ can be given as $h = u + RT$. As $R$ is a constant and $u = f(T)$, therefore, $h = f(T)$. As internal energy and enthalpy of ideal gas do not depend upon specific volume and pressure, therefore, Eqns. (2.14) and (2.15) can be expressed as

$$C_{v0} = \frac{du}{dT} \quad (2.18)$$

$$C_{p0} = \frac{dh}{dT} \quad (2.19)$$

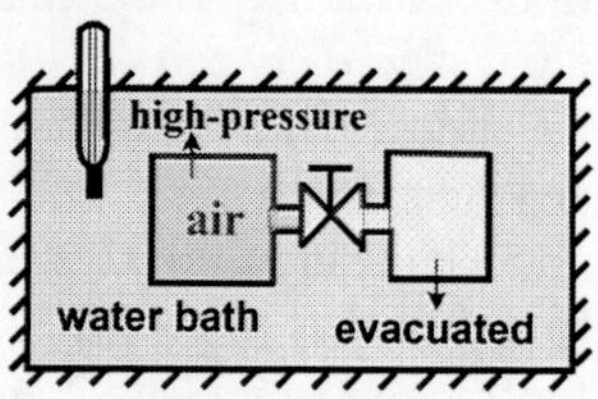


**Fig. 2.21:** Schematic depiction of Joule's experiment.

Here, $C_{v0}$ and $C_{p0}$ specific heats of an ideal gas at constant volume and pressure, respectively. As all the gases depict the ideal gas behavior when pressure approaches zero, therefore, the specific heat of a given ideal substance is also called the zero pressure specific heat and a subscript "0" is used to designate it. The physical interpretation of Eqns. (2.18) and (2.19) is presented in Fig. 2.22 which shows two constant temperature lines $T$ and $T + dT$ on a $P - v$ plot. As for ideal gases, internal energy and enthalpy are also functions of temperature only, therefore, these lines also represent constant internal energy and constant enthalpy. The state 1 having temperature $T$ can be converted to a state of higher temperature $T + dT$ via a number of paths as shown in Fig. 2.22 and the final state, i.e. $P$ and $v$ is different for different paths. As the lines of constant internal energy and constant enthalpy coincide with the constant temperature lines, therefore, change in $u$ and $h$ is the same for all processes. This establishes that irrespective of the pressure and volume, the specific heats of ideal gas ($C_{v0}$ and $C_{p0}$) are also the sole functions of temperature. Fig. 2.23 shows the variation of $C_{p0}/R$ as the function of temperature for a number of gases. These values are obtained using statistical thermodynamics which is beyond the scope of this book. The monoatomic gases in Fig. 2.23 don't show any dependence of temperature on their specific heat. The molecular vibrations are mainly responsible for variation in specific heats with temperature. As the complex molecules show more numbers of vibrational modes, hence depict significant variation of specific heat with temperature.

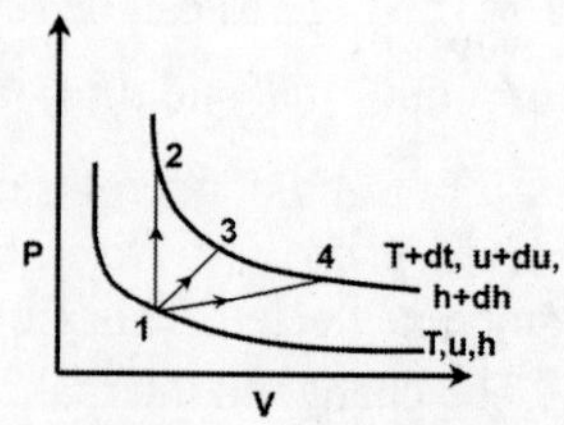


**Fig. 2.22:** Representation of isotherms on $P - V$ diagram for ideal gases.

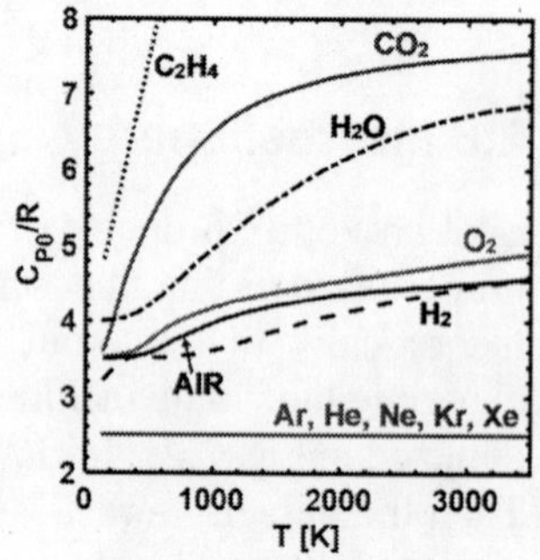


**Fig. 2.23:** Variation of constant pressure specific heat as a function of temperature for different gases.

The enthalpy relation can be employed to develop the relationship between constant volume specific heat and constant pressure specific heat for ideal gases. The differential form of $h = u + Pv = u + RT$ can be given as $dh = du + RdT$. If we substitute $du$ and $dh$ from Eqns. (2.18) and (2.19) and divide throughout by $dT$, we get

$$C_{p0} = C_{v0} + R \quad (2.20)$$

In Eqn. (2.20), $R$ is the characteristic gas constant. This explains that the difference between $C_{p0}$ and $C_{v0}$ is constant irrespective of their dependence on temperature. It also establishes that by studying the dependence of only one specific with temperature, another can be obtained using Eqn. (2.20). Another important parameter is the **ratio of specific heats** ($k$) which is defined as $k = C_{p0}/C_{v0}$. This ratio is constant for monoatomic gases and holds a value of 1.667. It shows mild dependence on temperature for other ideal gases and $k = 1.4$ for many diatomic gases including air at room temperature.

Using Eqns. (2.18) and (2.19), the internal energy and enthalpy in differential form can be written as $du = C_{v0}dT$ and $dh = C_{p0}dT$. If the state of a substance changes from 1 to 2, then changes in internal energy and enthalpy can be obtained using integration as

$$u_2 - u_1 = \int_1^2 C_{v0}(T)dT \quad (2.21)$$

$$h_2 - h_1 = \int_1^2 C_{p0}(T)dT \quad (2.22)$$

The functional relationship between specific heats and temperature is required to perform integration in Eqns. (2.21) and (2.22). If the integration is to be performed over a narrow range of temperature, then using the linear approximation (as shown in Fig. 2.24-a), $C_{v0,avg}$ and $C_{p0,avg}$ can be obtained at average temperature $\left(\frac{T_1+T_2}{2}\right)$ and Eqns. (2.21) and (2.22) can be expressed as

$$u_2 - u_1 = C_{v0,avg}(T_2 - T_1) \quad (2.21)$$

$$h_2 - h_1 = C_{p0,avg}(T_2 - T_1) \quad (2.22)$$

There also exists another way of estimating the average specific heats. In this method, first specific heats are obtained at $T_1$ and $T_2$ and then their average is taken (Fig. 2.24-b). Both these approaches of estimating average specific heats give reasonable accuracy. It is also to be noted that the relations of internal energy and enthalpy described here are applicable for any kind of process and their validity should not be confused with the definition of specific heats at constant volume and constant pressure.

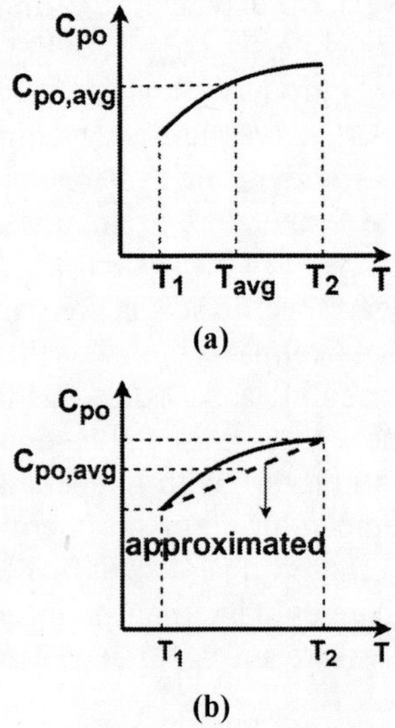


**Fig. 2.24:** (a) Estimation of specific heat at average temperature for small temperature range and (b) use of linear approximation for estimation of specific heat in the small temperature range.

## UNIT SUMMARY

This chapter introduced the concept of energy and explained energy in storage form and energy in transient. The energy stored in a system is present in the form of macroscopic and microscopic energies. The kinetic energy and potential energy are the constituents of macroscopic energy that depends upon the physical state of the system. The internal energy of the system is the microscopic form of energy that varies with change in thermodynamic state of the system. The internal energy is constituted by the molecular interactions and it mainly comprises of the sensible energy, latent energy, chemical energy and nuclear energy. The energy in transient cannot be associated with the thermodynamic state of the system and always occurs at the boundaries of the system while process takes place. Such an exchange of energy across the system boundary is possible in three forms, viz. heat transfer, work transfer and energy transfer with mass exchange. Heat transfer has further three modes which are conduction, convection and radiation. Similarly wrok transfer can also takes place in multiple forms. The most popular kinds of work transfer for thermodynamic systems are boundary/displacement work, shaft work, spring work and electric work, etc. The boundary work can be obtained with the help of polytropic relation ($PV^n = C$) between pressure and volume for a process. The polytropic index $n$ has values 0, 1, 1.4 and ∞ for isobaric, isothermal, adiabatic and isochoric processes, respectively.

Next we explained the first law of thermodynamics which is fundamentally the energy conservation principle. The applicability of this principle is demonstrated for different processes and systems. Then we introduced one more property which is formed by the combination of other properties (internal energy, pressure and volume) and is called enthalpy. This property is important for the analysis of thermodynamic systems. Next the specific heat which represents the amount of heat required to raise the temperature of unit mass of a substance by 1°C is elucidated. The specific heat is recognized as the property of the system which represents the heat storage capacity of the system. Two forms of specific heat, specific heat at constant volume and specific heat at constant pressure are explained and it is found that later has more value than former for any substance. Both the forms of specific heat become identical for solids and liquids due to their incompressible behaviour. The internal energy and enthalpy of the ideal gases is a sole function of temperature, so as the specific heats. Specific heats of monoatomic gases do not vary with temperature whereas for all other gases, these significantly change upon change of temperature. For the thermodynamic analysis of a system over a small temperature range, constant average value of specific heats can be obtained using interpolation and can be utilized for estimation of changes in internal energy and enthalpy. For ideal gases, the difference between specific heat at constant pressure and specific heat at constant volume is equal to characteristic gas constant.

## Solved Examples

**Example 2.1:** Consider a piston-cylinder system having gas at pressure $P_1$ and volume $V_1$ and the piston is held at this location by using a pin (Fig. E-2.1-a). The mass of the piston is $m_p$, its area is $A$ and pressure outside the piston is $P_0$ ($< P_1$). When the pin is released, the piston expands and finally comes to rest. Determine the work done during this expansion process.

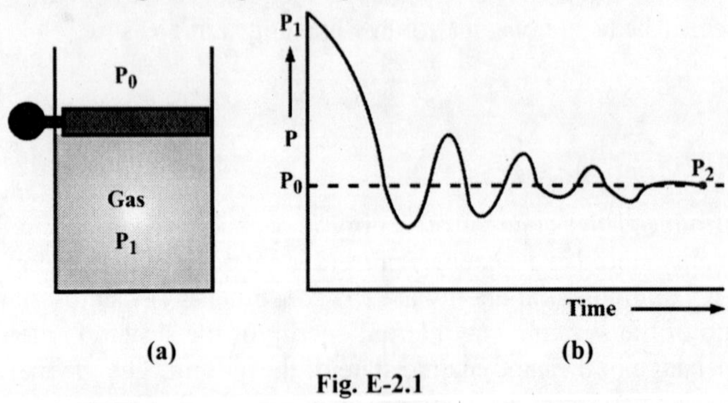


**Fig. E-2.1**

**Solution:**

**Assumptions:** During the expansion of the piston, the gas pressure will decrease, whereas the external force acting on it remains the same as it is exposed to constant surroundings pressure $P_0$ and the gravitational field $g$. Therefore, the external force can be given as $F_{ext} = P_0 A + m_p g$. This gives $P_{ext} = F_{ext}/A = P_0 + m_p g/A$ which is a constant parameter. This ascertains that for expansion to occur $P_1 > P_{ext}$ and piston will come to rest when $P_1 = P_{ext}$. As during the start of the expansion process, the gas pressure is not equal to the resisting pressure, thus the expansion will start at a finite rate and will occur in non-quasi-equilibrium manner. However, the piston expands against the constant external pressure, therefore, the work is done against this and it is to be used in the expression of boundary work.

**Analysis:** Work done in this process is $W = \int_1^2 P_{ext} dV = P_{ext} \int_1^2 dV = P_{ext}(V_2 - V_1)$.

It may be interesting to know the variation of gas pressure during this non-quasi-equilibrium process. If a pressure transducer is mounted on the cylinder and assumes that it measures the average pressure of the gas, it will show pressure variation with time as depicted in Fig. E-2.1-b.

**Example 2.2:** A gas expands in a piston cylinder system in a quasi-equilibrium process. Determine the work done during the expansion process if it is (a) isobaric, (b) isothermal, (c) adiabatic and (d) isochoric. Also, find the polytropic index for each process and represent the expansion processes on $P - V$ diagram.

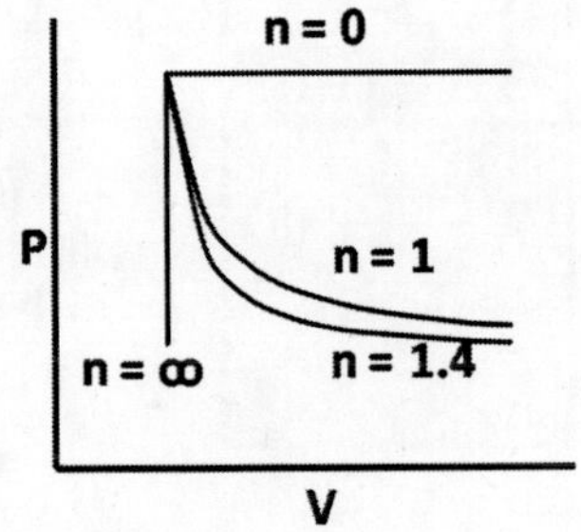


Fig. E-2.2

**Solution:** The expression of work and polytropic index are to be determined for given processes. Finally, all the processes are to be plotted on $P - V$ diagram.

**Assumptions:** Consider the ideal gas behaviour when required. As the process is expansion, therefore, $V_1 > V_2$ where $V_1$and $V_2$ represent the volume of gas at initial and final states, respectively. This assumption is not true for an isochoric process.

**Analysis: (a)** For an isobaric process, pressure remains constant, therefore, $n = 0$.

The work done for an isobaric process is

$W = \int_1^2 \ PdV = P(V_2 - V_1)$.

**(b)** For an isothermal process, temperature remains constant. If we assume the ideal gas behaviour $PV = mRT$, this gives $PV =$ constant. Now, by comparing it with polytropic equation $PV^n = C$, we can say that isothermal process can be represented using polytropic equation when $n = 1$.

The work done for an isothermal process is

$W = \int_1^2 \ PdV = \int_1^2 \ CV^{-1}dV = C\ ln\frac{V_2}{V_1} = P_1V_1\ ln\frac{V_2}{V_1}$.

**(c)** For an adiabatic process, the polytropic index is $n = k$. The detailed elucidation on this concept will be dealt in Chapter 6.

The work done for adiabatic process is

$W = \int_1^2 \ PdV = \int_1^2 \ CV^{-k}dV = C\left[\frac{V^{-k+1}}{-k+1}\right]_1^2 = C\left[\frac{V_2^{-k+1} - V_1^{-k+1}}{1-k}\right]$.

Since $P_1V_1^k = P_2V_2^k = C$, therefore, $W = \frac{P_2V_2 - P_1V_1}{1-k}$.

**(d)** For an isochoric process, volume remains constant. Therefore, $dV = 0$ and boundary work becomes zero. The situation of zero work can be seen as the limiting condition of Eqn. (2.2) when $n \to \infty$.

All the processes are represented on $P - V$ diagram in Fig. E-2.2.

**Example 2.3:** Consider a piston-cylinder system filled with a gaseous substance. A linear spring of stiffness $k$ is attached to the piston as shown in Fig. E-2.3-a. The mass of the piston is $m_p$ and its area is $A$. A constant point force $F_E$ is also acting on the other side of the piston and it is exposed to pressure $P_0$. Determine the work done during the expansion of the gas.

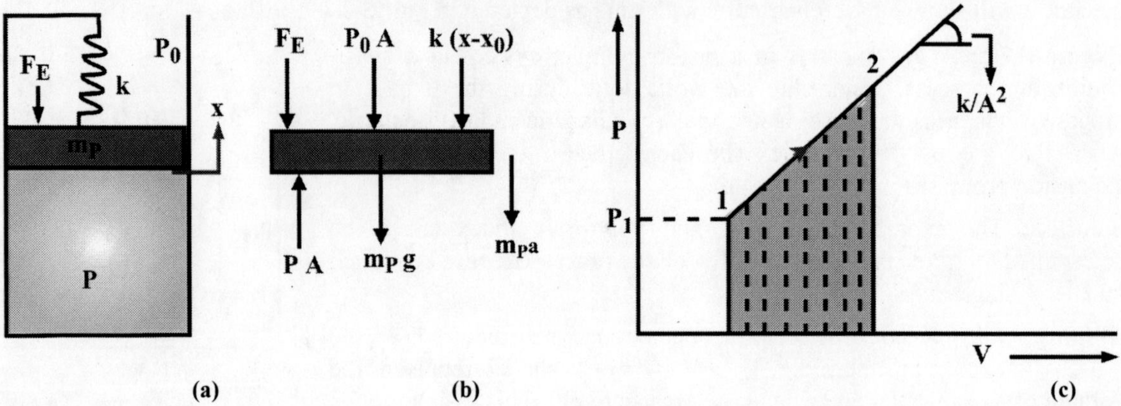


**Fig. E-2.3**

**Solution:** The expression of work is to be obtained for the system shown in Fig. E-2.3-a.

**Assumptions:** Consider the quasi-equilibrium expansion of the gas.

**Analysis:** The free body diagram of the piston is shown in Fig. E-2.3-b. The force balance for the piston results $m_p a = PA - F_E - P_0 A - m_p g - k(x - x_0)$, where $x_0$ is the position of the piston corresponding to the neutral position of the spring. As the process is quasi-equilibrium, thus $a \approx 0$. Hence, $PA = P_0 A + F_E + m_p g + k(x - x_0)$. Further $P = P_0 + \frac{F_E}{A} + \frac{m_p g}{A} + \frac{k(x-x_0)}{A} \times \frac{A}{A}$. Since $V = Ax$, this results $P = P_0 + \frac{F_E}{A} + \frac{m_p g}{A} + \frac{k(V-V_0)}{A^2} = P_0 + \frac{F_E}{A} + \frac{m_p g}{A} - \frac{kV_0}{A^2} + \frac{kV}{A^2}$.

Consider $C_1 = P_0 + \frac{F_E}{A} + \frac{m_p g}{A} - \frac{kV_0}{A^2}$, it includes all the constant quantities and $C_2 = \frac{k}{A^2}$. This yields $P = C_1 + C_2 V$. Therefore, the relation between gas pressure and volume is linear with a slope $C_2 = \frac{k}{A^2}$ and is independent of the medium present in the piston cylinder system. This variation is shown for an expansion process 1-2 on $P - V$ diagram in Fig. E-2.3-c. The work done for this process is $W = \int_1^2 PdV$ = area under the process line between points 1 and 2. Thus, it becomes $W = \frac{1}{2}(P_1 + P_2)(V_2 - V_1)$.

Similarly for the compression of the gas, the process will occur on the same line but in the opposite direction.

**Example 2.4:** Explain in a free expansion process the work done is zero.

**Solution:** Work done is to be estimated for a free expansion process.

**Assumptions:** Consider a pressurized gas separated by a membrane from an evacuated compartment in a vessel as shown in Fig. 2.4-a. When the membrane ruptures, the gas fills the entire vessel and its a non-quasi-equilibrium expansion process. Such an expansion of gas in vacuum against the zero resistance is called free expansion.

**Discussion:** This process can be analyzed in two ways. First consider that only gas is the part of the system (Fig. E-2.4-b). In this situation, the volume of the gas increases upon its expansion. Therefore, one may be tempted to calculate $W = \int \quad PdV$. However, it is to be kept in mind that the process of expansion is non-quasi-equilibrium. Thus, the external force against which the system boundary has expanded is to be considered in the integral equation. For the present case, the system boundary expands against vacuum, hence, the work done is zero.

In the second approach, the entire vessel can be considered the system (Fig. 2.4-c). Here, there is no change in system boundary upon expansion of the gas. Thus, $dV = 0$, so as the work.

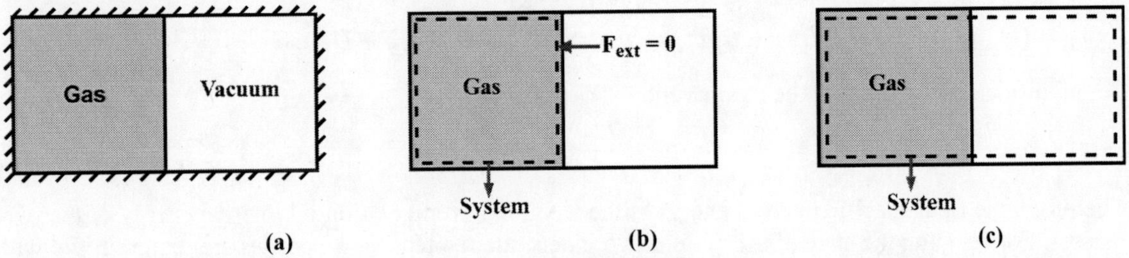


**Fig. E-2.4**

**Example 2.5:** Consider two different masses $m_A$ and $m_B$ of a substance are at different thermodynamic states and contained in two different compartments A and B as shown in Fig. 2.5. The compartments are connected by a pipe and a valve. Now the valve is opened and the two masses are allowed to mix and reach a uniform final state. Determine the final state of the system if the initial states of the two masses are known. What will happen if the two compartments and the additional fittings are insulated.

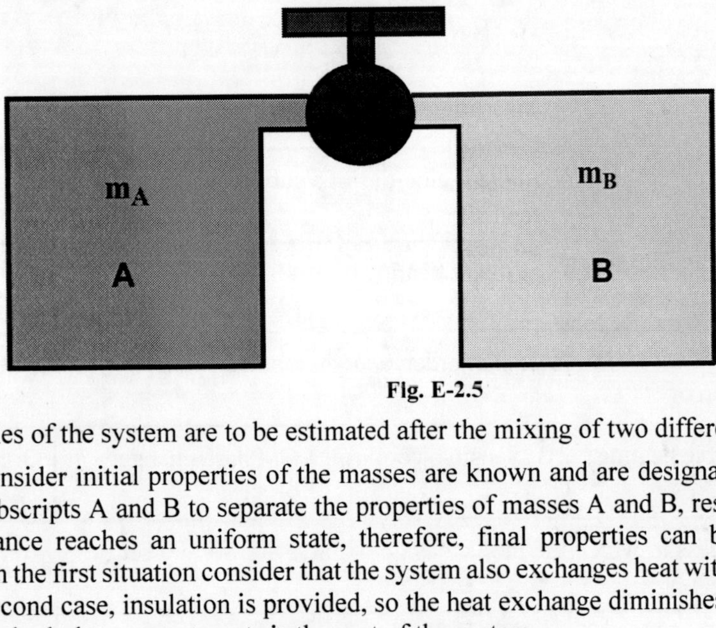


**Fig. E-2.5**

**Solution:** Properties of the system are to be estimated after the mixing of two different masses.

**Assumptions:** Consider initial properties of the masses are known and are designated by subscript 1. Let us also use subscripts A and B to separate the properties of masses A and B, respectively. As after mixing the substance reaches an uniform state, therefore, final properties can be designated with subscript 2 only. In the first situation consider that the system also exchanges heat with the surroundings, whereas for the second case, insulation is provided, so the heat exchange diminishes. Consider that the medium present in both the compartments is the part of the system.

**Analysis:** Applying conservation of mass and energy gives

$$m_2 = m_A + m_B$$

$$E_2 - E_1 = Q - W$$

Assume there are no changes in kinetic and potential energy of the system. Thus $\Delta KE = \Delta PE = 0$ and $\Delta E = \Delta U$. Also the boundary of the system is rigid, hence, $dV = 0$ which implies $W = 0$. Therefore, energy conservation becomes

$$U_2 - U_1 = Q$$

$$\Rightarrow m_2 u_2 - (m_A u_{1A} + m_B u_{1B}) = Q$$

$$\Rightarrow u_2 = (m_A u_{1A} + m_B u_{1B})/m_2 + Q/m_2 \Rightarrow u_2 = \frac{m_A}{m_2} u_{1A} + \frac{m_B}{m_2} u_{1B} + Q/m_2$$

For an insulated boundary of the compartments, $Q = 0$.

$$\Rightarrow u_2 = \frac{m_A}{m_2} u_{1A} + \frac{m_B}{m_2} u_{1B}$$

Therefore, the final specific internal energy is the weighted average of initial internal energy of the two masses. For defining the state of the system two independent intensive properties are required. We will attempt to determine the final specific volume by applying the volume conservation as

$$V_2 = V_{1A} + V_{1B} \Rightarrow m_2 v_2 = m_{1A} V_{1A} + m_{1B} V_{1B}$$

$$v_2 = \frac{m_{1A}}{m_2} V_{1A} + \frac{m_{1B}}{m_2} V_{1B}$$

Final specific volume is also the weighted mass average of the initial specific volumes.

## EXERCISES

### Multiple Choice Questions

| | | | | |
|---|---|---|---|---|
| 2.1 | During a process the molecular velocities of a substance have increased. It indicates that the process is | | | |
| | A. Sensible heating | B. Latent heating | C. Sensible cooling | D. Latent cooling |
| 2.2 | If in a process the molecular order and the spacing between molecules alters, it is an indication of | | | |
| | A. Sensible heating | B. Sensible cooling | C. Phase change process | D. None of these |
| 2.3 | The process in which the molecules dissociates and form a set of new molecules indicates the change of | | | |
| | A. Sensible energy | B. Latent energy | C. Nuclear energy | D. Chemical energy |
| 2.4 | All the forms of molecular kinetic energies contribute to | | | |

| | | | | |
|---|---|---|---|---|
| | A. Sensible energy | B. Latent energy | C. Nuclear energy | D. Chemical energy |
| 2.5 | The dissociation of an atom during a process indicates the change of | | | |
| | A. Sensible energy | B. Latent energy | C. Nuclear energy | D. Chemical energy |
| 2.6 | The binding potential of a monoatomic molecule is | | | |
| | A. Positive | B. Negative | C. Zero | D. Unpredictable |
| 2.7 | The $\oint \quad (\delta Q - \delta W)$ for a cyclic process is | | | |
| | A. Positive | B. Negative | C. Zero | D. Unpredictable |
| 2.8 | Air is contained in a rigid vessel of $2\ \square^3$ volume at 2 bar pressure. It is then heated to raise its pressure to 5 bar. What is the amount of work done during this process? | | | |
| | A. 600 $kJ$ | B. 0 $kJ$ | C. 6 $kJ$ | D. 1000 $kJ$ |
| 2.9 | A heat of 20 $kJ$ is added when a closed system undergoes a process 1-2 and it produces 50 $kJ$ of work. Now the system is returned from state 2 to state 1 and a heat of 10 $kJ$ is taken out of it. What will be the amount of work done during process 2-1? | | | |
| | A. +40 $kJ$ | B. +20 $kJ$ | C. -40 $kJ$ | D. -80 $kJ$ |
| 2.10 | A piston cylinder system is used to compress a gas to one half of its original volume and a heat of 200 $kJ$ is taken out of the gas during this process while its internal energy remains the same. How much amount of work is done on the gas? | | | |
| | A. 100 $kN - m$ | B. 200 $kN - m$ | C. 400 $kN - m$ | D. 0 $kN - m$ |
| 2.11 | A water heater of 5 $kW$ power is used to heat 10 Litres of water for 10 minutes. If the specific heat of water is 4.18 $kJ/kg - $°C, determine the rise in its temperature. | | | |
| | A. 1.43°C | B. 71.77°C | C. 1.20°C | D. 96.77°C |

| 2.12 | An identical mass of an ideal gas is heated at constant pressure and constant volume conditions between the same temperature limits. If the heat supplied at constant pressure and constant volume are $Q_p$ and $Q_v$, respectively, which of the following will be true? | | | |
|---|---|---|---|---|
| | A. $Q_p > Q_v$ | B. $Q_p < Q_v$ | C. $Q_p = Q_v$ | D. Cannot be determined |

**Answers of Multiple Choice Question**

**2.1 A, 2.2 C, 2.3 D, 2.4 A, 2.5 C, 2.6 C, 2.7 C, 2.8 B, 2.9 C, 2.10 B, 2.11 B, 2.12 A**

**Short and Long Answer Type Questions**

2.1 Consider a pan filled with water placed on a gas stove to heat the water. Determine the different energies involved with this process. Also comment on which constituents of the internal energy of water changes significantly in this process if water is heated before the limit of phase transformation.

2.2 What changes happen at the microscopic level to change the internal energy of the system during a chemical reaction.

2.3 What are the different forms in which energy can exchange across the boundary of the system?.

2.4 A perfectly insulated room is heated by an electric heater. What type of energy has exchanged across the system boundary if the heater is considered part of the system?

2.5 A refrigerator with its door open is kept for two hours in an insulated room. What will happen to the temperature of the room air at the end of the two hours duration?

2.6 Represent the isothermal and adiabatic expansion processes on $P - V$ diagram such that their final state is the same and operating between the same volume limits. Which of these processes have the highest initial pressure. Also, represent the displacement work done on $P - V$ diagram and comment which process will produce more work output?

2.7 Does the applicability of specific heats is only limited to constant volume and constant pressure processes? Explain.

2.8 Consider an identical mass of air heated at constant pressure in two situations. It is heated from 295K to 305K in the first case and secondly it is heated from 970K to 980K. Will the amount of heat required be the same in both cases?

2.9 If the working fluid is changed from air to helium in the previous question (1.8), how will the heat required affect in both the cases?

2.10 If an identical mass of an ideal gas is heated at two different constant pressure conditions of 1 atm and 5 atm between the same temperature limits. At which pressure, will it require more heat? Explain.

2.11 Can an ideal gas be expanded isothermally in an adiabatic piston cylinder system? Explain.

2.12 A balloon filled with air is placed in an evacuated and insulated room. When the balloon ruptured, air filled the entire volume of the room. What will happen to the internal energy of the air?

## Numerical Problems

2.1 Water is heated in a rigid tank and a stirrer is used within the tank to keep the water at uniform temperature during the heating process. Water receives a heat of 35 $kJ$ through the bottom surface of the tank and a total heat loss of 7 $kJ$ to the surroundings has taken place through the other surfaces of the tank. The stirrer does a work of 0.5 $kJ$. Calculate the final internal energy of the water if its initial internal energy is 25 $kJ$.

2.2 A 50 Litres vessel contains air at 25°C and 150 $kPa$. The vessel is connected to a piston cylinder system through a valve which is initially empty. The piston is frictionless and its mass is such that it floats corresponding to a pressure of 50 $kPa$. When the valve is opened, the air enters the piston cylinder system till it reaches an uniform state of 25°C and 50 $kPa$. Determine the work done in this process.

2.3 The isobaric expansion of air takes place in a piston cylinder system and gives 45 $kJ$ of work. Air was at 500 $kPa$, 25°C with 0.015 $m^3$ of volume at the start of the expansion process. Determine the final temperature and volume of the air. Also, calculate the amount of heat transfer involved in the process.

2.4 A steel block of 1.5 $kg$ is heated to a temperature of 700°C in a heat treatment process and then quenched in a 100 Litres pool of water which is initially at 25°C. If the entire heat from steel block enters into the water and there is no loss to the surroundings, find the final temperature. The specific heat of steel is 420 $J/kg - °C$ and that of water is 4182 $J/kg - °C$.

2.5 During the expansion of a gas, pressure varies with volume as $P = a + bV$ $(kPa)$ where $b =$ 5 $kPa/m^3$ and $a$ is a constant. If the pressure during the process changes from 25 $kPa$ to 150 $kPa$, determine the work done. The initial volume of the gas is 4 $m^3$.

2.6 Air at 20 bar and 300°C is contained in a piston cylinder device. The mass of the air is 0.12 $kg$ and it is expanded isothermally to a pressure of 5 bar. It is then compressed adiabatically to the initial pressure and further compressed in an isobaric manner to reach the initial state. Determine the work done in each process and the net work of the cycle.

2.7 A piston cylinder device contains 0.5 $kg$ of air at 2.5 bar and 20°C. The piston has cross-sectional area of 0.12 $m^2$. Now the air is heated and the piston encounters a linear spring of 90 $kN/m$ spring constant at volume of 0.22 $m^3$. The process of heating is continued till the piston rises 15 $cm$ distance after touching the spring. Determine the final pressure and temperature of the air. Also, calculate the work done during the process and represent it on $P - V$ diagram.

2.8 A room of 5 $m$ × 5 $m$ × 3 $m$ is provided with an electric resistance heater to raise its temperature from 2°C to 30°C in 20 minutes. The room is perfectly insulated and contains air at atmospheric pressure. Find the required power of the heater. Assume $C_{p0} = 1.005\ kJ/kg - °C$ and $C_{v0} = 0.718\ kJ/kg - °C$.

2.9 A well insulated rigid tank is divided in two parts having volume ratio of 1:5. The small volume contains 4 $kg$ of an ideal gas at 6 bar and 60°C and the large volume is evacuated. Now the partition between the two parts is ruptured and the gas fills the entire volume. Determine the final pressure and temperature of the tank. Also, comment on the work done during this expansion process.

2.10 A set of stops are connected to a piston cylinder device at its top. It initially contains 5 $kg$ of air at 2.5 bar and 20°C. Now the heat is supplied to the air and the piston rises until it touches the stops corresponding to a volume 1.5 times of the initial volume. Heat is further added until the pressures becomes 2.5 times of the initial pressure. Represent the process on $P - V$ diagram. Determine the work done and heat transfer for the process.

## PRACTICAL

**Activity 1:** Consider you are given a task to measure the specific heat of water and you have to design an experimental setup for the same using the easily available stuff. Make a list of the required instruments/components and prepare a detailed procedure for the measurement. Also, perform the detailed energy analysis of the proposed experimental setup and explain how primary measurements can be utilized to obtain specific heat.

## REFERENCES AND SUGGESTED READINGS

1. Bächtold, M., 2021. Introducing Joule's Paddle Wheel Experiment in the Teaching of Energy: Why and How?. *Foundations of Science*, *26*(3), pp.791-805.
2. Callendar, H.L., 1910. The caloric theory of heat and Carnot's principle. *Proceedings of the Physical Society of London (1874-1925)*, *23*(1), p.153.
3. Bejan, A., 2016. *Advanced engineering thermodynamics*. John Wiley & Sons.

# 3 Properties of Pure Substances

## UNIT SPECIFICS

*In this unit following aspects have been discussed:*

- *Definition of pure substances and different phases;*
- *Phase diagrams for representing the variation of phases with thermodynamic properties;*
- *Use of property tables for estimating the change in energy content and other properties;*
- *Definition of compressibility factor;*
- *Summary of the chapter and solved examples.*

*The practical applications of the topics are discussed for generating further curiosity and creativity as well as improving problem solving capacity.*

*Besides giving a large number of multiple choice questions as well as questions of short and long answer types marked in two categories following lower and higher order of Bloom's taxonomy, assignments through a number of numerical problems, a list of references and suggested readings are given in the unit so that one can go through them for practice. It is important to note that for getting more information on various topics of interest some QR codes have been provided in different sections which can be scanned for relevant supportive knowledge.*

## RATIONALE

*After getting the basic concepts of thermodynamic properties and first law of thermodynamics, the next step is the application of these concepts to real world applications. For these applications the properties of working fluids used in various thermodynamic systems and the variation of those properties with the thermodynamic state is very important. This chapter describes the general trends of variation of thermodynamic states of pure substances with the change in properties such as temperature, pressure, enthalpy, specific volume etc. The variation in these properties is further utilised for estimation of energy transformations. After theoretical description of these concepts the engineering approach of compiling the variation of properties by choosing convenient reference states is presented. Methods of estimation of phase transformations and energy flows with change in state variables is also illustrated through suitable solved examples.*

## PRE-REQUISITES

*Thermodynamic states, properties and first law of thermodynamics*

## UNIT OUTCOMES

*List of outcomes of this unit is as follows:*

*U3-O1: Appreciation of definition of pure substance and phase*

*U3-O2: Representation of phase transformations on phase diagrams*

*U3-O3: Use of property tables for estimation of energy and phase transformations*

*U3-O4: Learning the method of estimating the change in energy during phase transformations through solved examples*

| Unit-3 Outcomes | EXPECTED MAPPING WITH COURSE OUTCOMES (1- Weak Correlation; 2- Medium correlation; 3- Strong Correlation) | | | | | |
|---|---|---|---|---|---|---|
| | CO-1 | CO-2 | CO-3 | CO-4 | CO-5 | CO-6 |
| U3-O1 | 2 | 1 | 2 | 1 | 1 | 1 |
| U3-O2 | 3 | 2 | 1 | 1 | 1 | 1 |
| U3-O3 | 3 | 3 | 2 | 1 | 1 | 3 |
| U3-O4 | 2 | 3 | 2 | 1 | 1 | 3 |

## 3.1 Pure Substance and its Different Phases

A substance of homogenous and consistent chemical composition throughout is known as a **pure substance**. For example, water, nitrogen, carbon dioxide, oxygen and helium are pure substances. A mixture of different chemical compounds or elements can also be called a pure substance if it has homogenous and invariable chemical composition throughout. Air in gaseous phase is an example of a mixture that can be termed as a pure substance (Fig. 3.1-a). Whereas mixture of oil and water is not a pure substance as these components are insoluble and settle in different portions of the container as per their density, hence, results in variable chemical composition (Fig. 3.1-b).

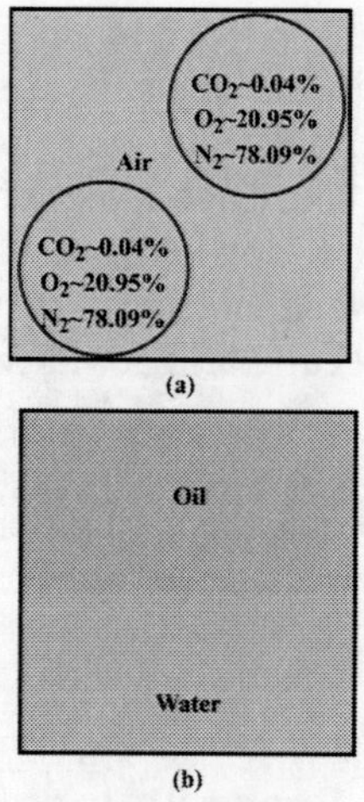


**Fig. 3.1:** (a) Gaseous air, a pure substance and (b) mixture of oil and water that is not a pure substance.

A pure substance can also exist in more than one phase, however, the chemical composition in all the phases should be the same. For instance, the mixture of ice and liquid water, liquid water and steam and a three phase mixture of ice, liquid water and steam are all pure substances (Fig. 3.2). On the other hand the mixture of liquid air and gaseous air is not a pure substance as the composition of liquid air is different from the gaseous air (Fig. 3.3). This is because different components of air condense at different temperatures for a given pressure. Therefore, it should be carefully noted that air, which is a mixture of different components, is a pure substance only if it remains in a single phase.

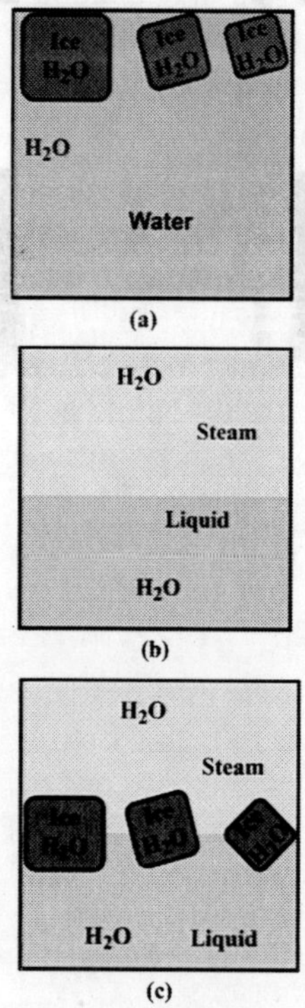


**Fig. 3.2:** Different phases of water in coexistence: (a) solid-liquid, (b) liquid-vapour and (c) solid-liquid-vapour.

We unusually see in our surroundings that iron exists as a solid, water exists as a liquid and oxygen stays as a gas at room temperature and pressure conditions. The same water exists in the gaseous phase as steam when its temperature is raised above 100°C at the same pressure condition. This signifies that all the substances can exist in different phases at different conditions. A **phase** is nothing but a distinct homogenous molecular arrangement that is easily distinguishable from other molecular structures by surface boundaries. The interfacial region between ice and liquid water is an example of separation between two phases. The three principle phases are solid, liquid and gas. In addition, there can be several other phases within a single principle phase depending upon the distinct molecular structures. For instance, ice shows seven different molecular structures at high pressures. Similarly carbon exists as diamond and graphite in the solid phase. In **solids**, the molecules are arranged in a three dimensional lattice structure and such structures are repeated throughout in all the directions. The molecules are closely spaced in solids, causing high attractive forces which bind them at fixed locations (Fig. 3.4-a). The molecules only oscillate at their positions and the velocity of oscillations is a strong function of temperature. When the temperature is increased such that the oscillation velocities are high enough to move a group of molecules relative to each-other, the transition of solid to liquid phase begins and is called the initiation of the melting process. In **liquids**, the molecular distances are comparable to solids, however, molecules are free to translate and rotate relative to each other, therefore, have slightly less intermolecular attractions than solids (Fig. 3.4-b). A slight increase in intermolecular distances is observed upon transition from solids to liquids for all the substances except water. In the **gas** phase, the molecular distances are large and attractions are weakest of all the phases (3.4-c). Such molecules move randomly and interactions between molecules are mainly in the form of molecular collisions. Gas molecules also collide prominently with the walls of the container where these are stored in. Specifically for low density gases, the molecular interactions are only due to intermolecular collisions. Usually the gaseous molecules have high energy compared to solids and liquids, therefore, they release a significant amount of energy before they condense or freeze.

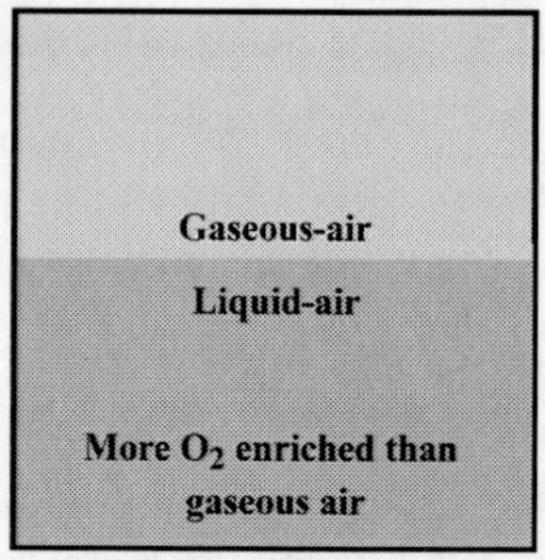


**Fig. 3.3:** Mixture of liquid air and gaseous air that is not a pure substance.

## 3.2 Phase Boundaries

Often a pure substance coexists in its different phases. For instance, a mixture of liquid and vapour exists in boilers, evaporators and condensers, etc. Therefore, it becomes essential to study the thermodynamic conditions and associated phases of a substance as it is useful for the design of thermodynamic systems. We will use water as an example case to demonstrate the effect of thermodynamic parameters on its phase transitions.

Consider water at 25°C and 1 atm pressure is contained in a piston-cylinder system such that the pressure is maintained constant (state 1 in Fig. 3.5). At this condition, water exists as a liquid. Now the heat is added to the water and its temperature is monitored continuously during the heat addition process. It is observed that the temperature of water increases upon addition of heat and its volume increases slightly.

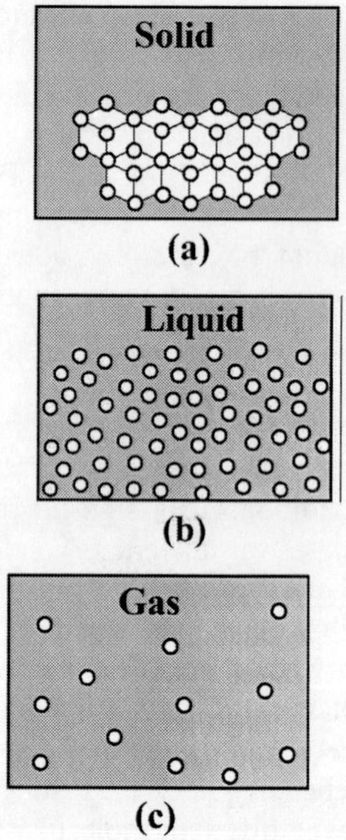


**Fig. 3.4:** Representation of molecular structure in different phases: (a) solid (b) liquid and (c) gas.

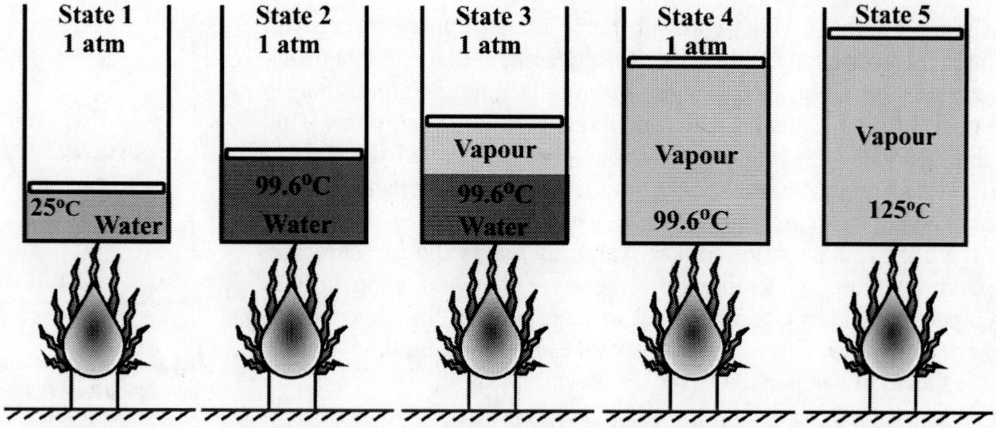


**Fig. 3.5:** Heating of water at constant pressure conditions depicting its phase transition from liquid to vapour phase.

To accommodate the increase of volume, the piston also displaces slightly whereas the pressure stays constant. The increase of temperature is manifested till it reaches 99.6°C and water remains as the liquid phase (state 2 in Fig. 3.5). When heat is added further, water starts converting to vapour phase, temperature remains the constant and volume increases considerably (state 3 in Fig. 3.5). This process of phase transfer continues upon addition of heat till the complete liquid is converted into vapour and a large increase in volume takes place during this transition (state 4 in Fig. 3.5). Further addition of heat results in an increase of both the temperature and volume (state 5 in Fig. 3.5). The complete process of this constant pressure heating is shown on $T - v$ diagram in Fig. 3.6.

State 2 represents complete liquid which is about to vapourize upon further addition of heat, therefore, it is termed the **saturated liquid state**. Similarly, state 4 represents complete vapour that is about to condense if heat is taken out of it, is called the **saturated vapour state**. The coexistence of liquid-vapour mixture in equilibrium between the saturated liquid (state 2) and saturated vapour (state 4) states is referred to as the **saturated liquid-vapour mixture** (state 3). When a substance is in pure liquid state and is not about to vapourize upon addition of heat, it is called the **subcooled** or **compressed liquid** (state 1). Contrary, a pure vapour phase that is not about to condense upon removal of heat is termed the **superheated vapour** (state 5). If the substance is cooled from state 5 at constant pressure condition, it will follow the same path on $T - v$ diagram in reverse direction till it reaches state 1 and the system will reject the same amount of heat as was added during the heating process.

In the previous illustration, the liquid to vapour transition started at 99.6°C. It is important to mention that it is the **boiling temperature** of water corresponding to 1 atm pressure. If the water is heated in a similar arrangement at 2 atm pressure, it will initiate the change of phase from liquid to vapour at 120.21°C. Thus the boiling temperature depends upon the pressure and for each pressure there is a unique value of boiling temperature. Therefore, the temperature at which pure substance changes phase for a given pressure is called the **saturation temperature**.

Similarly, a pressure at which pure substance changes phase for a given temperature is referred to as the **saturation pressure**. Figure 3.7 shows a plot between saturation pressure and saturation temperature that is obtained by heating water at different fixed pressure conditions. This curve that separates the liquid and vapour phases is called the **liquid-vapour saturation curve** or **vaporization line**. It is usual practice that if we cook food at high temperatures, it takes less cooking time. This is the reason for using the pressure cooker which increases the pressure and corresponding saturation temperature, therefore, allows adding heat at high temperatures. It is also important to mention that the atmospheric pressure decreases with elevation. Thus it takes less cooking time to cook food at sea level in comparison to higher elevations.

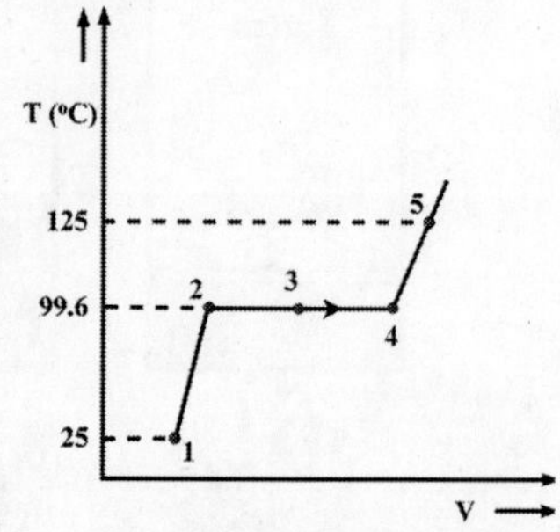


**Fig. 3.6:** Representation of constant pressure heating process on $T - v$ diagram (not to scale).

If in the previous experiment liquid water at state 1 is cooled under 1 atm pressure, first its temperature will decrease till it reaches 0°C and then liquid start changing into ice upon further cooling at constant temperature condition. This process continues till the whole liquid is converted into solid ice. In this process the volume of water expands upon its conversion from liquid to solid phase. Water is an exceptional substance which expands upon freezing whereas most pure substances contract upon freezing. This transitional state is another saturation state that represents the liquid to solid phase transition and vice-versa. The saturation temperature corresponding to solid-liquid transition is called the **freezing point**. If the cooling process is further continued after the complete ice formation, it results in decreasing the temperature of ice whereas its volume remains almost constant. When this experiment is repeated at different pressures, it gives a curve that separates the solid and liquid regions as shown in Fig. 3.8. This curve is called the **fusion line** and depicts that freezing temperature is not much influenced by the pressure. When the experiments are performed below the atmospheric pressure conditions for locating the boiling and freezing points, the fusion and vaporization lines extend toward low pressure and meet at a point. The further reduction of pressure below this point results in a single saturation curve, called the **sublimation curve** which represents the transition from solid to vapour phase and vice-versa (Fig. 3.8). The point at which all three phases coexist in equilibrium is called the **triple point**. The pressure and temperature of water corresponding to its triple point are 0.0061 bar and 0.01°C. This is the only condition at which all three phases can coexist in equilibrium and deviation from this condition gives coexistence of only two phases along the saturation lines. Also, the liquid phase cannot exist below the triple point. The diagram showing different saturation lines is called the **phase diagram** (Fig. 3.8). The region on the phase diagram other than the saturation lines shows the occurrence of a single phase. The vapourization curve on the phase diagram has the highest pressure of 220.9 bar and corresponding temperature is 374.14°C for water and it is called the **critical point**. The transition from liquid to vapour is not smooth above this point and liquid changes instantly to

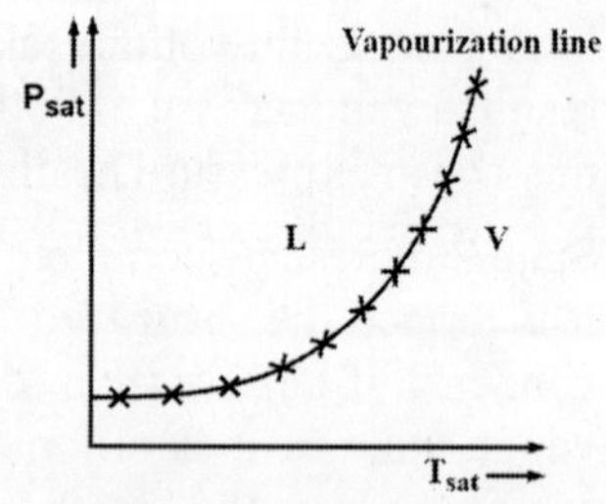


**Fig. 3.7:** Dependence between saturation pressure and saturation temperature.

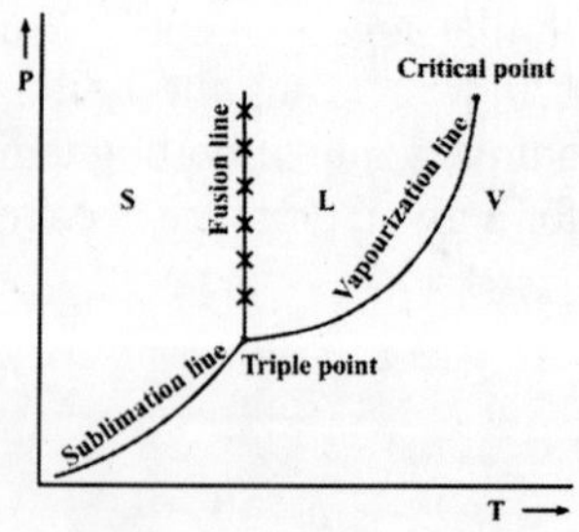


**Fig. 3.8:** Depiction of phase boundaries on $P - T$ curve.

vapour phase. Therefore, occurrence of two-phase mixture in equilibrium is not observed above critical point.

The complete phase diagram for water which covers a wide range of pressure and temperatures is shown on semi-log plot in Fig. 3.9. The logarithmic scale is chosen along the ordinate to cover a wide pressure range. This diagram shows seven different types of crystal structures for ice which are separated by different phase boundaries. Due to the occurrence of different solid phases, there is presence of multiple points where simultaneous presence of three phases can be observed. However, there is only a unique triple point where three principle phases: solid, liquid and gas can coexist. The general nature of the phase diagram remains the same for all other substances and there can be multiple phases under a single principle phase for other substances as well. Moreover, the specific features like triple point and critical point conditions are different for different substances. For instance, the triple point pressure of mercury is $1.3 \times 10^{-3}$ bar, which is the reason for its use as a barometric fluid. The triple point pressure of carbon dioxide is 5.208 bar, which is greater than atmospheric pressure. Therefore, at atmospheric pressure solid carbon dioxide converts directly to the vapour phase through a **sublimation process** (Fig. 3.10). The solid to vapour transition of carbon dioxide takes place at $\simeq -73°C$ for atmospheric pressure conditions. This is the reason solid carbon dioxide is used for shipping frozen food items instead of ice. Because even if it melts, it will keep the package dry unlike ice. Hence, solid carbon dioxide is also called **dry ice**.

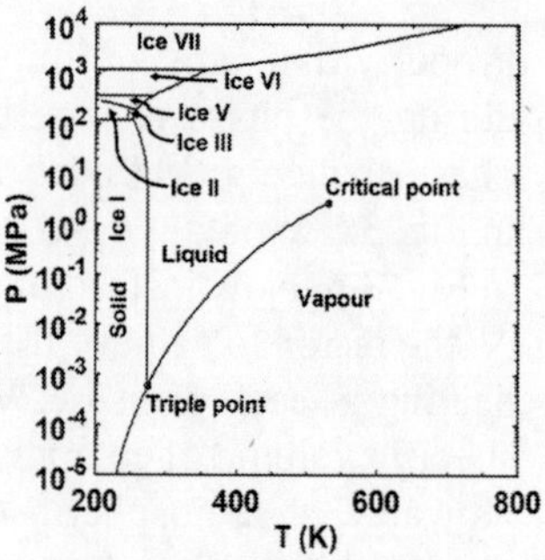


**Fig. 3.9:** Phase diagram of water.

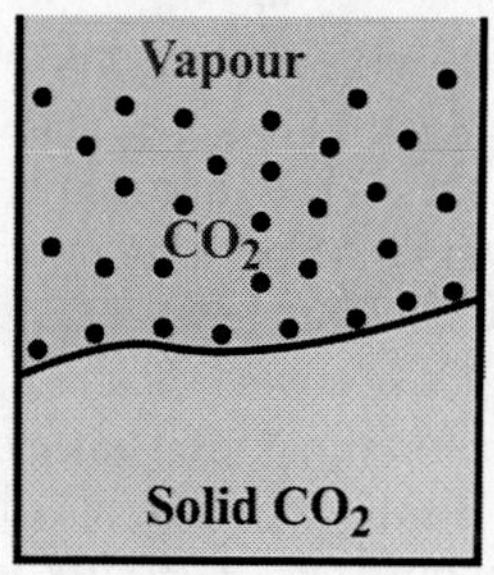


**Fig. 3.10:** Schematic representation of sublimation process.

## 3.3 Property Diagrams

Property diagrams are important to elucidate the variation of properties during the phase change processes. To demonstrate the liquid vapour phase change process, let us consider the piston-cylinder system as used in Fig. 3.5 to obtain the constant pressure heating at atmospheric pressure in Fig. 3.6. Now instead of atmospheric pressure, we will add the weights on the piston to achieve the high pressures and perform constant pressure heating at each pressure. When the pressure is increased to 1 MPa, the specific volume of compressed liquid decreases slightly in comparison to its value at atmospheric condition. During the heating process, it is observed that the change from liquid to vapour starts at a higher temperature of 179.9°C (Fig. 3.11). Therefore, this is the saturation temperature for 1 MPa pressure. The specific volume of saturated liquid is also slightly higher at this pressure in comparison to its counterpart at atmospheric pressure. When the complete liquid has converted into vapour, further its temperature and volume increase during the heating process. The saturated vapour state is found to have less specific volume than its corresponding value at atmospheric pressure. This signifies that the length of horizontal line between saturated liquid and saturated vapour states becomes shorter upon increase of pressure. Such a heating process is represented for other pressures as well on $T - v$ diagram in Fig. 3.11. It is found that the horizontal line between saturated liquid and saturated vapour states becomes a point for 22.09 MPa pressure and this point is called the critical point. The critical point represents the conditions at which saturated liquid and saturated vapour states are indistinguishable. At this point, heating results in a smooth transition from a saturated liquid state to a saturated vapour state without passing through the constant temperature vapourization process. The temperature and specific volume corresponding to critical point for water are 374.14°C and 0.003155 $m^3/kg$, respectively. Above the critical point, the two phase mixture does not coexist. In the region above the critical point where a substance is clearly distinguishable to change from a liquid to a vapour phase, it is termed a **dense fluid**. However, there is no clear boundary between a compressed liquid region and superheated vapour region above the critical point. Often a substance is called superheated vapour at temperatures above the critical temperature and compressed liquid at temperatures lower than the critical temperature.

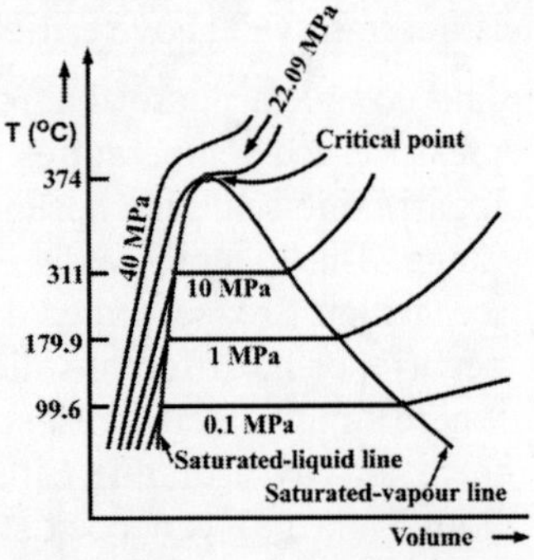


**Fig. 3.11:** □ – □ diagram for water in liquid-vapour regime.

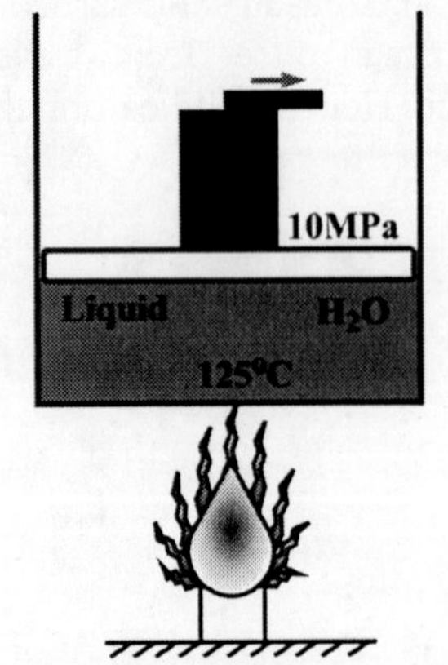


**Fig. 3.12:** Representation of pressure reduction by decreasing the weight acting on the piston.

The curve joined by the locus of saturated liquid states in Fig. 3.11 is called the **saturated liquid line**. Similarly a curve passing through saturated vapour states is termed as **saturated vapour line**. Both these lines meet at critical point and form a dome. The region within the dome represents liquid-vapour mixture in equilibrium and is called the **saturated liquid-vapour mixture region** or **wet region**. The region on the left of the saturated liquid line represents the **compressed liquid** or

**subcooled region**. Whereas the region on the right of the saturated vapour line depicts the **superheated region**. In both these regions, substance exists in a single phase, either a liquid or a gas. The amount of energy required in changing phase from saturated liquid to saturated vapour is called the **latent heat of vapourization**. When the state of a substance changes from saturated vapour to saturated liquid, then an equivalent amount of heat is released and the process is called the condensation. Similarly the amount of energy absorbed by a substance during its melting is called the **latent heat of fusion** and an equivalent amount of energy is released during the freezing process.

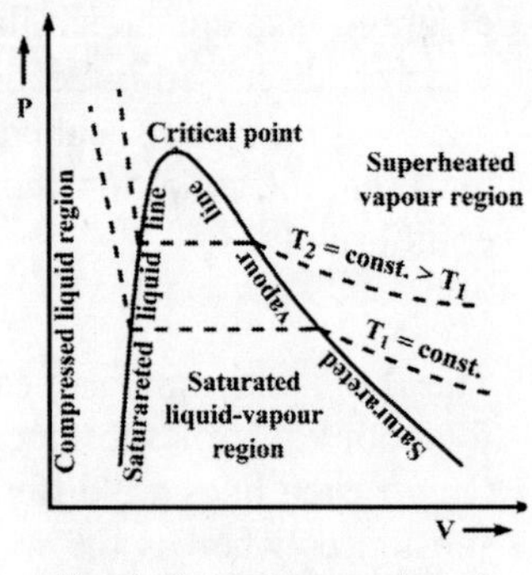


**Fig. 3.13:** Representation of constant temperature lines on $P-v$ diagram in liquid-vapour regime.

The $P-v$ diagram for a pure substance also has the similar general shape as that of the $T-v$ diagram whereas the major difference comes in the shape of constant temperature lines. To demonstrate the constant temperature line on a $P-v$ diagram, consider a compressed liquid at pressure of 10 MPa and 125°C is contained in a piston-cylinder arrangement as shown in Fig. 3.12. Now slowly the weights are removed and pressure is decreased gradually. In this process, the specific volume of the liquid increases slightly and heat is added to the liquid in order to maintain the constant temperature condition (Fig. 3.13). When the pressure reaches to 0.23 MPa, liquid starts converting into vapour. This is the saturation pressure corresponding to 125°C temperature. During the phase change process both the temperature and pressure are maintained constant and the volume increases considerably. Therefore, weights are not removed till liquid is converted completely into vapor. If the weights are removed during the phase change process, this will further reduce the pressure and hence the corresponding saturation temperature, therefore, isothermal conditions will not be maintained. After the saturated vapour state is achieved, further reduction of pressure results in considerable decrease of volume at the constant temperature. When the process is repeated at higher temperatures, the boiling starts at high pressures and the length of horizontal line between saturated liquid state and saturated vapour state becomes shorter. The saturated liquid and saturated vapour lines are obtained upon joining all the saturated liquid and saturated vapour states, respectively. The $P-v$ diagram for a pure substance with superimposed constant temperature lines is shown in Fig. 3.13.

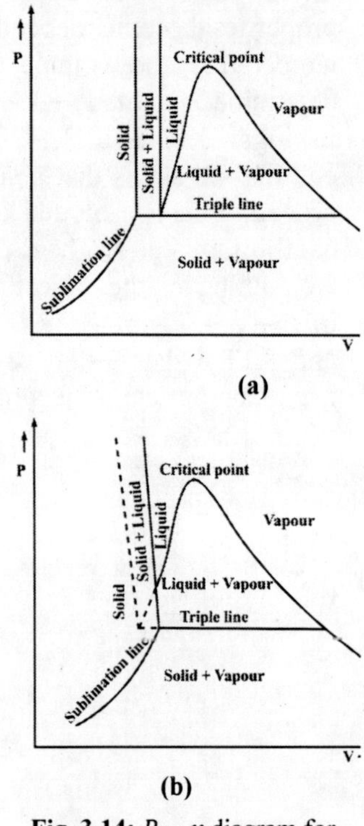


**Fig. 3.14:** $P-v$ diagram for a substance that: (a) contracts upon freezing and (b) expands upon freezing.

The property diagrams in the preceding discussion are shown for liquid-vapour phase change transitions. Similar experiments can be performed for solid-liquid and solid-vapour transformations to identify these two-phase regions on property diagrams. Majority of the substance contracts upon freezing, therefore their specific volume decreases in the solid phase. However a few substances such as water expands upon freezing. The $P-v$ diagrams including the solid-liquid and solid-vapor two phase regions are shown in Fig. 3.14 for both kinds of substances. The $T-v$

diagrams also appear similar to that of $P - v$ diagrams. The triple point, which is discussed in Section 3.2, appears as a line on $P - v$ (Fig. 3.14) and $T - v$ diagrams, called the **triple line**. Triple line on these diagrams represent the conditions corresponding to coexistence of three phases in equilibrium.

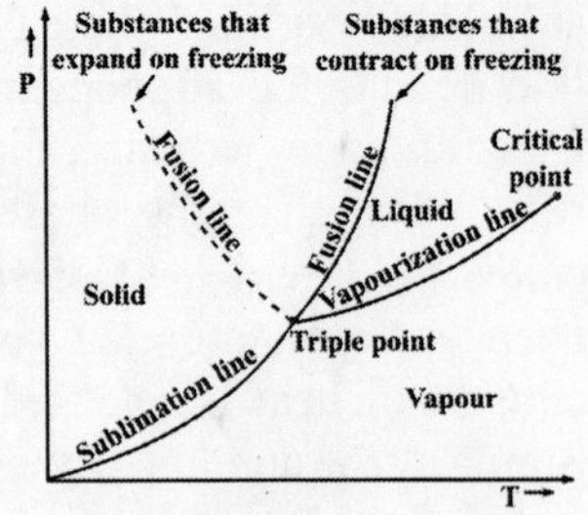


**Fig. 3.15:** Representation of $P - T$ diagram for substances that contract and expand upon freezing.

The $P - T$ diagram, also called the phase diagram, is shown in Fig. 3.15 for both kinds of substances that expands and contracts upon freezing. Only fusion lines are different on this diagram for two different kinds of substances whereas the vapourization and sublimation lines remain the same. The property diagrams shown here are only applicable for a simple compressible pure substance. For such substances, the state of the substance can be defined by two independent intensive properties. It means that if two independent intensive properties are fixed, all other properties become dependent on these. Using this idea, if we take temperature and volume as independent properties, pressure can be determined and the $P - v - T$ surface diagrams can be obtained as shown in Figs. 3.16 and 3.17. The $P - v - T$ surface represents all the equilibrium states that a substance can exist. All the quasi-equilibrium processes can be plotted on these diagrams as these pass through the equilibrium states. The 2-dimensional property diagrams shown earlier are nothing but the projections of $P - v - T$ surface. One such projection of $P - v - T$ surface in terms of $P - T$ diagram is shown in Figs. 3.16 and 3.17. Other 2-dimensional property diagrams can also be plotted using appropriate projections.

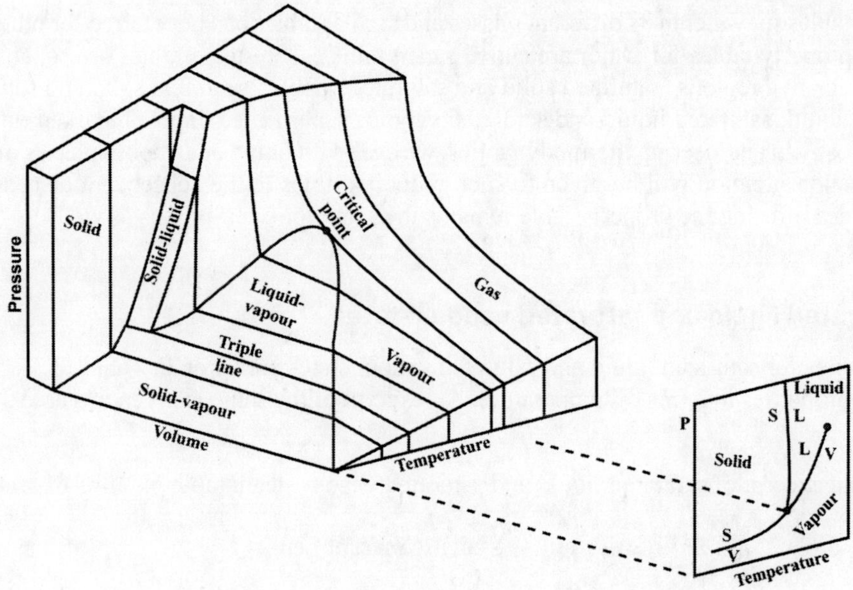


**Fig. 3.16:** The three dimensional $P - v - T$ surface plot for substances that contract upon freezing.

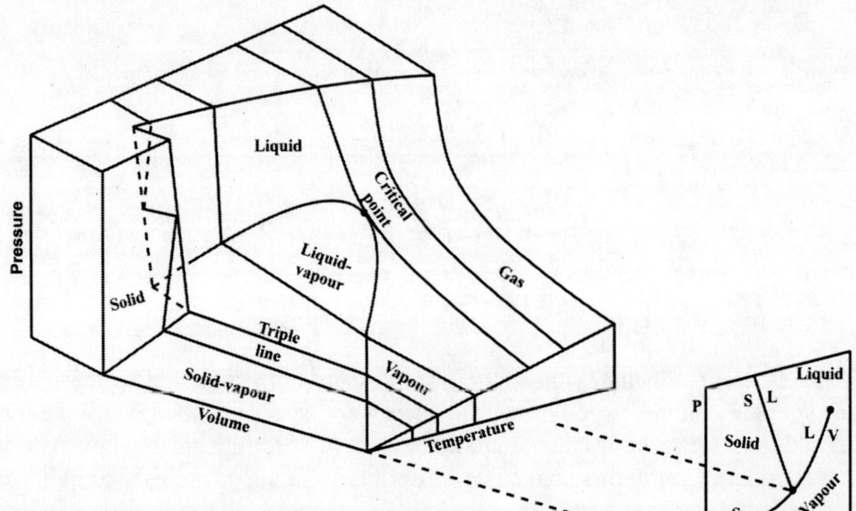


**Fig. 3.17:** The three dimensional $P - v - T$ surface plot for substances that expand upon freezing.

## 3.4 Property Tables

In general, the thermodynamic properties of most substances are presented in tabular form, as due to the complex behaviour of their variation, it is not possible to express in the form of simple equations. Also, the form of property tables is the same for all the substances. In the present section, we will introduce the property tables for water in its different phases and similar tables can be referred for other substances. Usually the property tables for water are called steam tables. The steam tables are separately given in Appendix B for five regions (saturated solid and saturated vapour; saturated solid and saturated liquid; compressed liquid; saturated liquid and saturated vapour; superheated vapour) marked on the property diagrams. Mostly in engineering thermodynamics, we deal with liquid and vapour phases of a substance. Therefore, major attention will be given to such property tables in the subsequent discussion whereas the general idea of using the property table remains the same for other phases as well.

### 3.4.1 Saturated liquid and saturated vapour states

The steam tables for saturated liquid and saturated vapour states are given in Tables B.1 and B.2. Table B.1 lists the properties in terms of temperature. An excerpt of the same is given in Table 3.1.

**Table 3.1:** An excerpt of saturated liquid and saturated vapour steam table in terms of temperature.

| **Temp.** (°C) | **Sat. Press.** $(kPa)$ | **Specific volume** $(m^3/kg)$ | | **Specific internal energy** $(kJ/kg)$ | | | **Specific enthalpy** $(kJ/kg)$ | | |
|---|---|---|---|---|---|---|---|---|---|
| $T$ | $P_{sat}$ | $v_f \times 10^3$ | $v_g$ | $u_f$ | $u_{fg}$ | $u_g$ | $h_f$ | $h_{fg}$ | $h_g$ |
| 50 | 12.352 | 1.012 | 12.026 | 209.33 | 2233.4 | 2442.7 | 209.34 | 2382.0 | 2591.3 |
| 55 | 15.763 | 1.015 | 9.5639 | 230.24 | 2219.1 | 2449.3 | 230.26 | 2369.8 | 2600.1 |
| 60 | 19.947 | 1.017 | 7.6670 | 251.16 | 2204.7 | 2455.9 | 251.18 | 2357.7 | 2608.8 |

In this table, the first entry includes the saturation pressure corresponding to the given temperature condition. The specific volume, specific internal energy and specific enthalpy are the other properties listed here. In addition, steam tables also include the data of specific entropy, however, it is not listed here as it is not introduced yet in this book. The subscripts "$f$" and "$g$" are used to indicate the property of saturated liquid and saturated vapour, respectively, whereas the subscript "$fg$" represents the difference of saturated vapour and saturated liquid values of the same property. For instance, $h_f$ denotes the enthalpy of saturated liquid and $h_g$ denotes the enthalpy of saturated vapour. The term $h_{fg} = h_g - h_f$, represents the amount of heat required to change the phase of 1 $kg$ of a substance from liquid to vapour at given temperature or pressure, called the **enthalpy of vapourization** or **latent heat of vapourization**. Similarly, an excerpt of Table B.2 is given in Table 3.2 which lists the properties in terms of pressure. It is important to note that both the tables list the same properties of substance at saturated states, the only benefit of two tables is to use an appropriate table based on the known property i.e. temperature or pressure.

It is also to highlight that for saturated states only a single thermodynamic property can give us the values of all other dependent properties. This is because the term saturated itself is the indication of the physical state of the system. For example, if one has to know the enthalpy of saturated liquid at 55°C, one can use saturation temperature table and refer to value given in $h_f$ column corresponding to 55°C temperature. From Table 3.1, it comes out to be 230.26 $kJ/kg$. If a property is to be determined at a given condition that is not listed in the table, then linear interpolation can be used between the closest given conditions to determine the property. For instance, if internal energy of saturated vapour ($u_g$) is required at 28 $kPa$ ($= P$). This value of pressure is not listed in Table 3.2. The closest given pressures are 25 $kPa$ ($= P_1$) and 30 $kPa$ ($= P_2$) and corresponding values of $u_g$ are 2462.4 $kJ/kg$ ($= u_{g1}$) and 2467.7 $kJ/kg$ ($= u_{g2}$), respectively. Using linear interpolation between these two pressures we can obtain

$$u_g - u_{g1} = \frac{u_{g2}-u_{g1}}{P_2-P_1}(P - P_1) = 2465.58\ kJ/kg.$$

**Table 3.2:** An excerpt of saturated liquid and saturated vapour steam table in terms of pressure.

| **Press.** | **Sat. Temp.** | **Specific volume** | | **Specific internal energy** | | | **Specific enthalpy** | | |
|---|---|---|---|---|---|---|---|---|---|
| $(kPa)$ | (°C) | $(m^3/kg)$ | | $(kJ/kg)$ | | | $(kJ/kg)$ | | |
| $P$ | $T_{sat}$ | $v_f \times 10^3$ | $v_g$ | $u_f$ | $u_{fg}$ | $u_g$ | $h_f$ | $h_{fg}$ | $h_g$ |
| 20 | 60.06 | 1.017 | 7.6481 | 251.40 | 2204.6 | 2456.0 | 251.42 | 2357.5 | 2608.9 |
| 25 | 64.96 | 1.020 | 6.2034 | 271.93 | 2190.4 | 2462.4 | 271.96 | 2345.5 | 2617.5 |
| 30 | 69.09 | 1.022 | 5.2287 | 289.24 | 2178.5 | 2467.7 | 289.27 | 2335.3 | 2624.6 |

## 3.4.2 Saturated liquid-vapour mixture

The saturated liquid-vapour mixture exists during the vapourization process. Though it is called mixture, liquid and vapour coexist as separate identities at the same temperature and pressure conditions and don't physically mix. A representation of such a two-phase mixture is shown in Fig. 3.18. Here liquid and vapour stays in different portions of the container and are clearly separated by a surface, called **interface**. In this condition, the entire volume of liquid has the properties of the saturated liquid and the whole vapour volume has an equivalent saturated vapour state. To define the combined property of the mixture, an additional independent parameter, called **quality** is used. It is defined as the ratio of mass of vapour to the total mass of the liquid-vapour mixture, i.e. $x = \frac{m_g}{m_f+m_g}$. The value of quality is zero for pure saturated liquid and it is equal to one for pure saturated vapour. Therefore, it varies from zero to one in the two phase region. The total volume of mixture can be given as the sum of liquid volume and vapour volume. Therefore, $V = V_f + V_g$. Using the definition of specific volume, it can be expressed as $mv = m_f v_f + m_g v_g$ where $m$ is the total mass of the mixture $(m_f + m_g)$ and $v$ is the specific volume of the mixture. Specific volume of mixture can be expressed as $v = \frac{m_f}{m} v_f + \frac{m_g}{m} v_g$. Now $x = \frac{m_g}{m}$ and $\frac{m_f}{m} = \frac{m - m_g}{m} = 1 - x$. Therefore, $v = (1 - x)v_f + xv_g = v_f + x(v_g - v_f) = v_f + xv_{fg}$. The values of $v_f$ and $v_g$ can be obtained from the saturated steam tables (Tables B.1 and B.2) based on the given temperature or pressure condition. It is important to mention that there is no mass of water which has the value of specific volume equals to $v$. It is just an average representative property of the mixture based on the fraction of liquid and vapour present in the mixture. Thus the value of $v$ will be always between $v_f$ and $v_g$ for a two-phase mixture. Like volume other properties such as internal energy and enthalpy can also be computed using mass weighted average of saturated liquid and saturated vapour values. Therefore, in general any property of the two-phase mixture can be expressed as $z = z_f + xz_{fg}$. The way saturated liquid-vapour mixture is dealt, other two-phase mixtures such as saturated solid-liquid mixture and saturated solid-vapour mixture can also be treated. The property tables for saturated solid and saturated liquid states and saturated solid and saturated vapour states are given in Tables B.5 and B.6, respectively.

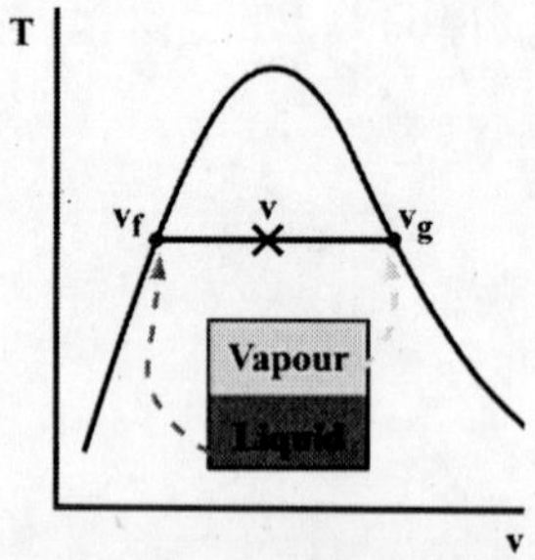


**Fig. 3.18:** Representation of saturated liquid-vapour mixture on $T - v$ diagram.

Scan Me

For more details on saturated steam tables.

Scan Me

For more details on superheated steam tables.

### 3.4.3 Superheated vapour states

The single phase region on the right of the saturated vapour line and the region above the critical point for temperature greater than the critical temperature is called superheated region. In this region, the pressure and temperature are independent parameters and no more depend upon each other as in the case of saturated states. Therefore, using these two parameters, other properties of the substance can be determined in this region. The steam tables for superheated states are given in Table B.3. An excerpt of superheated tables is given in Table 3.3. In Table 3.3, properties are listed against temperature for specific pressure conditions. The first entry shows the data of the saturated vapour state corresponding to given pressure and the value of saturated temperature is mentioned after pressure in small brackets. By knowing the value of pressure and temperature, all other properties can be obtained from superheated tables. The necessary conditions for superheated vapour states are: $T > T_{sat}$ at given $P$; $z > z_g$ at a given $T$ or $P$ where $z \in v, u, h$; $P < P_{sat}$ at a given $T$.

**Table 3.3:** An excerpt of superheated vapour steam table.

| $T$ (°C) | $v$ $(m^3/kg)$ | $u$ $(kJ/kg)$ | $h$ $(kJ/kg)$ | $v$ $(m^3/kg)$ | $u$ $(kJ/kg)$ | $h$ $(kJ/kg)$ |
|---|---|---|---|---|---|---|
| | $P = 6.0\ MPa\ (275.59°C)$ | | | $P = 8.0\ MPa\ (295.01°C)$ | | |
| Sat. | 0.03245 | 2589.9 | 2784.6 | 0.023525 | 2570.5 | 2758.7 |
| 300 | 0.03619 | 2668.4 | 2885.6 | 0.024279 | 2592.3 | 2786.5 |
| 350 | 0.04225 | 2790.4 | 3043.9 | 0.029975 | 2748.3 | 2988.01 |
| 400 | 0.04742 | 2893.7 | 3178.3 | 0.034344 | 2864.6 | 3139.4 |

### 3.4.4 Compressed or subcooled liquid states

When the pressure of a liquid is greater than the saturated pressure at the given temperature, it is called the **compressed liquid state**. Contrary, if we compare the same state to a saturated liquid state at the same pressure, it is observed that temperature is less than the saturation temperature, then it is referred to as **subcooled liquid state**. The compressed or subcooled liquid region is also a single phase regime and its property table is similar to that of superheated steam tables. Table B.4 represents the property values of compressed liquid states and its excerpt is given in Table 3.4. The necessary conditions for compressed liquid states are: $T < T_{sat}$ at given $P$; $z < z_f$ at a given $T$ or $P$ where $z \in v, u, h$; $P > P_{sat}$ at a given $T$. Therefore, the temperature entries of Table 3.4 include values less than saturation temperature at the given pressure. The first entry of the table shows the values of saturated liquid states. The compressed liquid property tables are very rare and if available includes a limited range of conditions. This is because properties of such liquids are not much influenced by pressure. Therefore, the usual practice is to use the corresponding saturated liquid properties at the given temperature for compressed liquid states as the properties of liquids are more influenced by temperature. It is evident from Table 3.4 that the properties show a sufficiently small variation upon increasing pressure from 10 $MPa$ to 15 $MPa$ whereas the change of properties with temperature is considerable. Therefore, in general, $z_{comp} \simeq z_{sat}$ at a given $T$. Usually, this formulation works well for specific volume and specific internal energy, however, it may give considerable error in enthalpy. Therefore, enthalpy can be corrected as $h_{comp} = h_f(@T) + v_f(@T)(P - P_{sat}(@T))$. This correction gives good accuracy for low to moderate pressures and temperatures, however, it may result in large error at very high pressures and temperatures due to overcorrection.

**Table 3.4:** An excerpt of compressed liquid steam table.

| $T$ (°C) | $v$ $(m^3/kg)$ | $u$ $(kJ/kg)$ | $h$ $(kJ/kg)$ | $v$ $(m^3/kg)$ | $u$ $(kJ/kg)$ | $h$ $(kJ/kg)$ |
|---|---|---|---|---|---|---|
| | $P = 10\ MPa\ (311°C)$ | | | $P = 15\ MPa\ (342.16°C)$ | | |
| Sat. | 0.0014522 | 1393.3 | 1407.9 | 0.0016572 | 1585.5 | 1610.3 |
| 20 | 0.0009973 | 83.31 | 93.28 | 0.0009951 | 83.01 | 97.93 |
| 40 | 0.0010035 | 166.33 | 176.37 | 0.0010013 | 165.75 | 180.77 |
| 60 | 0.0010127 | 249.43 | 259.55 | 0.0010105 | 248.58 | 263.74 |

### 3.4.5 Reference states for developing steam tables

The pressure, temperature and specific volume are measurable properties whereas the internal energy, enthalpy and entropy cannot be measured directly, therefore, these are obtained using the thermodynamic relations between the properties. Such property relations specify the change in property between the two state points rather than specifying the absolute value of property at a given thermodynamic state. Thus it becomes important to choose a reference state for a substance where zero value can be assigned to a few properties. The triple point is the reference state for water where its specific internal energy and specific entropy are assigned zero value. The properties at other state points are calculated with reference to this point. Hence, for some states of a substance below its triple point, the values of a few properties can also be negative. Generally, for water the reference point is the same in almost all the steam tables available in different literature. Whereas the reference states for other fluids such as ammonia and other refrigerants are differently selected in different sources of property tables. Therefore, we may find a situation where different values of the same property are given in different tables at the same states. As in thermodynamic analysis, we are mainly concerned with change of properties between two state points. Hence, it does not make any difference unless or until a consistent set of tables are referred.

## 3.5 Ideal Gas States

Majority of the time, it is convenient to use simplified relations between the properties that are more general and accurate instead of using the property tables. Such relations are mainly available for gaseous states under some prescribed conditions of pressures and temperatures. Any equation that relates the properties of a substance at equilibrium states is termed as **equation of state**. The most common equation of state that relates pressure, temperature and specific volume is the **ideal gas equation of state**. By combining the experimental observations of Robert Boyle, J. Charles and J. Gay-Lussac, the equation of state can be expressed for gases at low pressures as $Pv = RT$. Here, $P$ is absolute pressure, $T$ is absolute temperature and $v$ is the specific volume. The constant $R$ is called the characteristic gas constant and it is unique for each gas. It is stated as $R = \frac{\underline{R}}{M}$ where $\underline{R}$ is the universal gas constant and it value is 8.314 $kJ/kmol - K$. Here $M$ is the molecular mass of the gas. The relation $Pv = RT$ is called the **ideal gas equation** and is followed by most gaseous substances at low densities. Density of gases is low at high temperatures and low pressures. Thus these are the favorable conditions where ideal gas behaviour is shown by most gases. A gas following this behaviour is usually termed an **ideal gas**. In most engineering applications, the majority of gases such as air, oxygen, nitrogen, hydrogen, helium and carbon dioxide, etc. can be treated as ideal gas with very less errors. Whereas the dense gases such as water vapour and vapour of other common refrigerants show ideal gas characteristics in a very limited range of pressures and temperatures, therefore, property tables should be used for such substances instead of the ideal gas equation of state. A $T - v$ diagram is shown for water in Fig. 3.19 which shows the percentage error $\left(\frac{|v_{table} - v_{ideal}|}{v_{table}} \times 100\right)$ associated upon use of the ideal gas equation in the vapour region. The zone where error is less than 1% is suitable for use of ideal gas equations and mostly associated with high temperatures and low pressures. Usually air conditioning applications have low pressure of water vapours and use of ideal gas equation for such applications is justified. However, the steam used in power applications is maintained close to the critical point, therefore, using the ideal gas equation is not accurate for such situations.

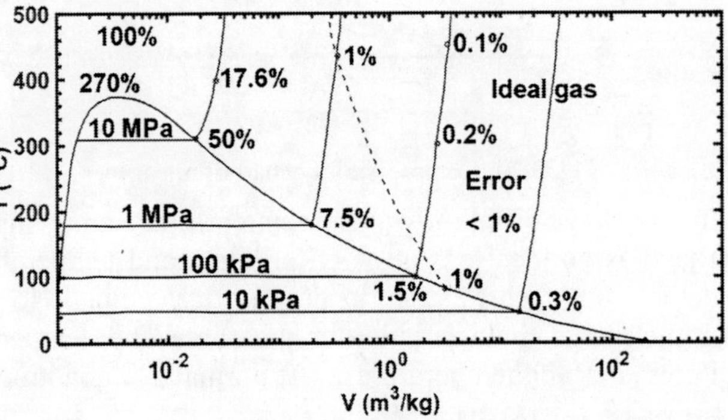


**Fig. 3.19:** Error associated upon using the ideal gas formulation on $T - v$ diagram for water. The dashed line demarcates the region having error less than 1%.

## 3.6 Compressibility Factor

It is evident from Fig. 3.19 that ideal gas behaviour is not accurate for vaour near the saturated vapour region and the critical point. However, the ideal gas equation of state is very handy and convenient to use. The deviation of a gaseous substance from ideal gas characteristics at a given temperature and pressure can be corrected by introducing a correction factor, called the **compressibility factor** which is defined as $Z = \frac{Pv}{RT}$. The term $\frac{RT}{P}$ is usually termed the ideal specific volume ($v_{ideal}$), hence, $Z$ can also be treated as the ratio of actual volume to the ideal volume. $Z$ is equal to one for an ideal gas and it can be positive and negative for real gases at the given conditions. The more is the deviation of $Z$ from unity, greater is the difference from the ideal gas behaviour. Usually for moderate density gases, the forces of attraction between molecules pull them close and reduce the actual volume in comparison to ideal volume. Thus $Z < 1$ for such conditions. On the other hand, for high density gases, the molecules repel each-other, occupy more volume than ideal behaviour and results in $Z > 1$. The compressibility chart of nitrogen is shown in Fig. 3.20. It is evident from Fig. 3.20 that when pressure approaches absolute vacuum, $Z$ becomes unity at all the temperatures. Compressibility factor deviates most at low temperatures and very high pressures. However, low temperatures and high pressures are relative to the critical point of a substance. For example, a temperature value of -120°C at atmospheric pressure is not considered low temperature for nitrogen whose critical temperature is -147°C whereas other substances like water exist as a solid at these conditions.

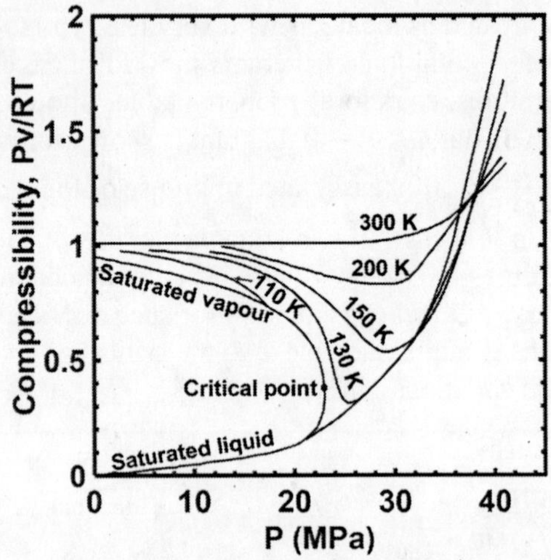


**Fig. 3.20:** Compressibility chart of nitrogen.

Different gaseous substances behave differently at the same pressure and temperature conditions. This means that a gas can behave as an ideal gas at given pressure and temperature whereas some other gaseous substance can deviate due to difference in their critical points. However, the most gaseous substances show almost the same characteristics at pressures and temperatures normalized with reference to their critical pressures and temperatures. The normalized quantities are called the **reduced pressure** and the **reduced temperature**. These are defined as $P_r = \frac{P}{P_{cr}}$ and $T_r = \frac{T}{T_{cr}}$, respectively. The subscript "$cr$" refers to the state of the substance corresponding to the critical point. The compressibility factor is approximately the same for all gaseous substances at the given reduced temperature and pressure. This behavioural similarity is termed as the **principle of corresponding states**. A common compressibility chart can be obtained for all substances based on the principle of corresponding states. Such a **generalized compressibility chart** shows the variation of compressibility factor in terms of reduced pressure and reduced temperature and is shown in Fig. 3.21. It is evident from Fig. 3.21 that maximum deviation from the ideal gas characteristics is observed near the critical point. This chart is obtained by curve fitting the data from experimental measurements for a large number of substances and gives reasonable accuracy.

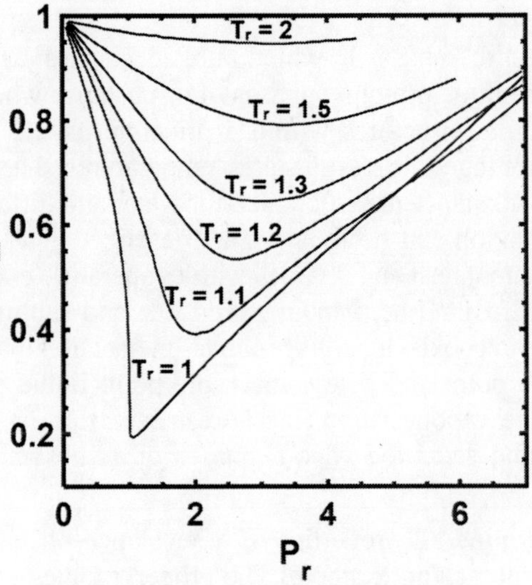


Fig. 3.21: Generalized compressibility chart.

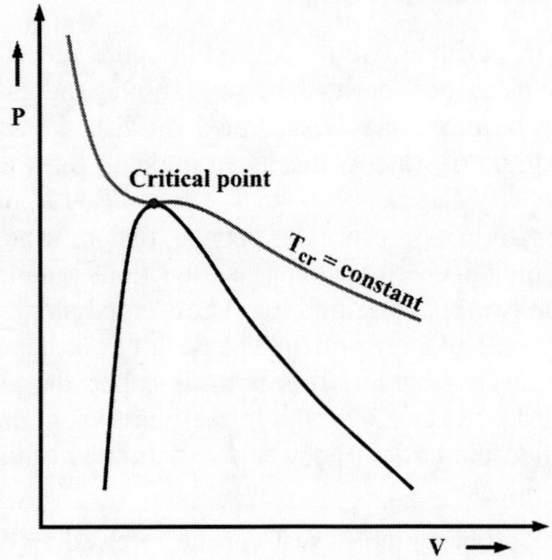


Fig. 3.22: Isotherm on $P - V$ diagram passing through the critical point.

## 3.7 Other Commonly Used Equations of State

The range of applicability of the ideal gas equation is very limited. Therefore, it is desirable to have equations of state which can represent the $P - v - T$ characteristics of a gas over the entire superheated region. Such equations are more complex to use. Many such equations have been proposed to describe the behaviour of gases with a very few limitations. In this section, we will introduce only a few of these equations. One of the earliest equations of state is the **Van der Waals equation of state** and it is expressed as

$$P + \frac{a}{v^2} = \frac{RT}{v-b} \tag{3.1}$$

In this equation, the term $a/v^2$ is included to account for intermolecular attractions and $b$ includes the volume occupied by the gas molecules. These effects are not considered in the ideal gas equation. The constant temperature line passing through critical point on $P - v$ diagram (Fig. 3.22) has a horizontal inflection at this point. Therefore, $\left(\frac{\partial P}{\partial v}\right)_{T=T_{cr}} = 0$ and $\left(\frac{\partial^2 P}{\partial v^2}\right)_{T=T_{cr}} = 0$. These conditions, when applied on Van der Waals equation, give the values of coefficients $a$ and $b$ as $a = \frac{27R^2T_{cr}^2}{64P_{cr}}$ and $b = \frac{RT_{cr}}{8P_{cr}}$. The parameters corresponding to critical point of a substance can be used to determine the values of $a$ and $b$. This equation is not very good in terms of accuracy, however, it is one of the initial attempts to describe behaviour of real gases. Another relatively simple equation of state is expressed as

$$P = \frac{RT}{v-b} - \frac{a}{v^2+cbv+db^2} \tag{3.2}$$

This is known as the **cubic equation of state**. The parameters $a, b, c$ and $d$ becomes zero for an ideal gas. In some of the equations of state, the parameters are also considered temperature dependent. Some of the recent equations of state are very complicated and contain a large number of empirical constants.

## UNIT SUMMARY

In this chapter, we introduced the pure substances, the phases in which pure substances can exist and the phase boundaries. The solid, liquid and gas are three principle phases of a substance whereas there can be other phases separated by distinct molecular structures within a principle phase. The phase diagram of water is discussed in detail for a wide range of pressures and temperatures. The important property diagrams ($P - v$, $T - v$ and $P - T$) of a substance are elucidated for water and different phase transformation processes such as fusion, vapourization and sublimation are described. The associated amount of energy with phase transitions is called latent heat and it varies with temperature and pressure. The two-phase equilibrium is observed along the fusion line, vapourization line and sublimation line whereas all three principle phases of a substance can coexist in equilibrium at intersection of these lines on phase diagram. This point is called the triple point and below the triple point liquid phase of a substance cannot exist. On the phase diagram, the vapourization line is restricted to a unique point, called the critical point where saturated liquid and saturated vapour phases of a substance become identical.

The property tables that can be referred to determine all properties of a substance once its state is designated, are also elucidated with the help of water as an example. The property tables of water are usually called steam tables and have different forms for saturated states and single phase states. In general two independent intensive properties are required to designate the state of a substance, whereas the saturated liquid and saturated vapour states can be designated by a single property. The properties of saturated two-phase mixture are dependent on a dimensionless parameter, called the quality. The properties of the two-phase mixture are obtained using the weighted average of its constituents. The properties of compressed liquid can be equated to saturated liquid state at the given temperature with negligible error. The superheated vapour region of a substance can be approximated as an ideal gas at low pressures and high temperatures. The deviation from ideal gas behaviour can be quantified using an important parameter, called compressibility factor. The deviation from ideal gas behaviour is highest for a substance near the saturated vapour line and the critical point. In addition to the ideal gas equation, other more complicated equations of state can be used to describe the complete behaviour of a superheated vapour.

### Solved Examples

**Example 3.1:** Explain why water expands upon freezing.

**Solution:**

**Discussion:** Water consists of oxygen and hydrogen atoms. Oxygen is more electronegative than hydrogen due to which it attracts the shared electrons more than hydrogen. This leads to formation of partial negative charge on oxygen and partial positive charge on hydrogen. In solid form, the water molecules are arranged in a crystalline structure. While doing so the positively charged hydrogen atom on one molecule shares a bond with negatively charged oxygen atom on other molecules, called the hydrogen bond (Fig. E-3.1-a). This results in a huge void space between the molecules and takes more space than the corresponding volume of liquid water. In liquid form of water, molecules have more energy, therefore, these do not stay in a structure and are randomly distributed (E-3.1-b). This is the reason for expansion of water upon freezing.

One interesting fact is that due to the lighter weight of ice than its liquid form it floats on water. That's why aquatic flora and fauna survive. If this is not the case, then aquatic life would have been completely different.

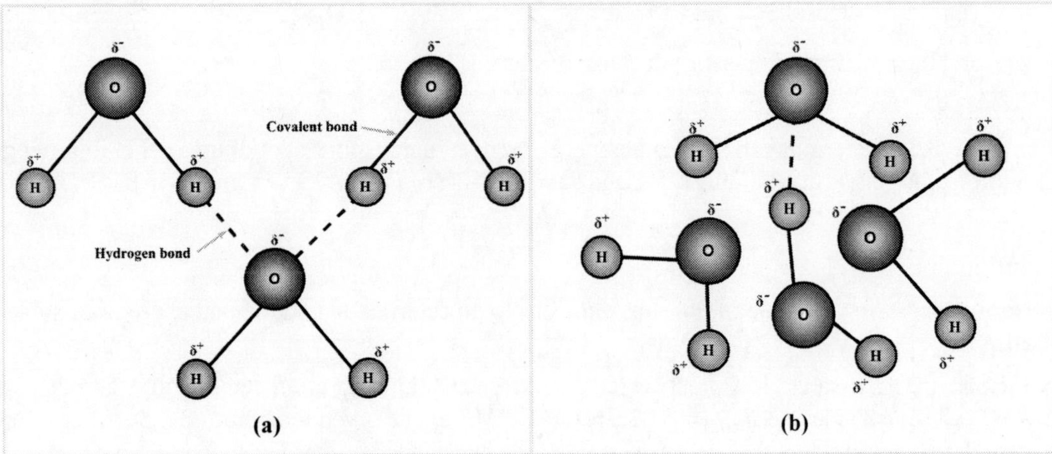


**Fig. E-3.1**

**Example 3.2:** Determine the saturated state of water based on the following given parameters and also find the corresponding saturation temperature: (a) $u = 1475\ kJ/kg$, (b) $h = 2727.9\ kJ/kg$, (c) $v = 0.001114\ m^3/kg$, (d) $P = 47.416\ kPa$ and (e) $u = 1260\ kJ/kg$.

**Solution:**

**Assumptions:** As the properties are mentioned corresponding to the saturation states, Tables B.1 and B.2 are to be referred. Also, the saturation state can be located by a single property value on steam tables.

**Analysis: (a)** If we refer Table B.1, $u = 1475\ kJ/kg$ is mentioned in the saturated liquid column against a temperature of 325°C. Therefore, the given state of the substance is saturated liquid and saturation temperature is 325°C.

**(b)** Similarly from Table B.1, $h = 2727.9\ kJ/kg = h_g$@310°C.

Therefore, the given state of the substance is saturated vapour and saturation temperature is 310°C.

**(c)** From Table B.1, $v = 0.001114\ m^3/kg = v_f$@170°C.

Therefore, the given state of the substance is saturated liquid and saturation temperature is 170°C.

**(d)** From Table B.1, $P = 47.416\ kPa = P_{sat}$@80°C. Since pressure and temperature both stay constant in the saturated liquid vapour region. Therefore, at the given saturation pressure, the substance can exist in either saturated liquid or a saturated vapour state. It can also stay as a saturated liquid-vapour mixture in equilibrium, however, in that case any other property than temperature is needed to determine its state. Because pressure and temperature are dependent parameters in this region. The saturation temperature corresponding to given pressure is 80°C.

**(e)** From Table B.1, it can be observed that $u = 1260\ kJ/kg$ is not available in Table B.1 whereas the closest values of 1253.7 $kJ/kg$ and 1279.7 $kJ/kg$ are available in saturated liquid column. Therefore, the state of the substance is saturated liquid and to find the corresponding saturation temperature, interpolation is required.

$u_1 = 1253.7\ kJ/kg = u_f@285°C$ and $u_2 = 1279.7\ kJ/kg@290°C$. Using linear interpolation,

$$T - T_1 = \frac{T_2 - T_1}{u_2 - u_1}(u - u_1) \Rightarrow T = 285 + \frac{290 - 285}{1279.7 - 1253.7}(1260 - 1253.7) = 286.212°C.$$

Therefore, the corresponding saturation temperature is 286.212°C.

**Example 3.3:** Determine the internal energy and enthalpy corresponding to the following given conditions for water: (a) $P = 1.2\ MPa$ and $T = 250°C$, (b) $P = 0.9\ MPa\ and\ T = 815°C$.

**Solution:**

**Assumptions:** First the state of the substance has to be determined and then an appropriate table can be used.

**Analysis: (a)** Use saturation pressure table and check the saturation temperature corresponding to $1.2\ MPa$. Using Table B.2, $T_{sat} = 187.96°C@1.2\ MPa$. The given temperature (250°C) is greater than $T_{sat}$ at the given pressure, therefore, the state of the substance is superheated vapour. Now, we can refer to superheated steam table. From Table B.3, $u = 2704.7\ kJ/kg$ and $h = 2935.6\ kJ/kg$.

**(b)** Using Table B.2, $T_{sat} = 175.35°C@0.9\ MPa$. The given temperature (815°C) is greater than $T_{sat}$ at the given pressure, therefore, the state of the substance is superheated. From Table B.3, it is observed that none of the values (pressure and temperature) are directly available in Table B.3. Therefore, interpolation needs to be performed using the closest values. The pressure conditions available in superheated steam tables are $0.8\ MPa$ and $1.0\ MPa$ and the temperatures available are 800°C and 900°C.

The corresponding values of internal energy and enthalpy are given as

| | $P = 0.8\ MPa$ | | $P = 1.0\ MPa$ | |
|---|---|---|---|---|
| $T$ (°C) | $u\ (kJ/kg)$ | $h\ (kJ/kg)$ | $u\ (kJ/kg)$ | $h\ (kJ/kg)$ |
| 800 | 3662.5 | 4157.0 | 3661.7 | 4156.1 |
| 815 | **3691.3** (using interpolation) | **4192.78** (using interpolation) | **3690.53** (using interpolation) | **4191.90** (using interpolation) |
| 900 | 3854.5 | 4395.5 | 3853.9 | 4394.8 |

The linear interpolation gives the values of internal energy and enthalpy at 815°C. Using the previous table, the data is now available at pressures of 0.8 $MPa$ and 1.0 $MPa$ for temperature of 815°C as given below:

| $T = 815$°C | | |
|---|---|---|
| $P\ (MPa)$ | $u\ (kJ/kg)$ | $h\ (kJ/kg)$ |
| 0.8 | 3691.3 | 4192.78 |
| 0.9 | **3690.8** (using interpolation) | **4192.34** (using interpolation) |
| 1.0 | 3690.53 | 4191.90 |

Further, interpolation along pressure gives the desired values of internal energy and enthalpy. Hence, $u = 3690.8\ kJ/kg$ and $h = 4192.34\ kJ/kg$.

**Example 3.4:** Find the specific volume, internal energy and enthalpy of water at $P = 5\ MPa$ and $T = 100$°C. Also, use the saturated liquid steam tables at given temperature to compute the above parameters and comment on the differences with the previous values.

**Solution:**

**Assumptions:** First the state of the substance has to be determined and then an appropriate table can be used.

**Analysis:** Use saturation pressure table and check the saturation temperature corresponding to 5 $MPa$. Using Table B.2, $T_{sat} = 263.94°C@5\ MPa$. The given temperature (100°C) is less than $T_{sat}$ at the given pressure, therefore, the state of the substance is compressed liquid. Hence, we can refer to compressed liquid steam tables. Using Table B.4, $v = 0.0010410\ m^3/kg, u = 417.65\ kJ/kg$ and $h = 422.85\ kJ/kg$.

Now, let's use the saturated liquid steam table (Table B.1) and find the values corresponding to 100°C. Using this table, we obtain $v = 0.001043\ m^3/kg, u = 419.06\ kJ/kg$ and $h = 419.17\ kJ/kg$.

The percentage error associated with values estimated from saturated steam table are:

$$\%\text{ error in specific volume} = \frac{|0.0010410 - 0.001043|}{0.0010410} \times 100 = 0.1921\%$$

$$\%\text{ error in specific internal energy} = \frac{|417.65 - 419.06|}{417.65} \times 100 = 0.3376\%$$

$$\%\text{ error in specific enthalpy} = \frac{|422.85 - 419.17|}{422.85} \times 100 = 0.8703\%$$

Among all parameters, the error observed in estimation of enthalpy is highest. Let's use the pressure correction and again compute enthalpy as

$$h = h_f@100°C + (P - P_{sat}@100°C) \times v_f@100°C = 424.28\ kJ/kg$$

$$\text{Now, }\%\text{ error in specific enthalpy} = \frac{|422.85 - 424.28|}{422.85} \times 100 = 0.3382\%$$

This shows that error has improved upon using the correction. However, this correction may give huge over estimation for moderate to high pressures.

Finally, it can be concluded that the compressed states can be assumed equivalent to saturated liquid at the given temperature for estimation of their properties with negligible error.

**Example 3.5:** Using the following parameters, find the state of a substance and estimate the corresponding enthalpy: (a) $P = 500\ kPa$ and $v = 0.265\ m^3/kg$ and (b) $T = 305°\text{C}$ and $x = 0.67$. Also, locate the given conditions on $P - v$ diagram. Here, the considered substance is water.

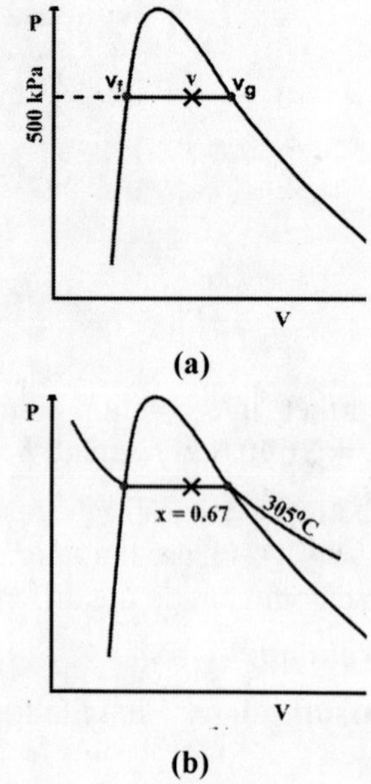


Fig. E-3.5

**Solution:**

**Analysis: (a)** Use saturation pressure table and check the corresponding $v_f$ and $v_g$ to determine the state of the substance. Referring to Table B.2 gives, $v_f = 0.001093\ m^3/kg$ and $v_g = 0.37483\ m^3/kg$ at $500\ kPa$ pressure. This show that $v_f < v < v_g$. It means that the substance exists as a saturated liquid-vapour mixture at the given conditions. Therefore, first quality has to be determined to calculate the enthalpy of the substance.

Using $v = \left(v_f + xv_{fg}\right)_{@500\ kPa} \Rightarrow x = 0.706.$

Therefore, $h = \left(h_f + xh_{fg}\right)_{@500\ kPa} \Rightarrow h = 2129.148\ kJ/kg.$

**(b)** It is already clear from the given conditions that the substance consists of the saturated liquid-vapour mixture. Use the saturated temperature table (Table B.1) to find required values at 305°C. This gives,

$h = \left(h_f + xh_{fg}\right)_{@305°\text{C}} \Rightarrow h = 2288.521\ kJ/kg.$

The corresponding states are shown on $P - v$ diagrams in Figs. E-3.5-a and E-3.5-b. It is always useful to plot the state of a substance on property diagrams. It helps in easy identification of the state of a substance and estimation of corresponding properties.

## EXERCISES

### Multiple Choice Questions

| 3.1 | A pure substance must have homogeneous | | | |
|---|---|---|---|---|
| | A. Molecular structure | B. Chemical composition | C. Phase | D. All of the above |

| | | | | |
|---|---|---|---|---|
| 3.2 | It takes more time to cook food at higher elevation because | | | |
| | A. Saturation pressure increases | B. Saturation pressure decreases | C. Saturation temperature increases | D. None of these |
| 3.3 | The latent heat of vapourization of water ________________ upon decrease of saturation pressure or temperature. | | | |
| | A. Remains the same | B. Can increase or decrease | C. Decreases | D. Increases |
| 3.4 | Which of the following phase does not exist below the triple point of a substance | | | |
| | A. Solid | B. Liquid | C. Gas | D. Mixture of solid and gas |
| 3.5 | A vapour state that is about to condense is called | | | |
| | A. Subcooled vapour | B. Superheated vapour | C. Saturated vapour | D. Compressed vapour |
| 3.6 | A liquid state that is about to boil is called | | | |
| | A. Subcooled liquid | B. Saturated liquid | C. Superheated liquid | D. Compressed liquid |
| 3.7 | A pure substance having temperature greater than its critical temperature is usually called | | | |
| | A. Superheated vapour | B. Critical vapour | C. Supercritical vapour | D. Superheated liquid |
| 3.8 | The number of independent properties required to designate the compressed liquid state of a substance on steam tables is | | | |
| | A. 0 | B. 1 | C. 2 | D. 3 |
| 3.9 | The set of independent intensive properties that cannot be used to designate the state of a saturated liquid-vapour mixture is | | | |

| | | | | |
|---|---|---|---|---|
| | A. $P$ and $v$ | B. $P$ and $T$ | C. $P$ and $x$ | D. $T$ and $v$ |
| 3.10 | Which of the following parameters cannot be used to designate the saturated liquid or saturated vapour states on steam tables? | | | |
| | A. $P$ | B. $T$ | C. $v$ | D. $x$ |
| 3.11 | At the critical point of a substance which of the following phases cannot exist? | | | |
| | A. Solid | B. Liquid | C. Gas | D. Mixture of liquid and vapour |
| 3.12 | The value of compressibility factor near the critical point of a pure substance is | | | |
| | A. $Z > 1$ | B. $Z < 1$ | C. $Z = 1$ | D. $Z = 0$ |

**Answers of Multiple Choice Question**

**3.1 B, 3.2 B, 3.3 D, 3.4 B, 3.5 C, 3.6 B, 3.7 A, 3.8 C, 3.9 B, 3.10 D, 3.11 A, 3.12 B**

**Short and Long Answer Type Questions**

3.1 Explain why water boils at low temperature at higher elevations.

3.2 In the cold regions, a long spindle is used to operate the valve of an external tap and usually the valve is positioned inside the wall. Explain why?

3.3 What is the reason for using solid carbon dioxide instead of ice for preserving food items while shipping?

3.4 Why is solid $CO_2$ called dry ice?

3.5 How to decide the pair of two independent intensive properties to designate the state of a system? Explain with examples.

3.6 Why does the cooking time decrease upon using the pressure cookers?

3.7 How much latent heat of vaporization is needed to change the phase of a substance from liquid to vapour at supercritical conditions?

3.8 Can the vapour phase of water exist below 0°C temperature?

3.9 Why does a gaseous substance deviate most from ideal gas behaviour near the saturated vapour line and critical point?

3.10 What happens when the pressure of a saturated liquid goes below its saturation pressure at a given temperature?

3.11 A mixture of saturated liquid-vapour is heated in a closed container having rigid walls. The initial specific volume of the mixture is less than the critical specific volume. What will happen to the liquid level in the container upon heating of the mixture?

3.12 In the previous question (3.11), if the container is replaced by a piston-cylinder system. What will happen to the water level upon heating?

## Numerical Problems

3.1 Determine the state of water and estimate the corresponding enthalpy for the given parameters: (a) $P = 2.50\ MPa$ and $u = 3026.2\ kJ/kg$, (b) $P = 200\ kPa$ and $x = 0.76$, (c) $T = 265°C$ and $u = 1443.2\ kJ/kg$, (d) $P = 30\ MPa$ and $u = 875\ kJ/kg$ and (e) $T = 725°C$ and $P = 166.35\ kPa$.

3.2 Water comes out of a condenser at the rate of $0.015\ m^3/s$ at 150°C temperature and 30 $MPa$ pressure. Determine the mass flow rate of the water. Consider if the saturated liquid table is used at the given temperature conditions, how much change will occur in the mass flow rate of the water?

3.3 A vessel contains liquid water at 25°C and it is sealed with a lid of 5 $kg$ mass. Outside the vessel, pressure is atmospheric. Determine the temperature at which water starts boiling.

3.4 The lid of a pressure cooker has 50 $g$ petcock which provides an opening of 4 $mm$ diameter. The lid is tightly fitted to the cooker and is exposed to surrounding atmospheric pressure. Determine the temperature at which water boils inside the pressure cooker.

3.5 The saturated liquid-vapour mixture is contained in a rigid vessel at 125°C and 22% quality. The mass of the mixture is 5 $kg$ and it is heated till the pressure inside the container reaches 850 $kPa$. Represent the process on $T - v$ diagram. Determine the amount of heat added to reach the final state. What will happen to the water level inside the container during this process?

3.6 In the previous problem (3.5), if the rigid container is replaced with a piston-cylinder system that maintains the constant pressure of 300 $kPa$. Determine the amount of heat needed to convert the saturated-liquid vapour mixture to saturated vapour state. Also, plot the process on $T - v$ diagram.

3.7 10 $kg$ superheated steam at 16 $MPa$ pressure and 480°C is contained in a rigid vessel. The steam is then cooled till it reaches the saturated vapour state. Determine the amount of heat rejected to the surroundings during the cooling process. Show the process on $P - v$ diagram.

3.8 2 $kg$ water is contained in a piston-cylinder system at 25°C as shown in Fig. P-3.8. The cylinder is provided with set of stops at volume of 0.5 $m^3$ and the massless piston is exposed to atmospheric pressure. Water is heated till it reaches a temperature of 500°C. Estimate the amount of heat required to reach the final state. Also, find the associated boundary work during the process. Show the process on $P - v$ diagram.

3.9 A piston-cylinder system contains 1.5 $kg$ of water at 30°C as shown in Fig. P-3.9. The water is heated till it reaches pressure and temperature of 1.2 $MPa$ and 600°C. The other side of the piston is exposed to atmospheric pressure and a linear spring touches the piston at a volume of 0.01 $m^3$. Determine: (a) the amount of heat added during the process, (b) stiffness of the spring if the diameter of the piston is 150 $mm$ and (c) the boundary work associated with the process. Show the process on $P - v$ diagram.

3.10 Two insulated containers are connected together through a valve. Volume of first container is 1 $m^3$ and it contains water at 500 $kPa$ and $v = 0.8\ m^3/kg$. The second container contains 5 $kg$ of water at 0.8 $MPa$ and 450°C. Now the valve is opened and the water is allowed to reach a final uniform state. Determine the specific volume, internal energy and enthalpy at the final state.

3.11 Saturated vapour is contained in a piston-cylinder system at atmospheric pressure and volume of 0.048 $m^3$. The diameter of the piston is 0.35 $m$ and the steam is cooled inside the cylinder by spraying cold water over the cylinder. During the cooling process, the piston is stuck at its position. If the temperature of water inside the cylinder reaches 35°C, determine the amount of heat transfer and the frictional force acting on the piston.

3.12 A freely moving piston divides a well insulated rigid cylinder into two parts of 0.8 $m^3$ and 0.2 $m^3$ volumes. The 0.2 $m^3$ volume of the cylinder is filled with superheated steam at 100 $kPa$ and 115°C whereas the other compartment contains air at 100 $kPa$ and 20°C. Determine the final temperature of the system if the thermal equilibrium is established as a result of heat transfer through the piston. The piston does not allow the physical mixing of fluids in two compartments.

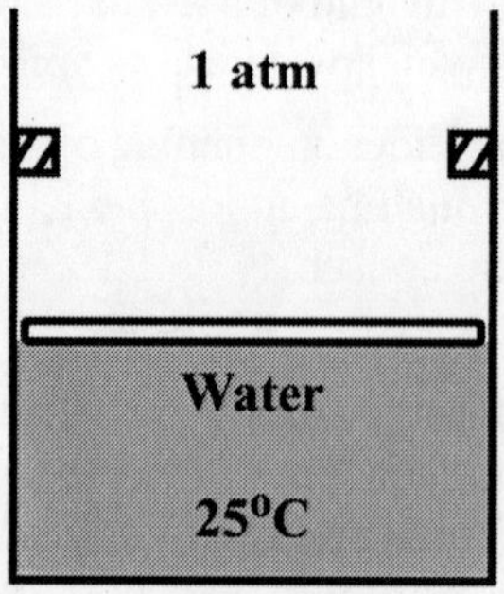

Fig. P-3.8

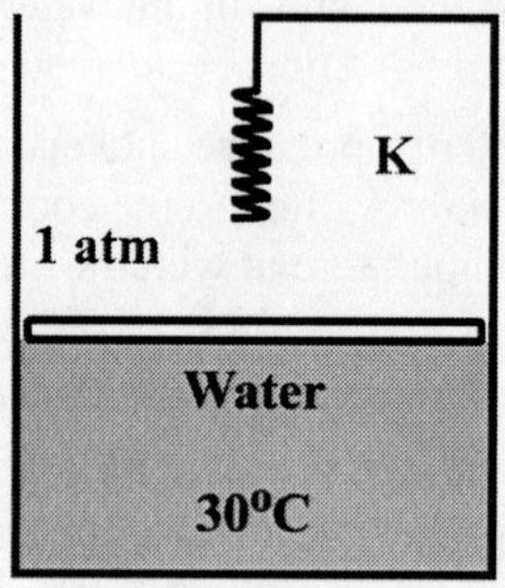

Fig. P-3.9

## PRACTICAL

**Activity 1:** Take an open pan filled with water and put it on the gas-stove. Put the gas-stove on a mild flame. During the heating process, continuously measure the temperature of the water using a thermometer and stir it using a wooden rod such that the temperature of the entire water volume is uniform. During this process, visually observe the condition at which vapour bubbles start appearing at the surface of the pan and note down the corresponding temperature. Use this information to calculate the atmospheric pressure at your location and relate it to elevation of your place.

## REFERENCES AND SUGGESTED READINGS

1. Parry, W.T., Bellows, J., Gallagher, J.S. and Harvey, A.H., 2009. ASME international steam tables for industrial use.
2. Eliezer, S., Ghatak, A.K., Hora, H. and Teller, E., 1986. *An introduction to equations of state: theory and applications* (pp. 153-164). Cambridge: Cambridge University Press.
3. Bejan, A., 2016. *Advanced engineering thermodynamics*. John Wiley & Sons.

# 4 Energy Analysis for Control Volumes

## UNIT SPECIFICS

*Through this unit we have discussed the following concepts:*

- *Application of first law of thermodynamics to open systems;*
- *Energy analysis through control volume approach;*
- *Law of mass conservation in open flow systems;*
- *Analysis of various open flow systems such as nozzles and diffusers, turbines, compressors, throttle valves, heat exchangers etc.;*
- *Energy analysis of transient systems;*
- *Solved examples on the analyses of open flow systems.*

## RATIONALE

*The practical application of the first law of thermodynamics includes the efficiency analyses of the power plants through the detailed analysis of flow analysis inside the individual components of the power plants. This chapter will help the learners to apply the basic concepts on the analysis of open flow systems such as heat exchangers, turbines, compressors, throttle valves etc. The application of the first law on the transient systems is also discussed. The Know More section discusses the Joule-Thomson coefficient.*

## PRE-REQUISITES

*First law of thermodynamics; Thermodynamic properties*

## UNIT OUTCOMES

*List of outcomes of this unit is as follows:*

*U4-O1: Application of mass conservation in open systems*

*U4-O2:* *Energy conservation principle for open systems*

*U4-O3:* *Thermodynamic analysis of practical steady flow devices*

*U4-O4:* *Thermodynamic analysis of transient systems*

*U4-O5:* *Introduction to Joule-Thomson coefficient*

| Unit-4 Outcomes | EXPECTED MAPPING WITH COURSE OUTCOMES (1- Weak Correlation; 2- Medium correlation; 3- Strong Correlation) | | | | | |
|---|---|---|---|---|---|---|
| | CO-1 | CO-2 | CO-3 | CO-4 | CO-5 | CO-6 |
| U4-O1 | 3 | 3 | 1 | 1 | 3 | 3 |
| U4-O2 | 3 | 3 | - | 1 | 3 | 3 |
| U4-O3 | 3 | 3 | - | 1 | 3 | 3 |
| U4-O4 | 3 | 3 | 1 | 1 | 3 | 3 |
| U4-O5 | 3 | 3 | 1 | 1 | 3 | 3 |

## 4.1 Conservation of Mass for a Control Volume

A control volume (or an open system) includes a region in space which is of interest and is separated from its surrounding by a boundary, called control surface. Such a system allows the exchange of heat, work and mass across its control surface as shown in Fig. 4.1.

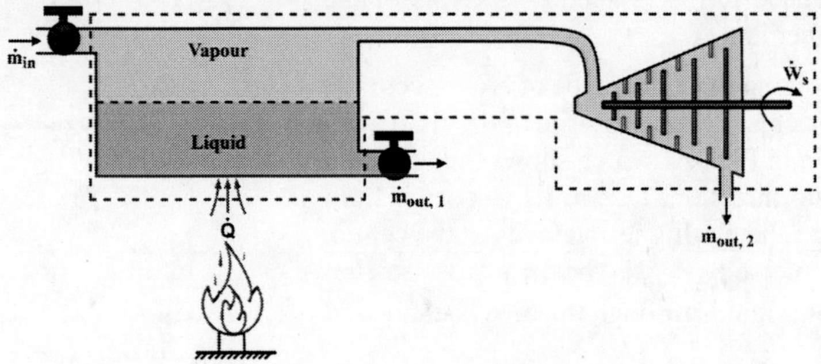


**Fig. 4.1:** A control volume depicting heat, work and mass interactions.

In a closed system where the quantity of matter remains fixed within the boundary of the system, mass conservation is inherently satisfied. Whereas for control volumes, depending upon the inflow and outflow of mass across its boundaries, the conservation of mass should be established. Like energy, fundamentally mass is also a quantity which cannot be created and destroyed. To mathematically explain this statement, let us consider a water tank in our home as a control volume. It usually has an inlet pipeline and an outlet that supplies water to different locations such as kitchen sink, bathroom,etc. A typical control volume representation of such a water tank is shown in Fig. 4.2 where the inner surface of the tank acts as the boundary of the control volume. Consider a water tank that is half filled with water and inlet pipeline is supplying water at the rate of 10 $kg/hr$. The water is being discharged through the tank outlet for different household requirements at the rate of 3 $kg/hr$. In this situation, we experience that the water level rises in the tank and the mass of water increases. The rate of increase in mass of water within the tank is 7 $kg/hr$. If we just take the opposite situation that mass is going through the discharge line at a higher flow rate than the supply flow rate, the water level, hence, the mass of water will decrease in the tank.

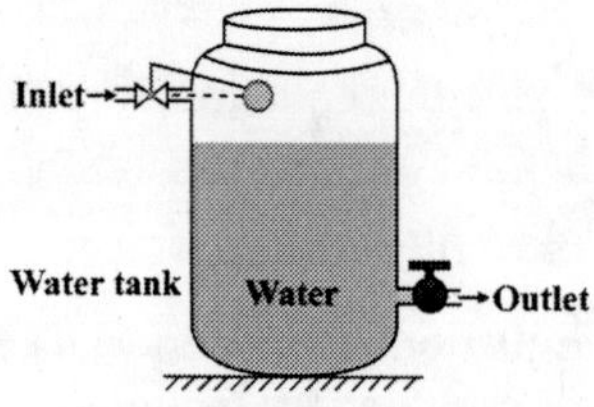


**Fig. 4.2:** Water tank, an example of a control volume.

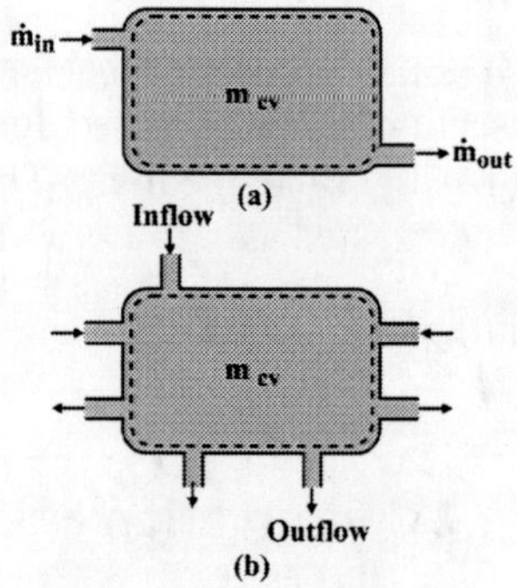


**Fig. 4.3:** Control volume having: (a) one inlet and one outlet and (b) multiple inlets and outlets.

This simple illustration explains that the rate of change of mass of a control volume is equal to the net rate of mass flow across its boundaries. Let $\dot{m}_{in}$ is the rate of mass inflow and $\dot{m}_{out}$ is the rate of mass outflow through the control surface (Fig. 4.3-a). Then, the time rate of change of mass of control volume can be expressed as

$$\frac{dm_{cv}}{dt} = \dot{m}_{in} - \dot{m}_{out} \tag{4.1}$$

For a control volume having multiple inlets and outlets (Fig. 4.3-b), Eqn. (4.1) can be expressed as

$$\frac{dm_{cv}}{dt} = \sum \dot{m}_{in} - \sum \dot{m}_{out} \qquad (4.2)$$

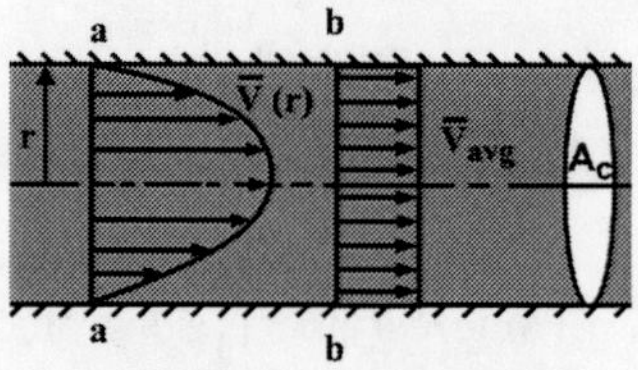


**Fig. 4.4:** Representation of flow through a pipe.

Equation (4.2) represents the conservation of mass and is usually called the **continuity equation**. This form of continuity equation often serves the purpose for most thermodynamic systems. However, the mass flow rate across the control surface can be further extended in terms of the properties of the associated fluid medium.

Consider a fluid of density, $\rho$ is flowing through a pipe of cross-sectional area, $A_c$ as shown in Fig. 4.4. The actual velocity profile of the fluid at an arbitrary cross-section a-a is shown in Fig. 4.4 which shows the variation in velocity of fluid along the radial direction. Due to its viscosity, fluid has zero velocity at the pipe surface and has highest velocity at the centre. The average representative of fluid velocity $\underline{V}_{avg}$ is shown at cross-section b-b. The volumetric flow rate of the fluid is through the pipe is expressed as

$$\dot{V} = \int_{A_c} \underline{V}(r) dA_c = \underline{V}_{avg} A_c \qquad (4.3)$$

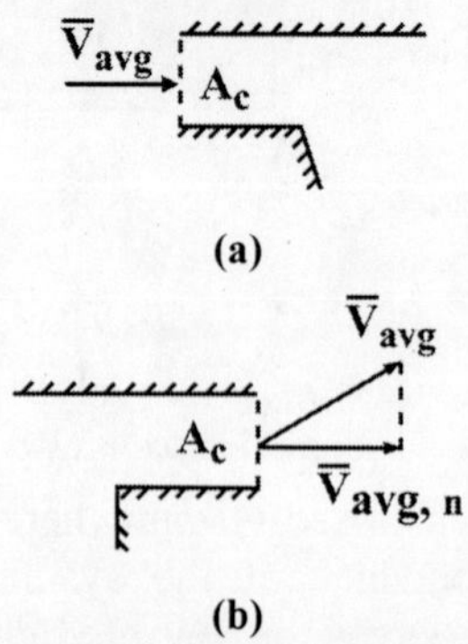


**Fig. 4.5:** Representation of direction of fluid velocity at a control surface: (a) parallel to surface normal and (b) at an angle to surface normal.

Using this equation, for a fluid of density, $\rho$, the mass flow rate is obtained as

$$\dot{m} = \rho \dot{V} = \rho \underline{V}_{avg} A_c \qquad (4.4)$$

Equations (4.3) and (4.4) are applicable for a stationary control surface (Fig. 4.5-a) for which flow direction is normal to the surface. In case the flow approaches or leaves a surface at an angle, the component of average velocity normal to the surface (Fig. 4.5-b) is used to calculate the mass flow rate as

$$\dot{m} = \rho \underline{V}_{avg,n} A_c \qquad (4.5)$$

It is also to be noted that the average density of the fluid at the given cross-section should be used for estimation of mass flow rate for a variable density fluid medium. The mass of the control volume can also be represented as

$$m_{cv} = \int \rho dV \qquad (4.6)$$

## 4.2 Conservation of Energy for a Control Volume

The fundamental principle of energy conservation is discussed in detail in Chapter 2. For a control mass system in rate form, energy conservation is expressed as $\frac{dE_{cm}}{dt} = \dot{Q} - \dot{W}$. Whereas in case of control volume, there is inflow and outflow of mass across its boundary (Fig. 4.6) in addition to heat and work interactions.

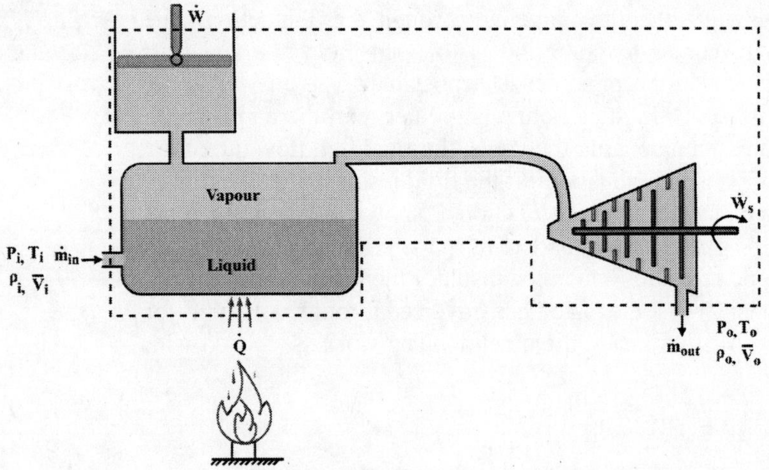


**Fig. 4.6:** A control volume involving heat, work and mass exchange with surroundings.

Mass also possesses energy that further leads to one more energy interaction associated with mass exchange across the control surface. Therefore, the energy conservation equation for a control volume is expressed as

$$\frac{dE_{cv}}{dt} = \dot{Q} - \dot{W} + \dot{E}_{mass,in} - \dot{E}_{mass,out} \qquad (4.7)$$

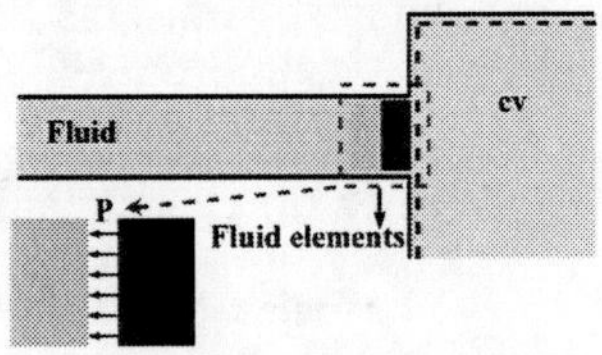


**Fig. 4.7:** Depiction of inflow of fluid through a control surface.

The energies possessed by mass are internal energy, associated with its thermodynamic state and kinetic and potential energies linked to the physical state of the mass. This gives total specific energy associated to mass flow as $e = u + \frac{1}{2}\underline{V}^2 + gz$. We have elaborated all these constituents of energy in Chapter 2. In addition for a flowing fluid stream, there is a requirement of some work to push fluid mass into or out of the control volume. This additional work is known as **flow work** or **flow energy** and it is necessary to maintain the continuous flow of fluid stream across the boundaries of the control volume. To understand this work, let us consider a pipe through which fluid is flowing and entering a control volume as shown in Fig. 4.7. If we consider a thin fluid element in the vicinity of the control surface, it is about to enter the control volume. It exerts a pressure on its immediate neighbouring fluid element upstream. This neighboring fluid element has to push against the pressure in order to make the flow happen. Likewise it can be said that from the source of the flow each upstream fluid element has to push its immediate neighbouring fluid element. Therefore, some energy is invested in making this flow happen. Commonly this energy is supplied to a flowing fluid stream by using a pump at the source of flow.

Let's consider another illustration in Fig. 4.8. It consists of a pipeline through which a fluid is flowing at pressure $P(> P_{atm})$ and velocity $\underline{V}$. This pipe has a T-junction and a valve as shown in Fig. 4.8-a. Now say we have $m$ mass of the same fluid at atmospheric pressure and we want to enter this mass into the flow pipeline through the T-junction. The immediate thought comes to our mind is to open the valve and start pouring the fluid inside the pipeline. Is it possible? The answer is no, because the pressure inside the pipeline is greater than the atmospheric pressure, therefore, upon opening the valve, the fluid will flow out of the pipe. So the solution is to put mass $m$ of the fluid inside a piston-cylinder arrangement and connect it to the T-junction as shown in Fig. 4.8-b. Now one has to push the piston with a force to raise the pressure of fluid inside the cylinder to pipe pressure $P$ and then displace the piston to insert fluid into the pipe. Consider that the piston has traversed through volume $V$ for inserting mass $m$ of fluid inside the pipeline. The work associated with this process is

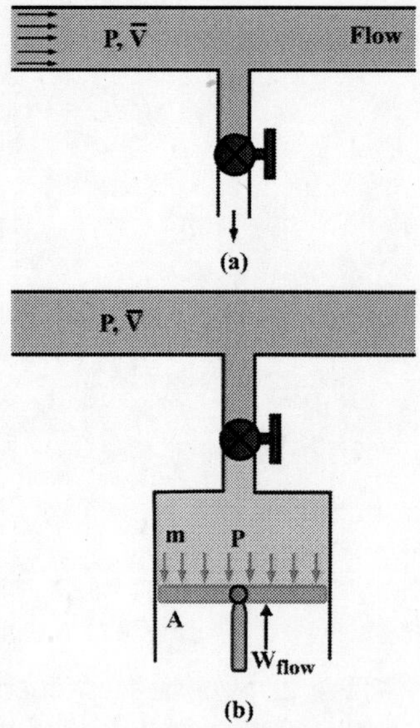


**Fig. 4.8:** (a) T-junction of a supply line provided with a valve and (b) a piston cylinder system used to inject water into the flow line.

$$W_{flow} = \int_V PdV = PV = Pmv \tag{4.8}$$

Here, $v$ is specific volume of the fluid at pressure $P$. Equation (4.8) shows the work required to make the flow of mass $m$ possible. In case the flow of fluid is to be maintained at mass flow rate of $\dot{m}$, the expression of work becomes

$$\dot{W}_{flow} = P\dot{V} = P\dot{m}v \tag{4.9}$$

In Eqn. (4.9), $\dot{V}$ is the volumetric flow rate and $\dot{m}$ is the mass flow rate of the flowing stream. This flow energy must be added to a flowing stream which it inherently exchanges with the control volume when it passes across the control surface. Therefore, the expression of total energy carried by a flowing stream can be written as

$$\dot{E}_{mass} = \dot{m}e + \dot{W}_{flow} = \dot{m}e + P\dot{m}v = \dot{m}(e + Pv)$$

$$\Rightarrow \dot{E}_{mass} = \dot{m}(u + \frac{1}{2}V^2 + gz + Pv)$$

$$\Rightarrow \dot{E}_{mass} = \dot{m}(h + \frac{1}{2}V^2 + gz) \tag{4.10}$$

Here, $h = u + Pv$ is enthalpy of the flowing stream at the given thermodynamic conditions. When we introduced enthalpy in Chapter 2, only it was mentioned that it is a property which is a combination of other thermodynamic properties. However, from the above illustration, it is revealed that when we add flow energy to the internal energy of a substance, it becomes enthalpy. Therefore, the value of enthalpy represents the flow work potential of a substance along with its internal energy. Consider that total specific energy constituted by a flowing mass is $e_{flow} = h + \frac{1}{2}V^2 + gz$. Thus, $\dot{E}_{mass} = \dot{m}\, e_{flow}$. Here, the enthalpy and other properties are to be estimated at the thermodynamic states of the corresponding inflowing and outflowing streams.

Scan Me

For more details on control volumes and flow energy.

Using this information, the energy conservation for a control volume (Eqn. 4.7) can be expressed as

$\frac{dE_{cv}}{dt} = \dot{Q} - \dot{W} + \dot{m}_{in} e_{flow,in} - \dot{m}_{out} e_{flow,out}$. In case of multiple mass interactions, this expression can be written as

$$\frac{dE_{cv}}{dt} = \dot{Q} - \dot{W} + \sum \dot{m}_{in} e_{flow,in} - \sum \dot{m}_{out} e_{flow,out} \qquad (4.11)$$

The sign conventions for heat and work interactions remain the same as discussed in case of closed systems, $\dot{Q}$ and $\dot{W}$ represent the net heat and net work transfer, respectively. Usually the size of the control volume is fixed for majority applications, therefore, displacement work is often zero and work transfer mainly occurs in the form of shaft work.

## 4.3 Energy Analysis of Steady-Flow Processes

Some engineering systems such as steam power plants operate for long duration under the same conditions after their starting transient operation and before shutdown. A steam power plant comprises of multiple engineering devices such as boiler, turbine, condenser, etc. which operate under the same conditions during their steady operation period and are called the **steady flow devices**. The processes involving such devices can be referred to as **steady flow processes** if these follow a certain set of conditions. The required conditions for a steady flow process are as follows:

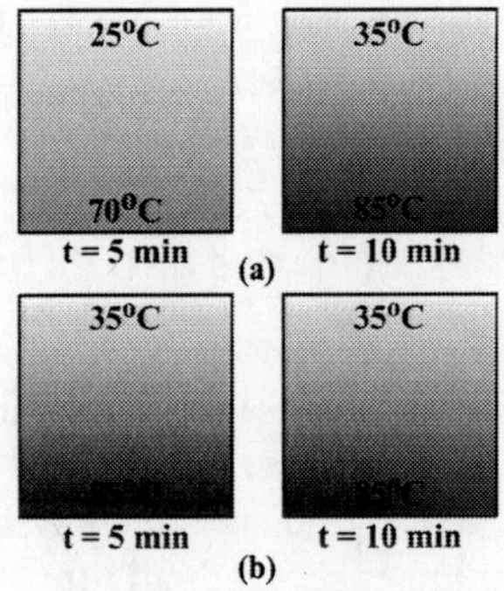


**Fig. 4.9:** (a) Unsteady control volume and (b) steady control volume.

- Any intensive or extensive properties of the control volume should not change with time. The properties can have different values at different locations within the control volume, however, at a given location these should not vary with time (Fig. 4.9).
- The mass flow rates and corresponding properties at each inlet and outlet should not change with time whereas the properties can have different values at different inlets and outlets (Fig. 4.10). Properties can also vary over the cross-section but should remain fixed at different locations of cross-section with time. Usually the average values of properties are specified for the entire cross-section at each inlet and outlet.
- The net rate of heat and work transfer between the control volume and its surroundings should not change with time.

**Fig. 4.10:** Control volume representation showing multiple inlets and outlets having invariable thermodynamic states with time.

Based on these assumptions, we obtain $\frac{dm_{cv}}{dt} = 0$ and $\frac{dE_{cv}}{dt} = 0$. Using this information, for a steady flow process, Eqn. (4.2) can be written as

$$\sum \dot{m}_{in} - \sum \dot{m}_{out} = 0 \Rightarrow \sum \dot{m}_{in} = \sum \dot{m}_{out} \qquad (4.12)$$

Similarly, energy conservation (Eqn. 4.11) is expressed as

$$\dot{Q} + \sum \dot{m}_{in} e_{flow,in} = \dot{W} + \sum \dot{m}_{out} e_{flow,out} \qquad (4.13)$$

If for a steady flow device, there is only one mass inlet and one outlet, then Eqn. (4.12) becomes $\dot{m}_{in} = \dot{m}_{out} = \dot{m}$ and Eqn. (4.13) becomes $\dot{Q} + \dot{m} \;\; e_{flow,in} = \dot{W} + \dot{m} \;\; e_{flow,out}$. Dividing both sides by $\dot{m}$ gives

$$q + e_{flow,in} = w + e_{flow,out} \qquad (4.14)$$

Here, $q = \frac{\dot{Q}}{\dot{m}}$ and $w = \frac{\dot{W}}{\dot{m}}$. A typical set of devices that follow the steady flow process are illustrated in the next section.

## 4.4 Examples of Steady Flow Devices

Here we present the flow and energy analysis of some common steady flow devices. These devices can have multiple inlets, outlets and multiple heat and work interactions.

### 4.4.1 Nozzles and diffusers

A steady flow device that is used to increase the velocity of a fluid stream at the expense of its pressure is called a **nozzle**. Nozzles are often used in garden hoses, rockets and jet engines, etc. The cross-sectional area of a nozzle gradually decreases in the direction of flow for subsonic flow and it increases gradually for supersonic flows. The profile of the nozzle is so designed that fluid expansion takes place smoothly to decrease its pressure and increase velocity. Usually the change in potential energy between inlet and outlet of a nozzle is negligible. Also, the velocity at the outlet of the nozzle is so high such that the inlet kinetic energy is negligible compared to outlet kinetic energy. Nozzle does not have any arrangement to produce work, therefore, its work interaction with surroundings is zero. Moreover, the velocity of fluid in a nozzle is sufficient to avoid the exchange of heat between flowing stream and outlet. It means that a fluid parcel entering the inlet leaves through the outlet before it can exchange heat with surroundings. Thus, the rate of heat transfer can be considered negligible. This particular assumption does not hold at very high temperature conditions such as the exit nozzle of a rocket and fuel injection nozzle of a furnace, etc. Such a control volume requires detailed heat transfer analysis and should be treated differently. A typical subsonic nozzle is shown in Fig. 4.11-a. It shows a fluid stream entering the nozzle with $h_i$ enthalpy, $\underline{V}_i$ velocity and leaving with $h_o$ enthalpy, $\underline{V}_o$ velocity. As there is no change of mass and energy within the control volume, therefore, Eqns. (4.12) and (4.13) gives $\dot{m}_i = \dot{m}_o = \dot{m}$ and $\dot{Q} + \dot{m} \;\; e_{flow,in} = \dot{W} + \dot{m} \;\; e_{flow,out}$. Using the conditions described above, energy equation becomes $\dot{m} \;\; \left(h_i + \frac{V_i^2}{2} + gz_i\right) =$

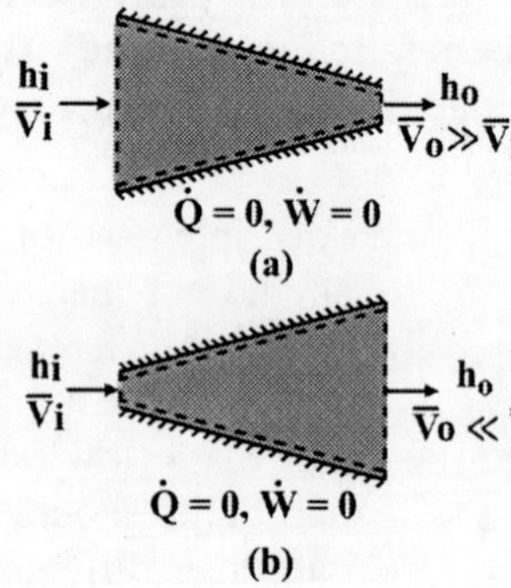


**Fig. 4.11:** Representation of a (a) nozzle and (b) diffuser.

$\dot{m}\ \left(h_o + \frac{V_o^2}{2} + gz_o\right)$. Now for a nozzle, $g(z_o - z_i) \simeq 0$ and $\frac{V_i^2}{2} \simeq 0$. Hence, $V_o \simeq \sqrt{2(h_i - h_o)}$. This depicts that velocity increases in a nozzle at the expense of enthalpy of the flowing stream.

A **diffuser** is a steady flow device that serves the purpose opposite to that of a nozzle. It increases the pressure of the flowing stream and decreases its velocity. Therefore, the enthalpy of the fluid stream increases in a diffuser at the expense of its kinetic energy. The profile of the diffuser is the reverse of the nozzle. All the assumptions of a nozzle are valid for diffusers except the negligible kinetic energy. Usually the velocity of fluid at the outlet of a diffuser is such that its kinetic energy cannot be neglected. A typical schematic representation of the diffuser is shown in Fig. 4.11-b.

### 4.4.2 Turbines and compressors

A rotary steady flow device that produces mechanical work at the expense of fluid's pressure is called a **turbine**. In a turbine, the fluid interacts with its blades and provides necessary torque to rotate the shaft and produce mechanical work. Turbines are used in thermal power plants, jet engines and hydroelectric plants. The turbines involving steam as the working medium are usually termed the steam turbines. Whereas turbines operated using high pressure gases (other than steam) are popularly called the gas turbines. In addition, the turbines used in hydroelectric plants and operated using water are termed the hydraulic turbines. In a turbine the change in potential energy is negligible. Whereas the velocities encountered in turbines are such that the changes in kinetic energy are considerable. However, the kinetic energy changes are very small compared to change in enthalpy, therefore, often neglected. Usually the turbines are well insulated and also the velocity of the fluid is such that the heat exchange with surroundings is negligible. Using these assumptions for single inlet and single outlet turbines, as shown in Fig. 4.12-a, the mass and energy conservation can be written as $\dot{m}_i = \dot{m}_o = \dot{m}$ and $\dot{Q} + \dot{m}\ \ e_{flow,in} = \dot{W} + \dot{m}\ \ e_{flow,out}$. Here $\dot{Q} \simeq 0$ and $\Delta PE \simeq 0$, therefore, $\dot{m}\ \left(h_i + \frac{V_i^2}{2}\right) = \dot{W}_s + \dot{m}\ \left(h_e + \frac{V_e^2}{2}\right)$. This gives $\dot{W}_s = \dot{m}\left(h_i - h_o + \frac{V_i^2}{2} - \frac{V_e^2}{2}\right)$. For a turbine, generally $\frac{V_i^2}{2} - \frac{V_e^2}{2} \ll h_i - h_o$. Hence, $\dot{W}_s = \dot{m}(h_i - h_o)$. It shows that the rate of shaft work done by a turbine per unit mass flow rate of working fluid is the difference of enthalpies of fluid medium at its inlet and outlet.

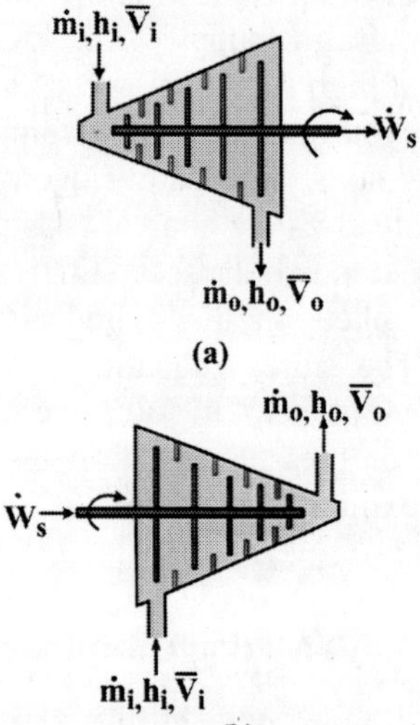

**Fig. 4.12:** Representation of a (a) turbine and (b) compressor.

The devices that are used to increase the pressure of fluid medium by taking shaft work as input are called **compressor**, **pump** and **fan**. A device that is majorly used to circulate the gaseous medium and slightly increases pressure is referred to as a fan. A compressor considerably increases the pressure of the gaseous medium, whereas a device that increases the pressure of the liquid significantly is called a pump. All these devices use some external source such as an engine or motor, etc. to supply the shaft work. The compressors/pumps are mainly of rotary type and reciprocating type. Rotary type devices are just the reverse of the turbine (Fig. 4.12-b) and the assumptions for energy analysis of such devices are the same as that of turbines. After applying the energy equation, it can be obtained that the rate at which shaft work is consumed by a compressor per unit mass flow rate of working fluid is equal to the difference of enthalpies at outlet and inlet of the compressor.

The reciprocating type compressors and pumps include a piston-cylinder arrangement to increase the pressure of the fluid medium. Such devices are mainly used for very high pressure requirements. In such a device, suction of the working medium takes place during downward movement of piston and it discharges high pressure fluid during the upward motion of the piston, as shown in Fig. 4.13. Such an arrangement results in a pulsating supply of the fluid. However, when the prime mover is rotated at high rotational speeds and flow is averaged for an integral number of cycles, steady state assumption holds valid. The properties of the control volume are also averaged for integral cycles in such systems to perform steady state analysis. Usually reciprocating compressors include either extended surfaces (fins) on the body of the cylinder or cooling water jackets in the cylinder body to dissipate the heat to surroundings. Therefore, adiabatic conditions do not work for reciprocating devices. Moreover, in our thermodynamic analysis we will consider the compressor operation as adiabatic unless the rate of heat transfer is explicitly mentioned.

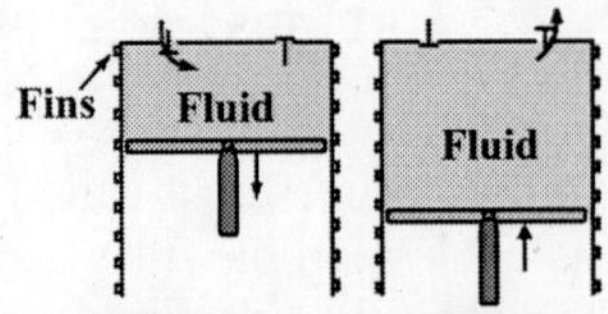

**Fig. 4.13:** A reciprocating compressor showing suction and discharge stroke.

### 4.4.3 Mixing chambers

Mixing of multiple streams has many occurrences in engineering applications. For example, in the faucet of a bathroom, hot and cold water mixes to supply outlet water at desired temperature. A physical arrangement that allows the mixing of two or more fluid streams is typically known as a **mixing chamber**. For instance, a T-junction in a pipeline can be called a mixing chamber. Usually the mixing chambers are insulated and do not exchange heat with the surroundings. Also, there is no work interaction in the mixing chambers. The changes in kinetic energy and potential energy are usually very small for mixing chambers compared to change in enthalpies. Therefore, $\Delta KE$ and $\Delta PE$ are also neglected. A schematic representation of a mixing chamber is shown in Fig. 4.14. It shows two streams 1 and 2 entering at different conditions and after mixing coming out at state 3. The mass and energy balance for such a control volume results in $\dot{m}_1 + \dot{m}_2 - \dot{m}_3 = 0$ and $\dot{m}_1 h_1 + \dot{m}_2 h_2 = \dot{m}_3 h_3$. This gives $\dot{m}_3 = \dot{m}_1 + \dot{m}_2$ and $h_3 = \frac{\dot{m}_1}{\dot{m}_3} h_1 + \frac{\dot{m}_2}{\dot{m}_3} h_2$. It shows that

**Fig. 4.14:** Mixing chamber as a steady flow device.

enthalpy at the outlet of a mixing chamber is the mass weighted average of the enthalpies at two inlets. Therefore, its values will be between $h_1$ and $h_2$. If there is heat addition to the control volume during the mixing process, then energy balance becomes $\dot{Q} + \dot{m}_1 h_1 + \dot{m}_2 h_2 = \dot{m}_3 h_{3a}$ which gives $h_{3a} = \frac{\dot{m}_1}{\dot{m}_3} h_1 + \frac{\dot{m}_2}{\dot{m}_3} h_2 + \frac{\dot{Q}}{\dot{m}_3} = h_3 + \frac{\dot{Q}}{\dot{m}_3}$. Here, $h_{3a}$ is the enthalpy of the fluid stream at the outlet of the mixing chamber considering heat addition. This shows that there will be a fixed addition of magnitude $\frac{\dot{Q}}{\dot{m}_3}$ in the enthalpy of the fluid stream at the outlet above the mass weighted average. Such a situation is graphically represented in Fig. 4.15. The point 3 in Fig. 4.15 represents the final state of the fluid stream for adiabatic conditions. Point 3a is the final state if heat is added at the rate of $\dot{Q}$ and 3b depicts the enthalpy of outlet stream when heat is rejected by the mixing chamber at the rate of $\dot{Q}$.

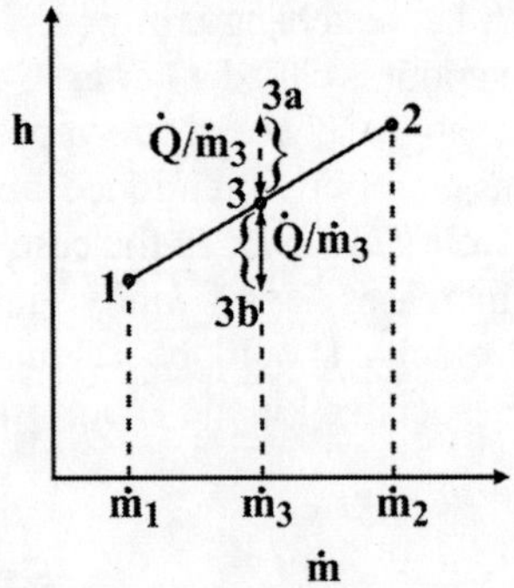


**Fig. 4.15:** Representation of enthalpy variation during mixing process.

### 4.4.4 Heat exchangers

A device that allows the exchange of heat between two or more fluid streams without their physical mixing is termed as a **heat exchanger**. The fluid streams participating in the heat transfer process may remain in a single phase or change phase from liquid to vapour or vapour to liquid depending upon the amount and direction of heat exchange. Heat exchangers are widely used in industries and have a variety of designs. A typical heat exchanger configuration is shown in Fig. 4.16-a which depicts a cold fluid stream passing through a tube that is surrounded by a larger diameter shell through which a hot fluid stream is flowing. The heat transfers from hot stream to cold stream through the thickness of the inner tube wall. If the complete heat exchanger is considered as a control volume shown by the dashed line in Fig. 4.16-a, it gives a simple control volume with four mass interactions as shown in Fig. 4.16-b. In a heat exchanger the mass flow rate of both the fluid streams remains the same at inlet and outlet. The heat exchange mainly takes place between the fluid streams and there is negligible heat interaction with surroundings ($\dot{Q} \simeq 0$). There is no work interaction in a heat exchanger. Also, the changes in kinetic and potential energies are also negligible. The energy balance for a heat exchanger shown in Fig. 4.16-b is $\dot{m}_h h_{h1} + \dot{m}_c h_{c1} = \dot{m}_h h_{h2} + \dot{m}_c h_{c2}$. This gives $\dot{m}_h (h_{h1} - h_{h2}) = \dot{m}_c (h_{c2} - h_{c1})$. Thus in a heat exchanger the amount of heat released by the hot stream is equal to the amount of heat gained by the cold stream. If only cold steam is considered the part of control volume, then there will be heat interaction between control volume and its surroundings (i.e. hot fluid stream). A control volume representation for such a situation is shown in Fig. 4.16-c. The energy balance for this control volume gives $\dot{Q} + \dot{m}_c h_{c1} = \dot{m}_c h_{c2}$. Thus $\dot{Q} = \dot{m}_c (h_{c2} - h_{c1})$.

Scan Me

For more details on steady flow processes.

A heat exchanger in which a cold stream changes its phase from liquid to vapour is called a **boiler**. Conversely, a heat exchanger where a hot stream changes phase from vapour to liquid is termed as a **condenser**. The assumptions mentioned above are valid for boilers and condensers as well. However, in the case of a boiler the specific volume of cold stream increases significantly during its phase change, therefore, its outlet velocity should be calculated using Eqn. (4.5) and the magnitude of associated kinetic energy should be considered if significant.

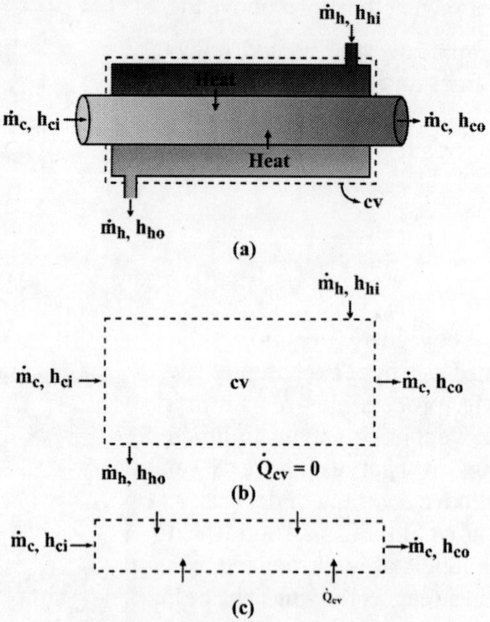


**Fig. 4.16:** (a) Schematic representation of a heat exchanger, (b) control volume representation of heat exchanger and (c) control volume representation of one fluid stream of a heat exchanger.

### 4.4.5 Throttle

Any kind of flow restriction in a fluid passage that results in significant pressure drop is termed as a **throttle** or **throttling device**. Typical examples (Fig. 4.17) of a throttle are a plate with a small hole, a capillary tube, an adjustable opening valve and a porous plug, etc. Such restrictions in the flow passage results in abrupt decrease of fluid's pressure and this expansion of fluid is completely different from the gradual expansion of fluid in a nozzle and a turbine. Throttle does not involve any work interaction and due to their smaller length heat transfer is also negligible. However, in case of long throttling devices such as a capillary tube, heat transfer should be estimated and checked for its significance. Usually the velocity of fluid increases at the outlet of a throttle, whereas the increase is not significant, hence the change in kinetic energy can be neglected.

The potential energy change is also negligible for a throttle. The mass of the fluid stream remains fixed during the expansion of a fluid through a throttle. Therefore, energy balance results in $h_i = h_o$. Thus the fluid expansion inside a throttle is an **isenthalpic process**. Typically $u_i + P_i v_i = u_o + P_o v_o$, therefore, if flow energy increases in a throttle ($P_o v_o > P_i v_i$), then the internal energy and the temperature of fluid decrease. On the other hand, when flow energy decreases ($P_o v_o < P_i v_i$), then internal energy increases during the throttling process and hence the temperature. For an ideal gas, $h = f(T)$, therefore, temperature remains constant during the throttling process. The magnitude of temperature rise or fall during a throttling process is governed by the **Joule-Thomson coefficient**. It is briefly described in the Know More section of this chapter.

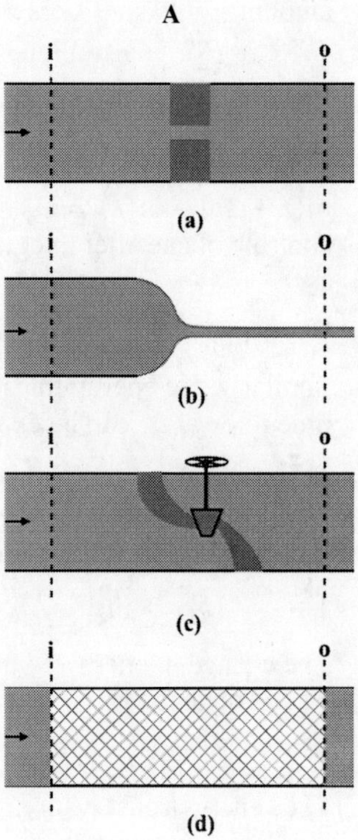


**Fig. 4.17:** Throttling device consisting of: (a) a plate with a small hole, (b) a capillary tube, (c) an adjustable opening valve and (d) a porous plug.

## 4.5 Energy Analysis of Transient Processes

In Sections 4.3 and 4.4, we described the processes for which there is no change of property within the control volume. There is another class of processes in which the properties also change inside the control volume. Such a process is called an **unsteady-flow process** or a **transient flow process**. Some examples of unsteady flow processes are charging of a tank by a pressurized fluid line, discharging of fluid from a pressurized vessel, inflating tires, cooking in pressure cookers, etc. In each of these processes, the mass and properties of the control volume change with time. A typical situation of charging a tank by a pressurized line is shown in Fig. 4.18. In order to analyse the transient flow processes, we consider a finite time interval $\Delta t$ and apply mass and energy conservation to find change of properties of control volume during this time interval. Most transient flow processes satisfy the following assumptions:

- The control volume can have a moving boundary, hence, associated boundary work.
- The properties of the control volume can change with time, however, these should be uniform throughout the control volume at a given time.
- The mass flow rate of inflowing and outflowing fluid streams can vary with time, whereas the state of the fluid stream should not change with time.

The conservation of mass and energy in rate form for a typical transient process are given in Eqns. (4.2) and (4.11), respectively. To thermodynamically analyse a transient process over the finite time interval $\Delta t$, let us consider a control volume having state 1 at time $t$. It has undergone a transient process involving heat, work and mass interactions during time interval $\Delta t$ and reached state 2 at time $t + \Delta t$. If $\sum \dot{m}_{in}$ is the total rate of mass flow entering the control volume, then amount of mass entered in time $\Delta t$ can be expressed as $\int_t^{t+\Delta t} (\sum \dot{m}_{in})dt = \sum m_{in}$. Similarly total mass leaving the control volume is expressed as $\int_t^{t+\Delta t} (\sum \dot{m}_{out})dt = \sum m_{out}$. The change in mass of the control volume is $\int_t^{t+\Delta t} \left(\frac{dm_{cv}}{dt}\right) dt = m(t + \Delta t) - m(t) = m_2 - m_1$. Therefore, the mass conservation for a control volume after integrating Eqn. (4.2) for a transient process is given as

$$\sum m_{in} - \sum m_{out} = m_2 - m_1 \quad (4.15)$$

Similarly, the energy transferred via heat, work and mass interactions in time interval $\Delta t$ can be expressed as

$\int_t^{t+\Delta t} \dot{Q}\, dt = Q$, $\int_t^{t+\Delta t} \dot{W}\, dt = W$,

$\int_t^{t+\Delta t} (\sum \dot{m}_{in} e_{flow,in})dt = \sum m_{in} e_{flow,in}$ and

$\int_t^{t+\Delta t} (\sum \dot{m}_{out} e_{flow,out})dt = \sum m_{out} e_{flow,out}$.

The change in energy content of the control volume for time interval $\Delta t$ can be obtained as

$\int_t^{t+\Delta t} \left(\frac{dE_{cv}}{dt}\right) dt = E(t + \Delta t) - E(t) = E_2 - E_1 = m_2 e_2 - m_1 e_1$.

Thus energy conservation for a transient process is expressed as

$$Q - W + \sum m_{in} e_{flow,in} - \sum m_{out} e_{flow,out} = E_2 - E_1 \quad (4.16)$$

The Eqn. (4.16) is derived considering the assumption that the state of the substance entering or leaving the system is not changing with time. However, for discharging of a tank, the leaving fluid stream from the control volume will have properties equal to that of the control volume at each instant during the process. Also, the properties of control volume, hence the leaving stream changes with time during a process. The energy balance for such a situation is explained in Example 4.4.

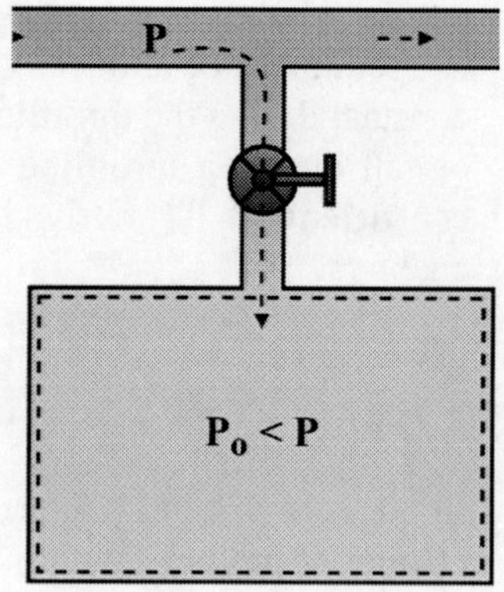


**Fig. 4.18:** Charging of a tank by a pressurized line.

## UNIT SUMMARY

In the present chapter, we introduced the mass conservation and energy conservation equations for control volumes. It is found that in addition to the internal energy, kinetic energy and potential energy, a flowing stream also carries flow energy. The sum of internal energy and the flow energy represents the enthalpy of a substance. A special kind of process in which the properties of a control volume does not change with time for longer durations is termed as a steady flow process. The mass and energy conservation are elucidated for steady flow processes and some practical examples of steady flow processes are also demonstrated. Finally the more generic processes in which the properties of control volume vary with time are discussed. Such a process is called the unsteady or transient flow process. The mass and energy balance equations are described for transient flow processes.

## Solved Examples

**Example 4.1:** Steam expands through an adiabatic nozzle from 0.8 $MPa$ and 250°C to 0.30 $MPa$. The velocity of steam at the inlet and outlet of the nozzles is 45 $m/s$ and 700 $m/s$, respectively. Determine the temperature and quality of steam at the nozzle outlet.

**Solution:** Given data: $P_i = 0.8\ MPa$, $T_i = 250°C$, $V_i = 45\ m/s$, $P_o = 0.30\ MPa$ and $V_o = 700\ m/s$.

**Assumptions:** The process inside the nozzle is a steady flow process. Therefore, $\dot{m}_i = \dot{m}_o = \dot{m}$. It is given that nozzle is adiabatic ($\dot{Q} = 0$). Work interaction for a nozzle is usually zero ($\dot{W} = 0$).

**Analysis:** Using energy balance $h_i + \frac{V_i^2}{2} = h_o + \frac{V_o^2}{2}$.

From superheated steam table B.3, $h_i = 2950.4\ kJ/kg$

$$\Rightarrow h_o = 2950.4 + \left(\frac{45^2}{2 \times 1000} - \frac{700^2}{2 \times 1000}\right) = 2706.41\ kJ/kg$$

At $P_o = 0.30\ MPa$, from saturated steam table B.2, $h_f = 561.43\ kJ/kg$, $h_g = 2724.9\ kJ/kg$ and $T_{sat} = 133.52°C$. Since $h_f < h_o < h_g$, thus the saturated liquid-vapour mixture exists at the nozzle outlet. $\Rightarrow h_o = h_f + xh_{fg} \Rightarrow x = 0.99$. Therefore, temperature at outlet of nozzle is 133.52°C and the quality of steam is 99 percent.

**Example 4.2:** A pump is used to lift water through a height of 20 $m$ to a tank maintained at 600 $kPa$ pressure. The water enters pump at 95 $kPa$ and 25°C. The discharge pipe has diameter of 50 $mm$ and the flow rate of water is 5000 $kg/h$. Determine the power required by the pump if the process is adiabatic. Also, the temperature of water does not change during the process.

**Solution:** The inlet of the pump is designated by subscript "$i$" and the point where discharge line puts water inside tank is designate by "$o$". Given data: $P_i = 95\ MPa$, $T_i = 25°C$, $d_o = 50\ mm$, $P_o = 600\ kPa$, $\dot{m} = 5000\ kg/h$ and $z_o - z_i = 20\ m$.

**Assumptions:** The process of pumping water is a steady flow process. Therefore, $\dot{m}_i = \dot{m}_o = \dot{m}$. It is given that process is adiabatic ($\dot{Q} = 0$). Also, consider that change in kinetic energy is negligible.

**Analysis:** Using energy balance $\dot{m}\left(h_i + \frac{V_i^2}{2} + gz_i\right) = \dot{m}\left(h_o + \frac{V_o^2}{2} + gz_o\right) + \dot{W}$.

From saturated steam table B.1, $h_i = h_f 25@°C = 104.83\ kJ/kg$, $v_i = v_f@25°C = 0.001003\ m^3/kg$. Since the temperature of the fluid remains constant during the pumping process, therefore, its internal energy will not change ($u_i = u_o$). Also, the specific volume is constant ($v_i = v_o$) due to the incompressible nature of fluid. Hence, $h_o = u_o + P_o v_o$ and $h_i = u_i + P_i v_i$ results in $h_o - h_i = (P_o - P_i)v_i$.

$$\Rightarrow \dot{W} = \dot{m}\left(h_i - h_o + \frac{V_i^2 - V_o^2}{2} + g(z_i - z_o)\right) = \frac{5000}{3600}\left((95 - 600) \times 0.001003 + \frac{9.81 \times (-20)}{1000}\right)$$

$\Rightarrow \dot{W} = -0.976\ kW$. Thus the required power input of the pump is 0.976 $kW$.

**Example 4.3:** Saturated liquid water enters a throttling device at 1500 $kPa$ and expands to a pressure of 500 $kPa$. Determine the quality and temperature of the water at the outlet of the throttling device.

**Solution:** Given data: Saturated liquid at $P_i = 1500\ kPa$ and $P_o = 500\ kPa$.

**Assumptions:** The process of expansion through a throttle is a steady flow process. Therefore, $\dot{m}_i = \dot{m}_o = \dot{m}$. For a throttling process, $\dot{Q} = 0$, $\dot{W} = 0$ and changes in kinetic and potential energy are negligible.

**Analysis:** Using energy balance $\dot{m}\left(h_i + \frac{V_i^2}{2} + gz_i\right) = \dot{m}\left(h_o + \frac{V_o^2}{2} + gz_o\right) \Rightarrow h_i = h_o$.

From steam table B.2, $h_i = h_f@1500\ kPa = 844.55\ kJ/kg = h_o$.

At $P_o = 500\ kPa$: $T_{sat} = 151.83°C$, $h_f = 640.09\ kJ/kg$ and $h_g = 2748.1\ kJ/kg$.

$h_o = h_f + xh_{fg} \Rightarrow x = 0.097$.

Therefore, the temperature at the outlet of the throttle is 151.83°C and the quality is 9.7 percent.

**Example 4.4:** A rigid insulated evacuated tank is connected to a steam line at a pressure of 1.2 $MPa$ and a temperature of 250°C through a valve. Now the valve is opened and steam is allowed to enter the tank till its pressure reaches 1.2 $MPa$. Considering the negligible kinetic and potential energies, determine the final temperature of steam within the tank.

**Solution:** Let the initial and the final states of the control volume are designated as "1" and "2", respectively. The state of incoming steam is designated by subscript "$i$". Given data: $P_i = 1.2\ MPa$, $T_i = 250°C$, $P_1 = 0\ kPa$ and $P_2 = 1.2\ MPa$.

**Assumptions:** The properties of the control volume are uniform at any instant. The properties of the inflowing steam are identical to the properties of the steam line. The work transfer is zero and the kinetic and potential energies are negligible. The tank is insulated, thus heat transfer is zero.

**Analysis:** Using mass and energy balance, $m_2 - m_1 = m_i$ and $m_i h_i = m_2 u_2 - m_1 u_1$.

From superheated steam table B.3, $h_i = 2935.6\ kJ/kg$.

For evacuated tank $m_1 = 0 \Rightarrow m_2 = m_i$. From energy balance, $h_i = u_2 = 2935.6\ kJ/kg$.

From saturation pressure table B.2 at $P_2 = 1.2\ MPa$, $u_g = 2587.8\ kJ/kg$. Since $u_2 > u_g$ at the given pressure, thus the final state of the steam in the tank is superheated. Now using the superheated steam table B.3 at $P_2 = 1.2\ MPa$,

$u = 2872.7\ kJ/kg@350°C$ and $u = 2955.5\ kJ/kg@400°C$. Hence, using interpolation, $T - 350 = \frac{400-350}{2955.5-2872.7}(2935.6 - 2872.7) \Rightarrow T = 387.98°C$.

Therefore, the final temperature of steam inside the tank is 387.98°C.

**Example 4.5:** A rigid tank of 80 $L$ volume contains air at 1.2 $MPa$ and 220°C. Now a valve is opened to release the air out till the pressure within the tank becomes 150 $kPa$ and heat interaction is allowed during the process such that the final temperature of air within the tank is 45°C. Find the amount of heat transfer.

**Solution:** Let the initial and the final states of the control volume are designated as "1" and "2", respectively. The state of outflowing air stream is designated by subscript "$o$". Given data: $P_1 = 1.2\ MPa$, $T_1 = 220°C$, $P_2 = 500\ kPa$, $T_2 = 45°C$ and $V_1 = V_2 = 80\ L$.

**Assumptions:** The properties of the control volume are uniform at any instant. The properties of the outflowing air stream are identical to the properties of the control volume at a particular time. The outflowing air stream has a constant mass flow rate throughout the process. The work transfer is zero and the kinetic and potential energy of the outflowing air stream are negligible. Consider air as an ideal gas.

**Analysis:** Using mass and energy balance, $m_1 - m_2 = m_o$,

$Q - m_o h_o = m_2 u_2 - m_1 u_1$.

Using ideal gas equation, $m_1 = \frac{P_1 V_1}{RT_1} = 0.6785\ kg$ and $m_2 = \frac{P_2 V_2}{RT_2} = 0.4383\ kg$, here $R = 0.287\ kJ/kg - K$ for air.

Therefore, the mass of the outflowing air is $m_o = 0.2402\ kg$.

From air property table A.1 at $T_1 = 220°C$, $u_1 = 354.62\ kJ/kg$, $h_1 = 496.125\ kJ/kg$.

At $T_2 = 45°C$, $u_2 = 227.26\ kJ/kg$, $h_2 = 318.49\ kJ/kg$.

The properties of the outflowing stream varied during the process because air within the tank was at different conditions during the process. If we consider the constant mass flow rate of air during the process as stated in assumptions, then properties of outflowing air stream can be considered as the average of initial and final conditions of the tank. Using this assumption, $h_o = \frac{h_1 + h_2}{2} = 407.31\ kJ/kg$.

Therefore, using energy balance $Q = m_o h_o + m_2 u_2 - m_1 u_1 = -43.16\ kJ$. Thus 43.16 $kJ$ heat is rejected during the process to maintain the desired temperature of air within the tank.

## EXERCISES

### Multiple Choice Questions

| 4.1 | A control volume allows interaction of_______________across its boundary. | | |
|---|---|---|---|
| | A. mass | B. work | C. heat | D. all of the above |
| 4.2 | The time rate of mass change within a control volume is equal to | | |
| | A. net mass transfer across control surface | B. total mass inflow across control surface | C. total mass outflow across control surface | D. none of these |
| 4.3 | The velocity of a fluid stream at a control surface should be _______________ to outward normal from the surface for computations of mass flow rate using $\dot{m} = \rho A_c \underline{V}$. | | |
| | A. parallel | B. perpendicular | C. at an angle | D. tangential |

| | | | | |
|---|---|---|---|---|
| 4.4 | The total specific energy associated to a flowing stream is equal to | | | |
| | A. $u + \frac{1}{2}V^2 + gz$ | B. $Pv + \frac{1}{2}V^2 + gz$ | C. $h + \frac{1}{2}V^2 + gz$ | D. $u + Pv$ |
| 4.5 | For a steady flow process,<br>(i) The thermodynamic state of the inflowing stream can change during the process.<br>(ii) The properties within the control volume can have different values at different locations.<br>(iii) The properties within the control volume can change with time.<br>(iv) The volume of the control volume should be fixed. | | | |
| | A. i and ii | B. ii and iii | C. iii and iv | D. ii and iv |
| 4.6 | A steady flow device has negligible kinetic energy at its inlet and is used to increase the velocity of fluid through an adiabatic expansion. The device is called | | | |
| | A. turbine | B. nozzle | C. diffuser | D. compressor |
| 4.7 | The expansion of a fluid through a throttling device can be<br>(i) isenthalpic<br>(ii) isothermal<br>(iii) isochoric<br>(iv) adiabatic<br>(v) isobaric | | | |
| | A. i, ii and iii | B. i, ii and iv | C. ii, iii and v | D. i, iv and v |
| 4.8 | A steady flow device that increases the pressure of a fluid stream by consuming external work is known as | | | |
| | A. compressor | B. diffuser | C. turbine | D. nozzle |
| 4.9 | A steady flow device that increases the pressure of a fluid at the expense of fluid's kinetic energy is called | | | |
| | A. compressor | B. diffuser | C. turbine | D. nozzle |
| 4.10 | In a transient process, the difference between amount of mass entering and amount of mass leaving the control volume in finite time $\Delta$□ can be | | | |
| | A. positive | B. negative | C. zero | D. all of the above |

| | | | | |
|---|---|---|---|---|
| 4.11 | The expansion of an ideal gas in a turbine results in _________________ of its temperature. | | | |
| | A. increase | B. decrease | C. no change | D. either increase or decrease |
| 4.12 | The expansion inside a throttle irrespective of the nature of working fluid is always | | | |
| | A. isenthalpic | B. isochoric | C. isothermal | D. isobaric |

**Answers of Multiple Choice Question**

| **4.1 D, 4.2 A, 4.3 A, 4.4 C, 4.5 D, 4.6 B, 4.7 B, 4.8 A, 4.9 B, 4.10 D, 4.11 B, 4.12 A** |
|---|

**Short and Long Answer Type Questions**

4.1 What is flow energy? Explain with the help of examples.

4.2 What are the necessary requirements of steady flow processes?

4.3 How the rate of work produced by a turbine will be influenced if the heat is transferred to surroundings during the expansion process?

4.4 During the expansion process of a saturated liquid through a throttling device, it changes to saturated liquid-vapour mixture. How does the flow energy change during the process? Explain.

4.5 A hydraulic turbine installed at a dam converts fluid power to mechanical work in terms of rotation of its shaft. Write the energy conservation for the same and identify the dominant terms.

4.6 Can cooking food in a pressure cooker be called a steady flow process? Elaborate with reasons.

4.7 Why kinetic energy of the cold stream at the outlet of a boiler can be significant?

4.8 A control volume has only one inlet through which a fluid enters during a process. Can the process be called a steady flow process? Explain.

4.9 A piston cylinder contains a fluid and the cylinder has one inlet and one outlet. The mass flow rate through the inlet and outlet are the same. Also, the thermodynamic state of the inlet stream and outlet stream do not vary with time. Is this process steady? Elaborate.

4.10 The boundary of a control volume expands during a process. Can it be called a steady flow process? Explain.

**Numerical Problems**

4.1 Superheated vapour at 200 $kPa$ and 350°C is discharged by a compressor to a pipe of 150 $mm$ diameter at the rate of 50 $kg/h$. Determine the average velocity of the vapour in the pipe.

4.2 Air flows through a circular pipe of 25 $cm$ diameter and enters with a velocity of 2.5 $m/s$ at 300 $kPa$, 30°C. Superheated steam flows outside the pipe and heats the air such that it leaves pipe at 55°C and 260 $kPa$. Determine: (a) volumetric flow rate at inlet, (b) velocity at outlet and (c) mass flow rate at outlet.

4.3 Steam at 2.2 $MPa$ and 280 $m/s$ enters to an adiabatic diffuser and leaves at 2.8 $MPa$ and 395°C with a velocity of 35 $m/s$. Determine: (a) the inlet to outlet area ratio of the diffuser. Also, find the inlet temperature of the steam.

4.4 Steam expands through a turbine under the adiabatic conditions from 8 $MPa$, 440°C to 5 $kPa$ and 88% quality. The velocity of steam at the turbine inlet is 70 $m/s$ and its mass flow rate is 10 $kg/s$. Find the inlet area of the turbine and the power produced by the turbine.

4.5 A compressore takes atmospheric air at 100 $kPa$ and 25°C, compresses it and supplies at 1500 $kPa$ and 750°C. Compressed air is further cooled in an isobaric cooler to 35°C. If the mass flow rate of the air is 0.01 $kg/s$, calculate the power rating of the compressor and the rate of heat transfer in the cooler.

4.6 Steam enters an adiabatic turbine at 3500 $kPa$ and 580°C at the rate of 10 $kg/s$. One fourth fraction of steam is extracted from the turbine at 950 $kPa$ and 330°C. The remaining portion of steam expands till 35 $kPa$ and 80°C. Determine the output power of the turbine.

4.7 An open feedwater heater is employed in a thermal power plant to heat the feedwater entering at 800 $kPa$ and 35°C using superheated steam at 800 $kPa$ and 195°C. Both the feedwater and superheated steam enter the feedwater heater and come out in a single stream as saturated liquid at the same pressure. Determine the required mass flow rate of the superheated steam when the feedwater enters at 10 $kg/s$.

4.8 A shell and tube heat exchanger is used to cool an oil from 170°C to 50°C at the rate of 3.5 $kg/s$. Water is used as the cooling medium which flows at the rate of 1.7 $kg/s$ and enters the heat exchanger at 27°C. Determine the exit temperature of water and rate of heat transfer in the heat exchanger. Consider $C_p = 2.18\ kJ/kg$ °C for oil and $C_p = 4.18\ kJ/kg$ °C for water.

4.9 Hot water at 1.8 $MPa$ and 170°C is throttled to a lower pressure of 450 $kPa$. Determine the rate of saturated liquid and saturated vapour formation at the outlet of throttle if the mass flow rate of hot water at the inlet of throttle is 10 $kg/s$.

4.10 If the saturated vapour produced in the previous question (4.9) is passed through an adiabatic turbine till its pressure becomes 15 $kPa$ and the quality at the exit of the turbine is 92%. Determine the output power of the turbine.

4.11 Air at 580 $kPa$ and 290°C is contained in a piston-cylinder arrangement at 0.22 $m^3$ initial volume. A valve connected to cylinder is opened till half mass of the air escaped the cylinder and corresponding volume of cylinder is 0.08 $m^3$. Determine the boundary work involved during this process and the final temperature of air in the cylinder.

4.12 An evacuated tank is connected to an air supply line which is at 220 $kPa$ and 130°C. The air is filled in the tank till it reaches 220 $kPa$ pressure and after that valve is closed to isolate it from the supply line. Determine the final temperature of air in the tank and the flow work associated with air flow at the supply line.

4.13 A 0.01 $m^3$ rigid tank contains saturated water at 350 $kPa$. Initially 75% of the volume is occupied by vapour and the rest is liquid. A valve connected at the top of the tank is opened to allow escaping the vapour out of the tank. Heat is added to the water during this process to maintain the pressure constant. The valve is closed when 5% of liquid is left in the tank. Determine the amount of heat added during the process.

4.14 A pressure cooker of 2 $L$ capacity operates at a pressure of 220 $kPa$. It is observed at an instant that 75% of the cooker volume is filled with water and remaining is vapour. The pressure cooker is kept on the gas stove for further 30 minutes. Determine the highest rate of heat transfer permissible if it should not run out of water till the end of 30 minutes.

4.15 A rigid tank of 35 $kg$ is made up of steel and has a volume of 0.8 $m^3$. It contains air at 450 $kPa$ and 25°C. It is then connected to an air supply line at 2.5 $MPa$ and 25°C. Air is filled in the tank till its pressure reaches 1.75 $MPa$ and then the valve is closed. The final temperature of both the tank and the air is 55°C. Also, both tank and air remain at thermal equilibrium during the entire process. Determine the amount of heat transfer involved in the process and final mass of the air in the tank.

## PRACTICAL

**Activity 1:** Use of centrifugal pumps for lifting water at home is very common. Measure the height through which it is being used to lift water at your home. Also, measure the flow rate of water being pumped using the bucket and stopwatch method. Using this information apply the steady flow energy equation and estimate the power rating of the pump. Compare it to the given power rating of the pump specified by the manufacturer and find the associated reasons for differences.

## KNOW MORE

### Joule-Thomson Coefficient

We observed that the expansion of a fluid through a throttling device is an isenthalpic process and the temperature of fluid in this process can increase or decrease or may remain unchanged. A important property of fluid called **Joule-Thomson coefficient**, governs the temperature behaviour of fluid in an isenthalpic process and is defined as $\mu_{JT} = \left(\frac{\partial T}{\partial P}\right)_{h=const.}$. This reveals that Joule-Thomson coefficient represents the slope of constant enthalpy process on a $T - P$ diagram and is a measure of change in temperature of a substance with pressure during the isenthalpic process. Temperature decreases for $\mu_{JT} > 0$ during the isenthalpic expansion whereas it increases for $\mu_{JT} < 0$ and remains unchanged when $\mu_{JT} = 0$. Consider expansion of a fluid through a valve from state 1 to state 2 as shown in Fig. K-4.1-a. Let us measure pressure and temperature at the exit of the valve during the throttling process. This expansion process is repeated for multiple openings of the valve for a fixed inlet state of the fluid, hence the fixed enthalpy. We will experience that the state of the fluid at the exit of the valve will change for its varying openings. The final states obtained in each experimental trial are plotted on a $T - P$ diagram as shown in Fig. K-4.1-b. If we join the locus of all the final states obtained in different experiments and the inlet state, a constant enthalpy line is obtained because the initial state was kept constant.

This experiment can be performed for different initial states, hence, different enthalpies of the fluid streams and a diagram, as shown in Fig. K-4.2, can be obtained. This diagram shows different constant enthalpy lines. For a given constant enthalpy line, the point where its slope becomes zero is called the

**inversion point**. The line joined by locus of different inversion points is termed as an **inversion line.** The region on the left of the inversion line shows the positive value of Joule-Thomson coefficient. On the other hand, its value is negative on the right of the inversion line. The point on the temperature axis where the upper portion of the inversion line crosses it, represents the maximum inversion temperature. The throttling devices used in refrigeration applications are operated on the left of the inversion line because this is the only region where isenthalpic expansion causes cooling. Isenthalpic expansion of any fluid above its maximum inversion temperature results in heating of the substance.

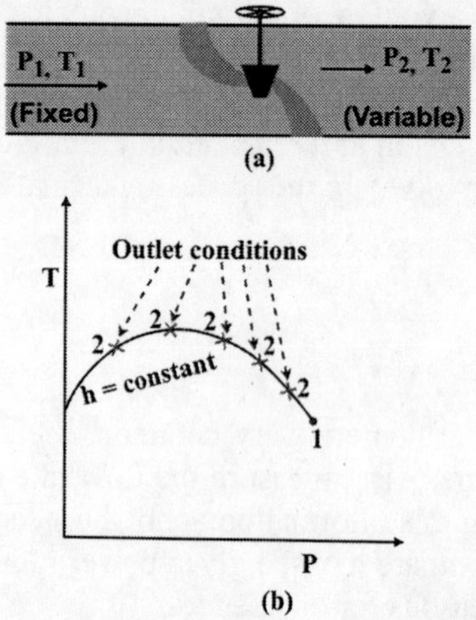


**Fig. K-4.1:** (a) Representation of throttling process and (b) $T - P$ diagram depicting constant enthalpy line.

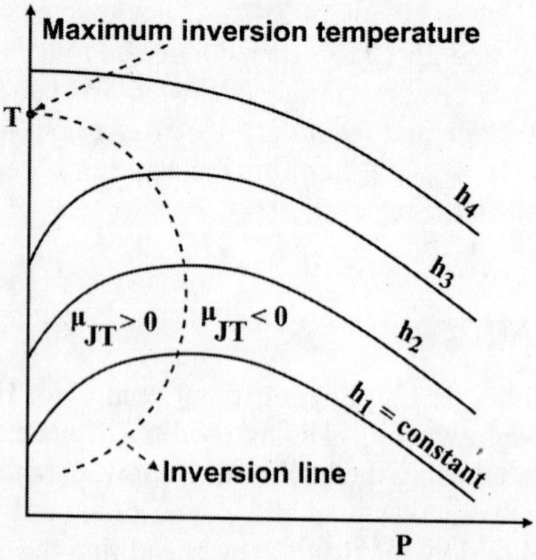


**Fig. K-4.2:** $T - P$ diagram showing isenthalpic lines and inversion curve.

Further, to represent Joule-Thomson coefficient in terms of specific heats, temperature, pressure and specific volume, general enthalpy equation can be used as $dh = c_p dT + \left(v - T\left(\frac{\partial v}{\partial T}\right)_P\right) dP$. For isenthalpic process, $dh = 0$, thus Joule-Thomson coefficient can be expressed as $\mu_{JT} = \left(\frac{\partial T}{\partial P}\right)_{h=const.} = -\frac{1}{c_p}\left(v - T\left(\frac{\partial v}{\partial T}\right)_P\right)$. It can be easily obtained from this expression that for an ideal gas Joule-Thomson coefficient is zero.

## REFERENCES AND SUGGESTED READINGS

1. Bier, K., Ernst, G. and Maurer, G., 1974. Flow apparatus for measuring the heat capacity and the Joule-Thomson coefficient of gases. *The Journal of Chemical Thermodynamics*, *6*(11), pp.1027-1037.
2. Yunus A. Çengel and John M. Cimbala, 2010. *Fluid Mechanics: Fundamentals and Applications*. McGraw-Hill Higher Education.

# 5 Second Law of Thermodynamics

## UNIT SPECIFICS

*In this unit we have discussed the following aspects:*

- *Need of second law of thermodynamics;*
- *Concept of irreversibility and sources of irreversibilities;*
- *Heat engines, refrigerators and heat pumps;*
- *Statements of second law of thermodynamics;*
- *Concepts of perpetual motion machines;*
- *Thermodynamic temperature scale;*
- *Ideal and real machines;*
- *Solved examples for illustrating the applications of the second law of thermodynamics.*

*After energy balance and energy efficiency analysis of the thermodynamic systems up to unit 4, this unit focuses on the feasibility prediction of the thermodynamic processes through the application of the second law of thermodynamics. The need for the second law of thermodynamics is established and various sources of irreversibilities are discussed. For illustration of various transformations of heat into work and vice versa, the concepts of heat engines and heat pumps along with refrigeration systems are discussed. Carnot cycle consisting of reversible thermodynamic processes is described for enabling us to calculate the limiting values of cycle efficiencies operating between specified temperature limits. Definition of absolute temperature scale which is independent of material properties is discussed based on the efficiency limit considerations of Carnot cycle.*

## RATIONALE

*This unit describes the second law of thermodynamics for determining the feasibility of thermodynamic processes and estimating the limit values of performance indicators such as efficiency and coefficients of performance. Need of second law of thermodynamics has been established and different causes of the irreversibilities have been analysed in detail. Equivalence between different statements of the second law of thermodynamics has been established in order to identify the feasible direction of the thermodynamic processes. Ideal and real processes have been discussed to further enhance the understanding of the concept of feasibility.*

## PRE-REQUISITES

*First law of thermodynamics; Thermodynamic properties; Basics of analysis of thermodynamic processes*

## UNIT OUTCOMES

*List of outcomes of this unit is as follows:*

*U5-O1:* *Appreciation of need of second law of thermodynamics*

*U5-O2:* *Identification of sources of irreversibilities*

*U5-O3:* *Statements of second law of thermodynamics and establishment of their equivalence*

*U5-O4:* *Appreciation of Carnot cycle, representation of thermodynamic systems for calculation of performance limits*

*U5-O5:* *Need and definition of thermodynamic scale of temperature*

| Unit-5 Outcomes | EXPECTED MAPPING WITH COURSE OUTCOMES (1- Weak Correlation; 2- Medium correlation; 3- Strong Correlation) | | | | | |
|---|---|---|---|---|---|---|
| | CO-1 | CO-2 | CO-3 | CO-4 | CO-5 | CO-6 |
| U5-O1 | 1 | 2 | 3 | 3 | 2 | 3 |
| U5-O2 | 1 | 2 | 3 | 3 | 2 | 3 |
| U5-O3 | 1 | 2 | 3 | 3 | 2 | 3 |
| U5-O4 | 1 | 2 | 3 | 3 | 2 | 3 |
| U5-O5 | 1 | 2 | 3 | 3 | 2 | 3 |

## 5.1 Need for the Second Law of Thermodynamics

We did elaborate discussion on the first law of thermodynamics or the principle of conservation of energy in the previous chapters. It was elucidated that energy is a conserved quantity and no process, taking place in nature, violates the first law of thermodynamics. However, now it must be emphasized that the satisfaction of only the first law of thermodynamics does not guarantee the actual occurrence of the process. Let us explain this point with the help of examples. Consider a hot water bucket placed in a room which is at ambient temperature (Fig. 5.1). Due to the difference of temperature, heat will transfer from the water bucket to the surrounding air till its temperature reaches the ambient condition. The amount of heat loss from the water bucket will be equal to the amount of heat gained by the surrounding air in this process. Hence, it satisfies the first law of thermodynamics. Now if we attempt to reverse this situation such that the hot water bucket receives the heat from the surrounding air and gets more warmer. Such a situation will also result in conservation of energy because the heat loss from air will be equal to heat gained by the water bucket. Hence, it will satisfy the first law of thermodynamics whereas this process will not be feasible as the spontaneous heat transfer takes place from high temperature to low temperature.

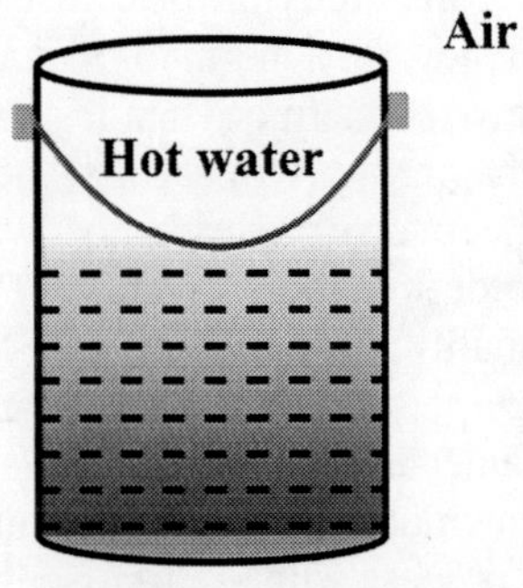


**Fig. 5.1:** A hot water bucket placed in a room.

In another situation, let us consider a rigid vessel containing a gas and a paddle wheel as shown in Fig. 5.2-a. The paddle wheel rotates and does work on the gas when the mass ($m$) is lowered through a certain height. This increases the internal energy of the gas. Now the gas is returned back to its original state by transferring heat equivalent to the amount of work done on the gas to the surrounding air. This heat transfer decreases the internal energy content of the gas and it reaches the initial state. The sequence of processes used here, form a cycle in which work is done on the system and is equal to the amount of heat released by the system to surroundings. This thermodynamic cyclic process has negative heat and work interactions and it ensures the applicability of the first law of thermodynamics. Now let us attempt to reverse this cycle. In this effort, first transfer the heat from surroundings to the gas, which will increase its internal energy content, then use this internal energy to raise the weight and bring the gas back to its initial conditions (Fig. 5.2-b). If we observe such a cyclic process it will satisfy the first law of thermodynamics, however, it is not practically possible to rotate the paddle wheel solely by the internal energy of the gas to raise the weight. This explains that it is not practically possible to achieve a cyclic process under the given conditions which can have positive heat and work interactions, i.e. heat addition to the system and work done by the system.

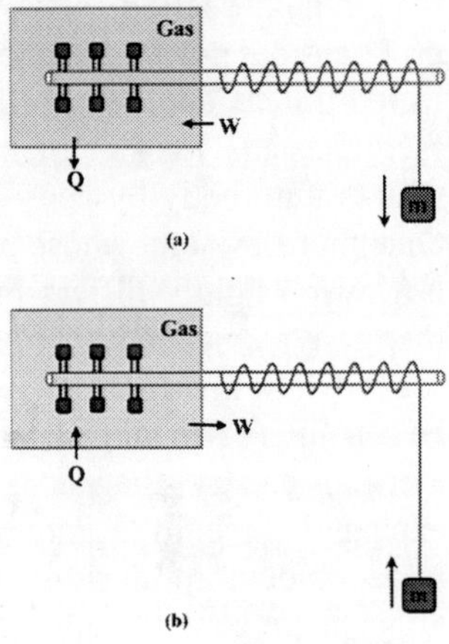


**Fig. 5.2:** A paddle wheel and mass arrangement that undergo an impossible cyclic process.

The above examples illustrate that the thermodynamic processes take place in a certain direction and it is not always possible to reverse the processes even if the first law of thermodynamics is satisfied. Therefore, the first law of thermodynamics cannot explain the feasibility of the process. We have to introduce another generic law, called the **second law of thermodynamics**, to explain the feasible direction of the process. Hence, for a thermodynamic process to occur both the first and the second laws of thermodynamics should be satisfied. In the later portions of this chapter, we will see that the violation of second law was taking place for the reverse processes that we discussed in the above examples. In addition to defining the direction of the process, the second law also explains the quality of the energy. For instance, the heat energy at the higher temperature has more potential to produce work in comparison to the same amount of energy available at low temperatures. The detailed elucidation regarding the quality of the energy and its degradation during a process will be discussed in later portions. Further, the second law also defines the theoretically possible maximum performance of the commonly used thermodynamic devices.

In the next section, we will discuss the thermodynamic system that produces net positive heat and work interactions during a cyclic process, called a **heat engine**. On the other hand, a thermodynamic system that facilitates the transfer of heat from a low temperature body to a high temperature body by consuming external work is called a **refrigerator**. One more system whose working is similar to that of a refrigerator, however, slightly differs in its usage, is termed as a **heat pump**. The thermodynamic analysis of such systems and associated discussion of the second law involve the use of a hypothetical body, called a **thermal reservoir**. A typical thermal reservoir is a body of very large capacity of thermal energy that does not have any temperature change upon exchanging the finite amount of heat. Such a body must have either a very large amount of mass or very high specific heat. For example, the atmosphere contains a huge amount of air and upon addition of a finite amount of heat its temperature does not change. Conversely, a fluid in the saturated two phase regime can be considered a thermal reservoir of infinite heat capacity whose temperature does not change upon energy interaction if its amount is less than the corresponding latent heat. It is important to mention that thermal reservoirs need not always be of infinite size. For instance, the air in a perfectly sealed room can act as a thermal reservoir for the amount of heat released by a tube light in the room. On the other hand, the temperature of air in the same room increases if a heater is switched on inside the room. Engineering devices such as a furnace can act as a thermal reservoir during their steady operation. A thermal reservoir that supplies the heat energy is referred to as a **heat source** or **source** and a reservoir that absorbs the heat energy is called a **heat sink** or **sink** (Fig. 5.3). The further discussion in this book predominantly uses these terms while describing different concepts.

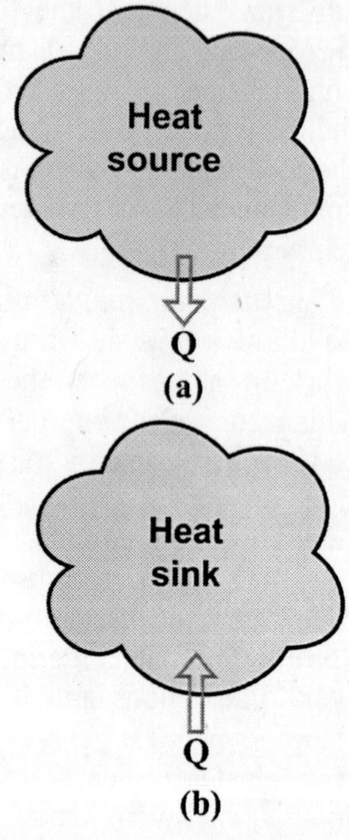


**Fig. 5.3:** A depiction of a thermal reservoir as a: (a) heat source and (b) heat sink.

## 5.2 Heat Engines, Refrigerators and Heat Pumps

It is established in the previous section (Fig. 5.2) that work can be completely converted into heat by operating in thermodynamic cyclic manner whereas the heat cannot be converted completely into an equivalent amount of work upon using the cyclic process. For example, the heat added to the gas during the reverse process (Fig. 5.2-b) could not raise the weight. Using this fact, we introduce here a class of devices that can have positive heat and work interactions by operating in a cyclic manner. Such a device which converts heat into work by operating in a thermodynamic cycle is called a heat engine. A heat engine can have plenty of engineering arrangements to achieve this task, however, it principally involves the following main processes:

- A heat engine receives heat from a high temperature thermal reservoir,
- It utilizes part of the received heat to produce work,
- It rejects the remaining amount of heat to a low temperature thermal reservoir and
- It includes any additional energy interaction if required to operate in a thermodynamic cycle.

Considering the above processes, we demonstrate here a piston-cylinder arrangement based device which works as a heat engine (Fig. 5.4).

Figure 5.4-a shows a cylinder including two sets of stops and a piston resting on the lower stops. A gas is contained between the piston and cylinder and a weight is placed on the piston rod. Now the system is made to contact a high temperature heat source that supplies heat to the gas and it expands till the piston reaches the top stops (Fig. 5.4-b). Let us consider the amount of heat added during this expansion in $Q_{in}$ and the gas has done $W_{out}$ amount of work. After this the high temperature heat source and the weight acting on the piston are removed (Fig. 5.4-c). Now the system is made to contact a low temperature heat sink which absorbs the heat from the gas and the piston displaces in downward direction during this process (Fig. 5.4-d). Let $Q_{out}$ is the amount of heat loss from the gas till piston reaches the lower stops and $W_{in}$ is the work done on the gas during this process. After this we can remove the low temperature heat sink and once again place the weight on the piston rod (Fig. 5.4-a). Here, it can be observed that the piston has reached its initial state and the same sequence of processes can be repeated to further continue the cycle. The net amount of heat transfer during the complete cycle is $\oint \delta Q = Q_{net} = Q_{in} - Q_{out}$. Similarly, the net amount of work interaction is $\oint \delta W = W_{net} = W_{out} - W_{in}$. The work output $W_{out}$ here is greater than the input work $W_{in}$ because the piston has lifted an external weight during the expansion process and this weight was not there during the compression process. Using the first law of thermodynamics for this cyclic process $Q_{net} = W_{net}$. Therefore, we obtained a piston-cylinder system that operates in a cyclic manner and produces a net positive work output by exchanging the heat between high temperature and low temperature thermal reservoirs. A simplified schematic representation of such a heat engine is shown in Fig. 5.5. All the heat engines, irrespective of their mechanical configuration, can be represented using the schematic shown in Fig. 5.5. The substance used in a typical heat engine to achieve different processes of heat and work transfer is known as **working fluid** or **working medium**.

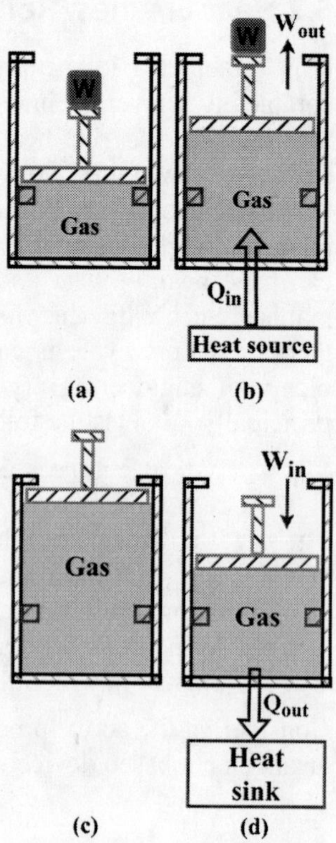


**Fig. 5.4:** A piston cylinder arrangement that operates as a heat engine.

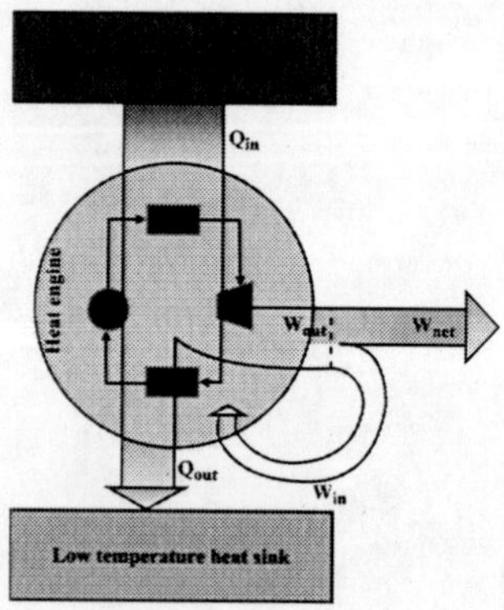


**Fig. 5.5:** Schematic representation of a heat engine.

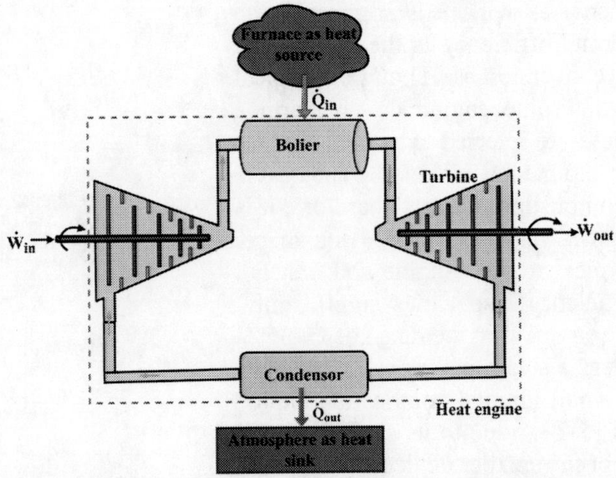


**Fig. 5.6:** Schematic representation of a steam power plant as a heat engine.

A steam power plant is another example of a heat engine which consists of an assembly of different steady flow devices. A typical representation of such a heat engine is shown in Fig. 5.6. The working fluid in a steam power plant is water. The high pressure water receives heat from a high temperature heat source which is typically a furnace in this case. The water undergoes a phase change process from liquid to vapour during the heat addition inside a boiler. The steam is then expanded over a turbine to obtain work output. The pressure of the steam decreases during the expansion inside the turbine and then it is passed through a condenser where it rejects heat to the atmosphere. After the condenser, the working fluid is pumped to its initial conditions in order to complete the cycle. The rate of heat transfer in boiler and condenser is $\dot{Q}_{in}$ and $\dot{Q}_{out}$, respectively. The rate of work done by the turbine is $\dot{W}_{out}$ and the rate of work input to the pump is $\dot{W}_{in}$. The net rate of heat transfer for the complete cycle is $\dot{Q}_{net} = \dot{Q}_{in} - \dot{Q}_{out}$ and net rate of work transfer is $\dot{W}_{net} = \dot{W}_{out} - \dot{W}_{in}$. For a thermodynamic cyclic process $\dot{Q}_{net} = \dot{W}_{net}$.

It is established in the previous discussion that a heat engine converts part of the input heat supply to work done and the remaining fraction is rejected to the heat sink in order to complete the cycle. A heat engine for which amount of heat rejection is less for the given heat input will have better performance. It becomes important now to define **thermal efficiency** as a parameter to gauge the performance of a heat engine. In general, efficiency is the ratio of desired output to the input required to produce that output. For a heat engine, the net amount of work done is desired and heat is supplied to achieve this work output. Therefore, thermal efficiency can be defined as

$$\eta_{th} = \frac{\dot{W}_{net}}{\dot{Q}_{in}} = \frac{\dot{Q}_{net}}{\dot{Q}_{in}} = \frac{\dot{Q}_{in} - \dot{Q}_{out}}{\dot{Q}_{in}} = 1 - \frac{\dot{Q}_{out}}{\dot{Q}_{in}} \qquad (5.1)$$

As the heat engine transforms heat energy to work, it is nothing but an energy conversion device and the thermal efficiency is the measure of how efficient the process of energy conversion is. Though a typical internal combustion engine and a gas turbine engine do not form a complete thermodynamic cycle, yet these are referred to as heat engines because these also convert heat energy into mechanical work. The heat is supplied to such heat engines by the combustion of fuel. Therefore, it is important to mention that the heat engines vary a lot in size and shape, ranging from a scooter engine to a typical marine engine and a steam thermal power plant. Hence, their efficiency also varies significantly. Typical thermal efficiency of a gasoline engine is in the range of 25-35%, for a diesel engine, it is 30-40% whereas a steam power plant can have efficiency in the range of 35-50%. A small internal combustion engine can have thermal efficiency as low as 15-20% due to its simple control system and more proportion of losses for the smaller devices.

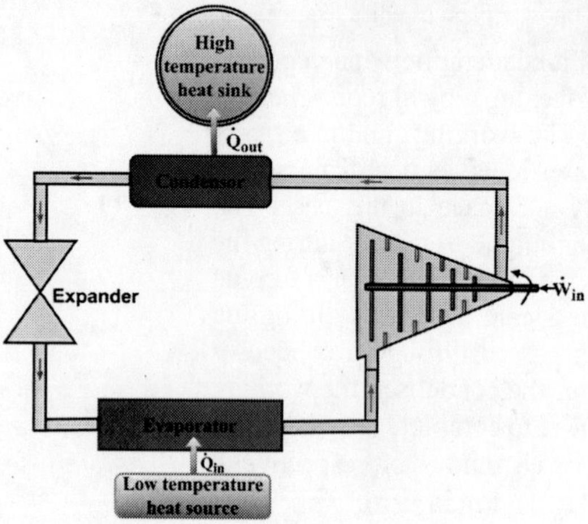


**Fig. 5.7:** Representation of a refrigerator.

We observed another impossible situation of transferring heat from a low temperature body to a high temperature body in the previous discussion. A device that facilitates such a heat transfer by consuming work and operating in a cyclic manner is termed as a refrigerator. The working fluid used in the refrigerator is termed as **refrigerant**. The schematic of a commonly used vapour compression refrigeration system is shown in Fig. 5.7. It consists of four main components: an evaporator, a compressor, a condenser and a throttling device. The refrigerant inside the evaporator receives heat from the refrigerated space and then it enters the compressor where its pressure and temperature is increased. The high temperature refrigerant then rejects heat inside the condenser and then it is passed through a throttling device where its pressure and temperature decreases. The rate at which heat is added to refrigerant in the evaporator is $\dot{Q}_{in}$ and rate of heat rejection from the refrigerant inside the condenser is $\dot{Q}_{out}$. The rate of work done required by the compressor is $\dot{W}_{in}$. Using the first law of thermodynamics for the complete system gives $\dot{W}_{in} = \dot{Q}_{out} - \dot{Q}_{in}$. The freezer of a typical household refrigerator is the place where refrigerant receives heat and it acts as an evaporator whereas the coil behind the refrigerator acts as a condenser where refrigerant dissipates heat to the surrounding air. The performance of a refrigerator is defined by a parameter called its **coefficient of performance (COP)** and it is represented by symbol $\beta$. It should be emphasized that the purpose of a refrigerator is to maintain the refrigerated space below the atmospheric temperature and it does this by absorbing the heat in the evaporator. Thus the desired output in case of a refrigerator is $\dot{Q}_{in}$ and it requires $\dot{W}_{in}$ amount of work to operate in a thermodynamics cycle. Therefore, $\beta$ is expressed as

$$\beta = \frac{\dot{Q}_{in}}{\dot{W}_{in}} = \frac{\dot{Q}_{in}}{\dot{Q}_{out}-\dot{Q}_{in}} = \frac{1}{\dot{Q}_{out}/\dot{Q}_{in}-1} \quad (5.2)$$

It can be observed from Eqn. (5.2) that the COP of a refrigerator can be greater than unity. This implies that the heat absorbed by the refrigerant can be greater than the amount of work required to do this. This is the distinction of COP from the thermal efficiency of a heat engine and it is the reason to mention the efficiency of a refrigerator in terms of COP to avoid the peculiarity of having an efficiency value greater than one. The COP of a typical household refrigerator ranges from 2-3.

Another device that operates on the same principle as that of a refrigerator is termed as a heat pump. The objective of a heat pump is to maintain a space above the atmospheric temperature. So in this case, the refrigerant receives the heat from the cold surroundings and then after the work of compression, it supplies the heat to the desired space. The schematic representation of a typical refrigerator and a heat pump is shown in Fig. 5.8 to show the distinction between the two. The performance of a heat pump is also measured in terms of its coefficient of performance and is denoted by $\beta'$. As $\dot{Q}_{out}$ is the rate at which heat is supplied by a refrigerant to the maintained space, therefore, the COP of a heat pump is expressed as

$$\beta' = \frac{\dot{Q}_{out}}{\dot{W}_{in}} = \frac{\dot{Q}_{out}}{\dot{Q}_{out}-\dot{Q}_{in}} = \frac{1}{1-\dot{Q}_{in}/\dot{Q}_{out}} \quad (5.3)$$

The relationship between COP of heat pump and COP of refrigerator can be obtained as

$$\beta' = \frac{\dot{Q}_{out}}{\dot{Q}_{out}-\dot{Q}_{in}} = \frac{\dot{Q}_{out}+\dot{Q}_{in}-\dot{Q}_{in}}{\dot{Q}_{out}-\dot{Q}_{in}} = 1 + \frac{\dot{Q}_{in}}{\dot{Q}_{out}-\dot{Q}_{in}} = 1 + \beta \quad (5.4)$$

The COP of a typical heat pump operating between moderate temperatures ranges from 3 to 4.

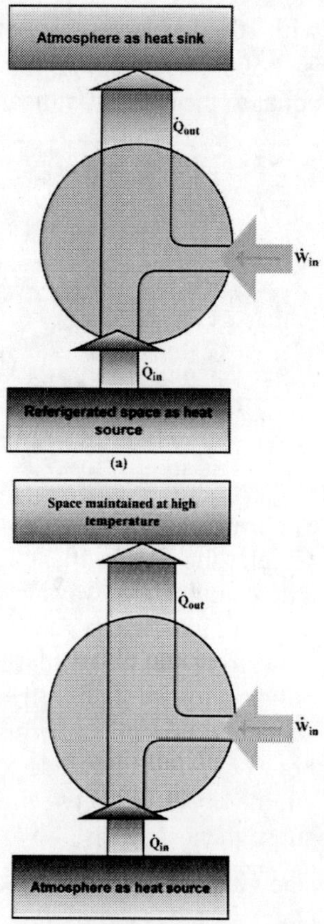


**Fig. 5.8:** Schematic representation of a: (a) refrigerator and (b) heat pump.

## 5.3 Second Law of Thermodynamics

The heat engine, refrigerator and the heat pump described in the previous section are useful in explaining the second law of thermodynamics. This law consists of two classical statements, referred to as the **Kelvin-Planck statement** and the **Clausius statement**. The Kelvin-Planck statement is related to the heat engines and it states that ***it is impossible to operate a device in a cycle that produces net work output by exchanging heat with a single thermal reservoir***. There are a number of statements that formulate the Kelvin-Planck statement in different words, however all these convey the same meaning. It basically explains that a heat engine with 100% efficiency is not possible (Fig. 5.9). In Fig. 5.9, a heat engine is shown for which $\dot{Q}_{out}$ is equal to zero and it produces net work output equal to the amount of heat added.

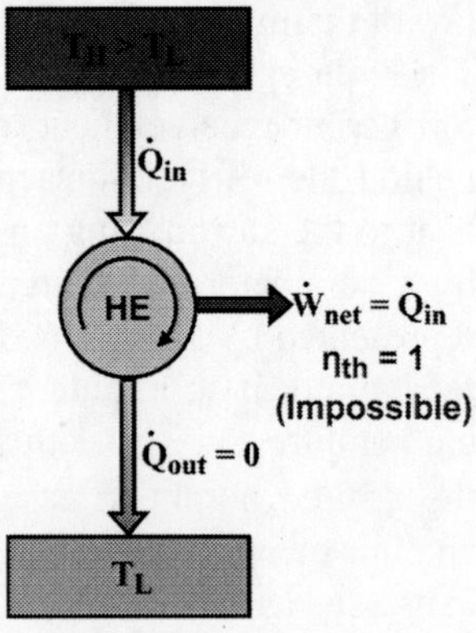


**Fig. 5.9:** Representation of Kelvin-Planck statement.

Such a heat engine with 100% thermal efficiency is not possible as it violates the Kelvin-Planck statement. Therefore, the thermal efficiency of a heat engine is always less than 100%. It is important to mention that a heat engine has to reject some heat in order to complete the cycle. Therefore, the reason for less than 100% efficiency of a heat engine is not only the losses such as friction or other dissipative effects, however, the statement is applicable for ideal cycles as well for which such losses do not exist. The magnitude of maximum limiting efficiency of an ideal heat engine operating between two temperature limits will be obtained in the subsequent sections.

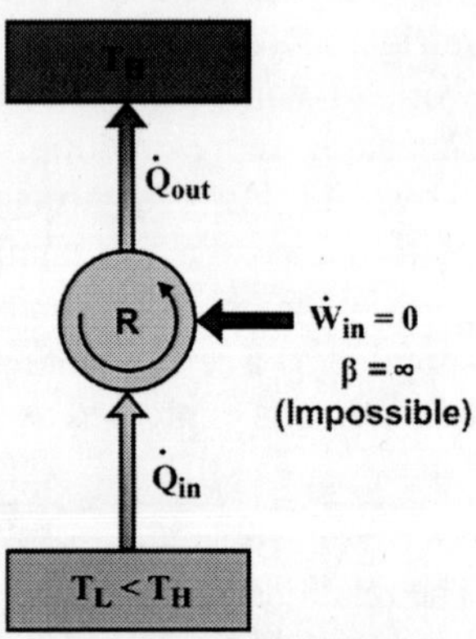


**Fig. 5.10:** Representation of Clausius statement.

The Clausius statement is related to refrigerators and heat pumps and it states that ***it is impossible to construct a device that produces no effect other than transfer of heat from a low temperature reservoir to a high temperature reservoir by operating in a cycle.*** In simple words, it states that no refrigerator or heat pump can operate without consuming work (Fig. 5.10). Therefore, the COP of a refrigerator cannot be infinite. An example of this, we can observe in our kitchen. If we switch off the electric supply of a refrigerator for longer durations, the temperature inside the refrigerator becomes equal to the room temperature. It means electric work should be supplied to a refrigerator to maintain the refrigerated space below the room temperature.

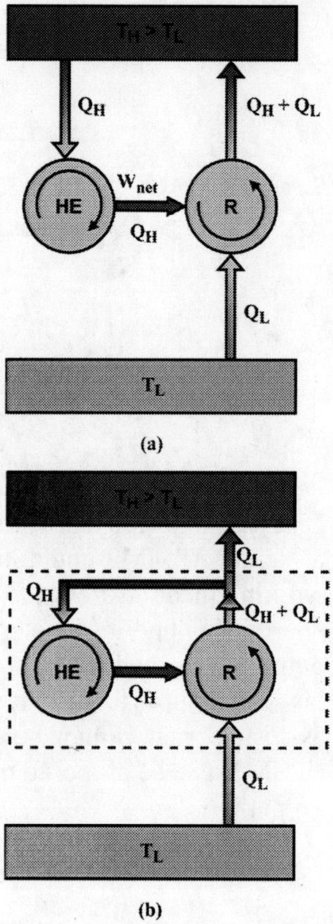


**Fig. 5.11:** Equivalence of the two statements of the second law of thermodynamics.

It is utmost important to highlight that both the statements of the second law of thermodynamics are the negative statements. Thus, these cannot be proved as it is not possible to prove a negative statement. Both the statements are based on the experimental evidence and to date no experiment has been performed which violates the second law of thermodynamics. Also, both the statements are equivalent, that means, the violation of one statement by any system results in violation of the other for the same system. Therefore, either of the statements can be considered the expression of the second law of thermodynamics as both carry the same consequences. To prove the equivalence let us consider a heat engine and a refrigerator operating between the same temperature limits $T_H$ and $T_L$ (Fig. 5.11-a). Let us assume that the heat engine violates the Kelvin-Planck statement, therefore produces work output equal to the heat obtained from heat source ($Q_H$). The work obtained from the heat engine is supplied to the refrigerator which gets the $Q_L$ amount of heat from the low temperature reservoir and by the principle of energy conservation, supplies a heat of $Q_H + Q_L$ amount to the high temperature reservoir. In this process, the net heat transfer associated with the high temperature reservoir is $Q_H + Q_L - Q_H = Q_L$. Therefore, both the heat engine and the refrigerator, we can put inside an equivalent refrigerator (Fig. 5.11-b) that transfers $Q_L$ amount of heat from a low temperature reservoir to a high temperature reservoir without involving any work interaction. This is the violation of the Clausius statement. This proves the equivalence of the two statements. Further, if we start with violation of the Clausius statement, it will result in violation of the Kelvin-Planck statement. Such a situation is demonstrated in Example 5.2.

As stated previously that for a process to occur, both the first and the second law of thermodynamics should be satisfied. Any device that violates any of the laws is termed as a **perpetual-motion machine**. To date, no perpetual-motion machine has been known to work despite various efforts. The perpetual-motion machine of the first kind (PMM1) would produce work without any energy input, thus violating the first law of thermodynamics. Consider a steam power plant which utilizes electric resistance heaters inside the boiler to produce steam as shown in Fig. 5.12. In this system, the part of electricity produced will be used for producing steam and running the pump whereas the rest of electricity can be supplied to the grid as the net work output. If such a system becomes possible, it will produce energy at the rate of $\dot{W}_{net} + \dot{Q}_{out}$ where $\dot{Q}_{out}$ is the rate of heat transfer to surroundings in order to operate the system in a cyclic manner. Such a device produces energy without any input, hence called the PMM1.

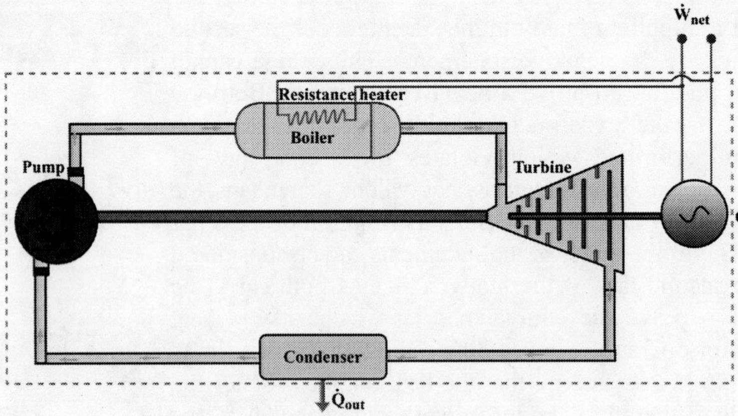


**Fig. 5.12:** Representation of PMM1.

Likewise a device that violates the second law of thermodynamics is referred to as PMM2. An attempt to achieve PMM2 could be a steam power plant that operates without a condenser and supplies the working medium from the exit of the turbine directly to the pump (Fig. 5.13). In such a device all the energy supplied to the working medium in the boiler will produce an equivalent amount of net work output under the ideal conditions. Such a device will have 100% efficiency. Though it satisfies the first law of thermodynamics, it violates the Kelvin-Planck statement of the second law, thus it is referred to as PMM2 and cannot be achieved practically. Similarly a perpetual-motion machine of third kind (PMM3) can be proposed that has no friction, hence would run indefinitely without producing any work output.

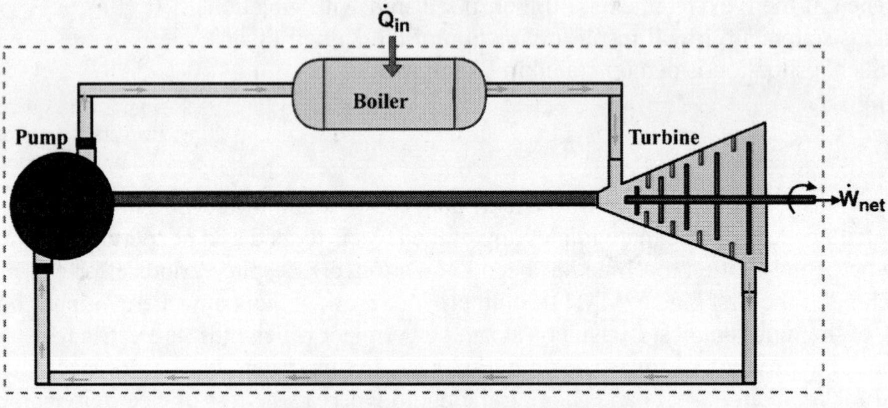


**Fig. 5.13:** Representation of PMM2.

## 5.4 Reversible Process

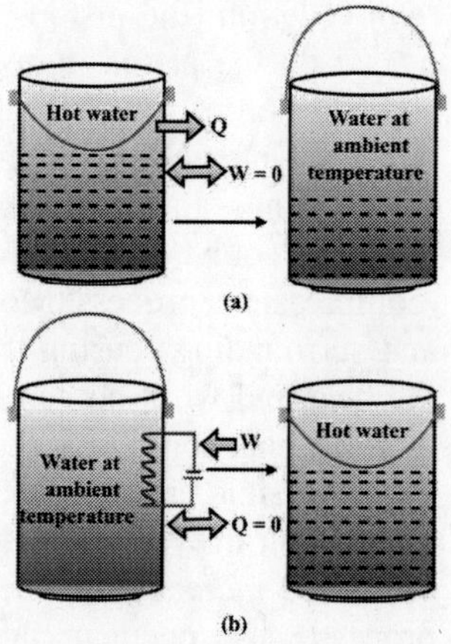


**Fig. 5.14:** Demonstration of irreversibility during a forward and a reverse process.

It is well established in the previous section that the efficiency of a heat engine cannot be 100%. This poses an important question: what is the theoretically possible maximum efficiency of a heat engine? The answer of this question lies with the understanding of the reversible process. A process that can be reversed without keeping any change on its surroundings is termed as a **reversible process**. It means that upon reversing the process, both the system and the surroundings should reach their initial conditions at the end of the reverse process. To understand this let us consider a simple example of a hot water bucket (Fig. 5.14). The hot water bucket transfers heat to the surroundings and reaches to ambient conditions (Fig. 5.14-a). If we have to heat the water in the bucket to reach its initial conditions, we need to heat it using an electric heater which will consume the electric work (Fig. 5.14-b). It means that to reverse this process, the surroundings have to supply electric work to heat the water. If we observe the water bucket has supplied, say $Q$ amount of heat to its surroundings during the forward process and there is no work interaction between the system (water bucket) and surroundings (Fig. 5.14-a). On the other hand, the surroundings has supplied, say $W$ amount of electric work to the system which is equivalent in magnitude to $Q$ during the reverse process (Fig. 5.14-b) to bring the water temperature to its initial conditions. Now the net heat transfer between system and surroundings during the forward and the reverse process is $Q$ which is a loss to the system and gain to surroundings. Similarly, the net work transfer is $W$ which is gain to the system and loss to the surroundings. As the magnitude of $Q$ and $W$ is equivalent, therefore, on the total energy basis the net balance of energy is satisfied for the system and surroundings. However, the system has gone through a gain of work and equivalent loss of heat. Contrary, the surroundings have gained heat and equivalent loss of work. As heat and work are different forms of energy and usually work is a high grade energy whereas heat is a low grade energy. Further, the restoration of heat in the form of work is not fully possible as per the second law of thermodynamics. Therefore, despite the balance of total energy, the quality of energy is not balanced in the present situation, thus this process is not a reversible process. This brings an important conclusion that the net heat and net work transfer between the system and the surroundings should be zero during the forward and the reverse process for a reversible process. Hence, to satisfy the condition of reversibility the quality of the energy should also be preserved along with its quantity.

To further understand the concept of a reversible process, let us consider an adiabatic piston cylinder system which contains a gas and a number of small weights are placed on the piston (Fig. 5.15). When each weight is removed from the piston it expands in a quasi-static manner and gas does work on the surroundings. This expansion continues till the last weight is removed from the piston. Now to reverse the process, the weights are placed on the piston in sequential manner and it results in quasi-static compression of the gas. In the forward and the reverse process, there is no heat interaction, whereas the work is done by the system on surroundings during the expansion process and the surroundings have done the equivalent amount of work on the system during the reverse (compression) process. Moreover, the net work transfer between system and surroundings during the combined process is zero. This means that the heat and work content of the system and surroundings has reached their original conditions once the process is reversed. This is an example of a reversible process. Contrary, the non quasi-equilibrium expansion of a gas is an irreversible process. The detailed illustration of the same is explained in Example 5.3. It is now important to highlight that all the processes that occur in nature are irreversible. Reversible processes are just the idealization of the actual processes for meeting the two purposes: 1. These are easy to analyze as the system undergoes a number of equilibrium states for a reversible process and 2. These form the basis for comparing the performance of the actual processes. As achieving a fully reversible process is not possible, therefore, attempts should be to operate a process with least possible deviation from a reversible process.

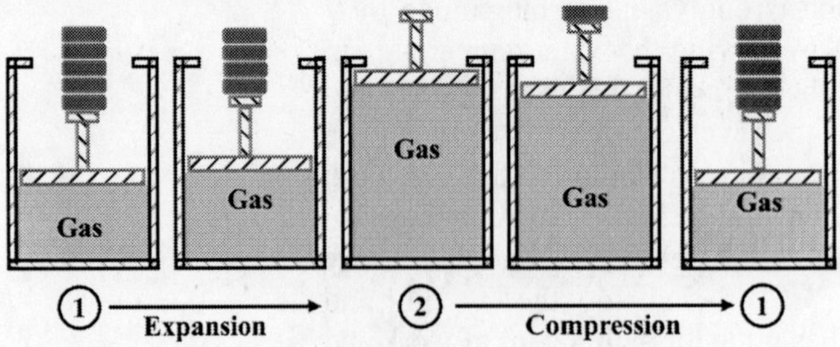


**Fig. 5.15:** Quasi-equilibrium expansion as a reversible process.

Scan Me

For more details on causes of irreversibility.

## 5.4.1 Factors responsible for irreversibility

There are many factors responsible for irreversibilities in a process. Some of these factors are friction, mixing of two or more substances, unrestrained expansion, heat transfer through a finite temperature difference, chemical reactions, inelastic deformation in solids and electric resistance, etc. The occurrence of any of the factors during a process makes it irreversible. We will illustrate the nature of irreversibility introduced by a few common factors in the following discussion.

Friction is the most common factor that comes into consideration when there is a relative motion between two surfaces. For example, in a piston cylinder system, during the expansion of the gas, frictional force opposes the motion of the piston. Therefore, some portion of work done by the gas is utilized in overcoming the frictional force and it produces heat at the piston cylinder interface. Now if we reverse the process and allow the piston to compress the gas in order to reach its initial conditions, the heat generated at the interface in the previous process will not convert into work, rather more amount of heat will be produced at the interface. Ultimately, it will increase the temperature of the system and energy equivalent to the amount of heat generated due to friction should be transferred to surroundings to fully recover the system to its original state. This results in gain of heat by the surrounding during the combined (expansion and compression) process. Thus, friction makes processes irreversible as net heat exchange during the combined process is not zero. The more is the amount of frictional force, the more will be the irreversibility associated with a process. Friction does not only occur between solids, but it is also encountered between a solid surface and a moving fluid and even between the fluid layers moving at different velocities due to the viscosity of the fluids. The shape of the aircraft is made streamline in order to decrease the friction and hence the loss of energy produced by its engine to overcome the friction. The energy produced at the interface of the aircraft body and the surrounding air due to friction ultimately increases the internal energy content of the air.

Unrestrained expansion is another cause of irreversibility that can be observed with the help of an example shown in Fig. 5.16. It shows the portion of a compartment filled with a gas and the remaining portion is in absolute vacuum (Fig. 5.16-a). The two portions are separated by a membrane. When the membrane is ruptured, the gas fills the entire volume (Fig. 5.16-b). If we have to reverse this process, the gas is to be compressed to its initial volume (Fig. 5.16-c). The process of compression requires $W$ amount of work to be done on the gas and an equivalent amount of heat $Q$ to be transferred to the surroundings to bring the gas to its initial conditions. During this process surroundings suffer a loss of work and gain of heat. Therefore, unrestrained expansion is an irreversible process. It is to highlight that the expansion of the gas associated with infinitesimal small differences in gas pressure and the restrained pressure causes the infinitesimal expansion and is a reversible process. However, the finite difference between the two makes the process irreversible as explained in Example 5.3. In general, a partly restrained expansion would be irreversible due to lack of mechanical equilibrium of which unrestrained expansion (or free expansion) happens to be a special case.

Heat transfer through a finite temperature difference is also a cause of irreversibility. A hot glass of water kept in a room gets cooled to room temperature. However, to bring back the water to its initial temperature, either a heat pump or an electric heater is required which will consume the work from the surroundings. During this process, the surroundings cannot be returned to their original state as there will be loss of work, therefore, this heat transfer associated with finite temperature difference is an irreversible process. As the necessary condition for heat transfer to take place is the temperature difference, therefore, all the real life processes involving heat transfer are irreversible in nature. The associated irreversibility can be decreased by decreasing the temperature difference. The only way to achieve the reversible heat transfer is to have infinitesimally small temperature difference between the participating media. However, such a process will require a very large time and surface area to achieve the finite amount of heat transfer. Therefore, reversible heat transfer is just an idealization of the actual process which is of relevance for the purpose of theoretical analysis.

Mixing of fluids is another cause of irreversibility. Consider hot water and cold water are mixed together to reach an equilibrium temperature. Now if we have to reverse the process and obtain the separate quantities of water at their respective initial temperatures, we have to use a heat pump between the two water volumes which will require work input from the surroundings. As surroundings cannot be restored to their original state in this process, therefore, mixing is an irreversible process.

### 5.4.2 Internal and external reversibility

For a reversible process, there should not be any irreversibility within the system and outside the system (in surroundings). However, we may sometimes find situations where there is no irreversibility within the system but presence of irreversibilities outside the system and vice-versa. Let us understand such situations with the help of an example shown in Fig. 5.17. It includes a constant pressure piston cylinder system which contains saturated two phase mixture of liquid and vapour of a substance at temperature $T$. This mixture is being heated with a heat source maintained at a temperature $T + dT$ in the first case (Fig. 5.17-a) and heat source is having temperature $T + \Delta T$ in the second situation (Fig. 5.17-b). Here, $dT$ represents the infinitesimally small temperature difference and $\Delta T$ is the finite temperature difference. During this heat addition process, there will be a constant temperature phase change process within the system and upon reversing the process, the mixture will pass through the same intermediate states to reach its initial state. Therefore, we can say that there is no irreversibility within the system. However, outside the system boundary, the first case is reversible and the second is irreversible as it involves heat transfer through finite temperature difference. Therefore, we can say that both the situations are **internally reversible** as these involve no irreversibilities within the system boundary. However, outside the system boundary, the first situation is **externally reversible** and the second is externally irreversible. A process is said to be reversible or totally reversible if both the internal and external reversibility conditions are satisfied. Hence, we can say that the first case is totally reversible and the second is only internally reversible.

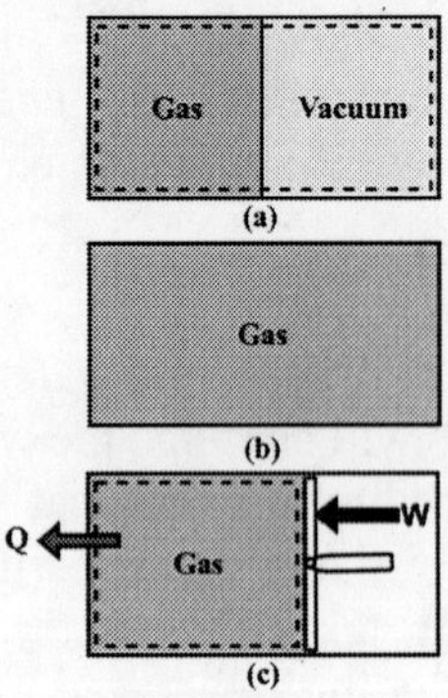


**Fig. 5.16:** Free expansion as an irreversible process.

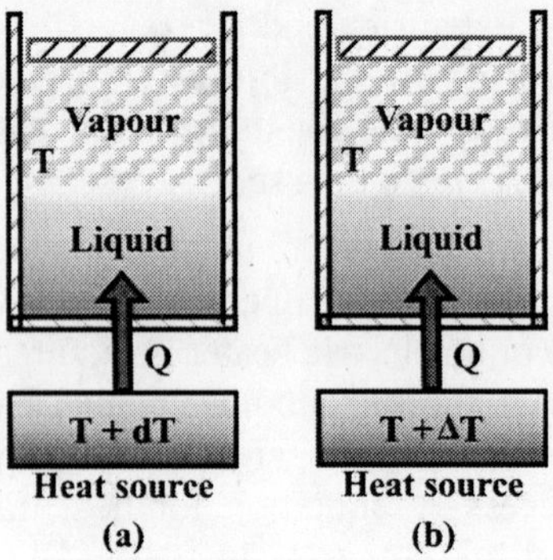


**Fig. 5.17:** (a) Total reversible process and (b) internally reversible and externally irreversible process.

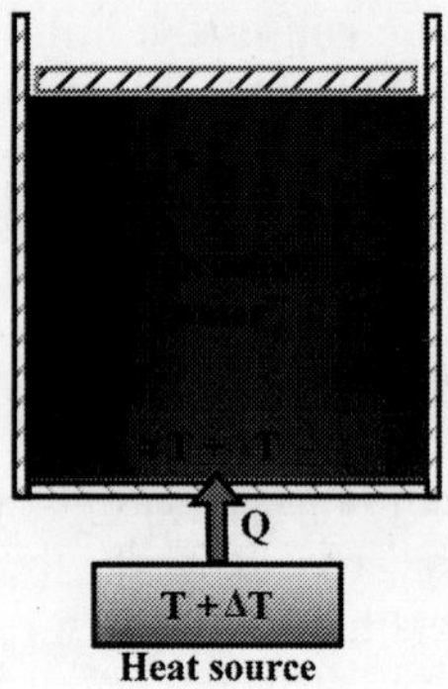


**Fig. 5.18:** Both internally and externally irreversible process.

Another illustration is shown in Fig. 5.18 which shows a piston cylinder system containing water at subcooled state for the given pressure. Here, the layer of water in the vicinity of the bottom wall of cylinder will acquire higher temperature close to $T + \Delta T$ whereas the farthest layer of water near the piston will be at lesser temperature ($\simeq T$). Under this situation, we will experience that there is a finite temperature difference within the system boundary during the heat transfer process. Hence, this process is both internally and externally irreversible.

## 5.5 Carnot Cycle

We explained the concept of reversible process in the previous section, however, the question pertaining to maximum efficiency of a heat engine is still unanswered. If we consider a thermodynamic cycle that consists of all the reversible processes, it is called a reversible cycle. The best known reversible cycle was proposed by a French engineer Nicolas Leonard Sadi Carnot in 1824 and is called the **Carnot cycle**. A heat engine that works on the Carnot cycle is called a **Carnot heat engine** and a refrigerator working on the same is termed a **Carnot refrigerator**. There are plenty of arrangements possible to operate a heat engine in a Carnot cycle. We will demonstrate here two such arrangements to achieve the Carnot cycle. The first arrangement includes a steam power plant in which the working medium is water and operates in the two-phase regime. Let us consider that a high temperature thermal reservoir is at temperature $T_H$ and the water is allowed to exchange heat with it while maintained at the same temperature condition ($T_H$). In this process, water must be at saturated conditions while it receives heat inside the boiler from the high temperature thermal reservoir (Fig. 5.19). Water receives $Q_H$ amount of heat during this process and gets converted to vapor. As there are no irreversibilities associated with this process, therefore, this is a totally reversible isothermal heat addition process. The next process in this sequence is adiabatic expansion of the working fluid in a turbine such that its temperature reaches to the temperature of the low temperature thermal reservoir ($T_L$).

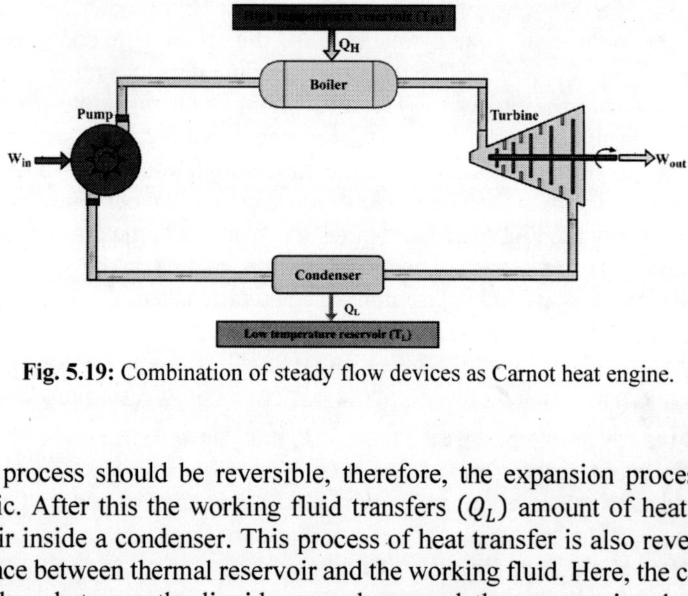


**Fig. 5.19:** Combination of steady flow devices as Carnot heat engine.

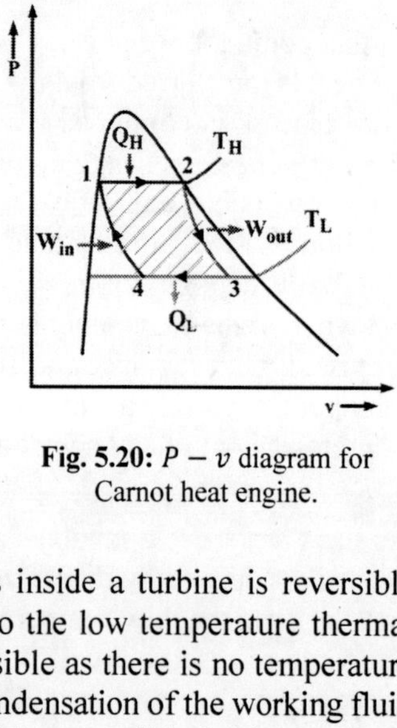


**Fig. 5.20:** $P - v$ diagram for Carnot heat engine.

As the process should be reversible, therefore, the expansion process inside a turbine is reversible adiabatic. After this the working fluid transfers ($Q_L$) amount of heat to the low temperature thermal reservoir inside a condenser. This process of heat transfer is also reversible as there is no temperature difference between thermal reservoir and the working fluid. Here, the condensation of the working fluid takes place between the liquid-vapor dome and the process involved is reversible isothermal heat rejection. After condensing, the working fluid is passed through a pump to compress it adiabatically such that it reaches the initial state to complete the cycle. The process inside the pump is reversible adiabatic. The combination of all these reversible processes forms a totally reversible heat engine, called the Carnot heat engine. The representation of different processes of Carnot heat engine on $P - v$ diagram is shown in Fig. 5.20. The shaded region on the $P - v$ diagram represents the net work produced by the cycle.

Another method to demonstrate the Carnot heat engine includes a piston cylinder system that contains a gaseous working medium at temperature $T_H$ as shown in Fig. 5.21. The piston is loaded with a number of small weights at the initial condition (state 1). Now a heat source having temperature $T_H$ is made to contact the cylinder and it supplies heat to the gaseous medium. The weights are also removed from the piston during the heat addition process such that the temperature of the gas remains unchanged. This process is continued till $Q_H$ amount of heat is supplied to the gas (state 2) and then the heat source is removed. The process 1-2 here is reversible isothermal heat addition. After that weights are removed in sequence to allow the further expansion of the gas under the adiabatic conditions. As the expansion process is quasi-equilibrium, therefore, process 2-3 is reversible adiabatic expansion. During this expansion, the temperature of the gas decreases and reaches to a value $T_L$ at state 3. Now the cylinder is made to contact a low temperature thermal reservoir having temperature $T_L$.

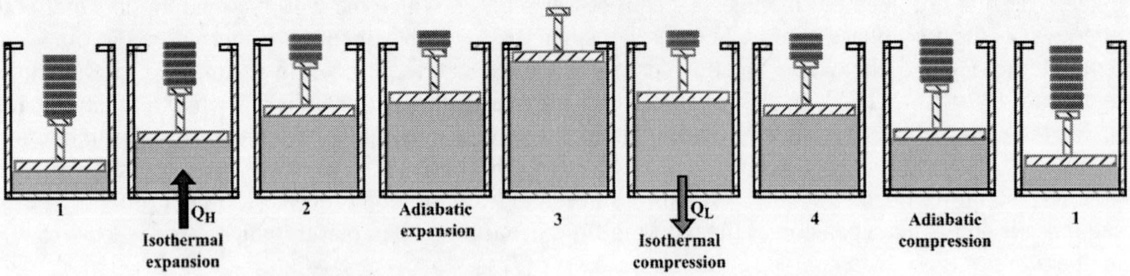


**Fig. 5.21:** Piston-cylinder arrangement based Carnot heat engine.

At this point, the weights are placed on the piston one by one to allow the quasi-static compression and the heat is transferred from the gas to the reservoir such that the temperature of the gas remains constant. This process is continued till state 4 and the process 3-4 is a reversible heat rejection process. The amount of heat exchange involved here is $Q_L$. Further, the piston is loaded with weights sequentially to allow reversible adiabatic compression till the gas reaches its initial conditions. It is important to mention that the working medium remains in the single gaseous phase during the complete cycle here. The $P - V$ diagram for this cycle is shown in Fig. 5.22. It involves work output in the processes 1-2 and 2-3 whereas the work input is required for processes 3-4 and 4-1. The shaded area on the $P - V$ diagram represents the net work done by the cycle. Now it must be emphasized that irrespective of the mechanical arrangement used to achieve the Carnot cycle, the sequence and nature of the processes remain the same. Therefore, it can be concluded that the Carnot cycle requires two reversible isothermal heat exchange processes, one reversible adiabatic expansion and one reversible adiabatic compression process.

As the Carnot cycle involves all the reversible processes, therefore, the Carnot refrigeration cycle can be easily achieved by reversing the processes performed for heat engine operation. The only difference will be that the low temperature thermal reservoir will now act as the heat source and the high temperature reservoir will be used as the heat sink. Moreover, the expansion and the compression processes will be swapped. The $P - V$ diagram for such a Carnot refrigeration cycle is shown in Fig. 5.23. Here, the shaded region shows the work input required by the refrigeration cycle.

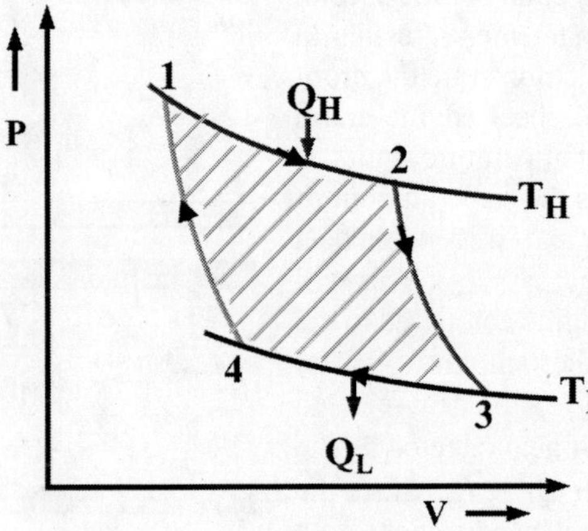


**Fig. 5.22:** $P - V$ diagram for closed system based Carnot heat engine.

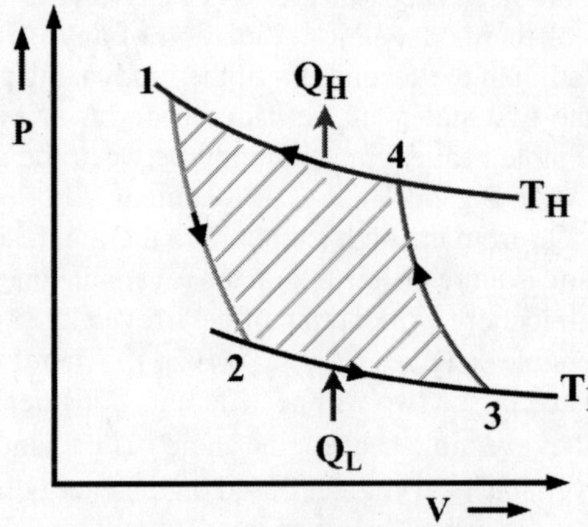


**Fig. 5.23:** $P - V$ diagram for closed system based Carnot refrigerator.

Due to the total reversible processes involved in the Carnot cycle, it is the most efficient cycle operating between the specified temperature limits. The magnitude of its efficiency will be determined in Section 5.7. It is quite evident that the Carnot cycle is not possible in reality because irreversibilities such as friction, etc. cannot be avoided in actual processes. However, it can be used as a benchmark to compare the efficiency of the actual cycles. Further, the efficiency of the actual cycles can be improved by attempting their close approximation to the Carnot cycle.

## 5.6 Propositions Regarding the Efficiency of Carnot Cycle

The two statements pertaining to the efficiency of the reversible and irreversible heat engines, known as **Carnot principles**, are as follows:

1. It is impossible to construct a heat engine operating between the two thermal reservoirs that has more efficiency than a reversible heat engine operating between the same temperature limits.
2. All the reversible heat engines operating between the two given thermal reservoirs have the same efficiency.

The proof of these statements is based on the fundamentals of the second law of thermodynamics. Violation of any of these statements results in violation of the second law of thermodynamics. To demonstrate the proof of the first statement, let us consider an irreversible heat engine and a reversible heat engine operating between the same temperature limits as shown in Fig. 5.24-a. Let us assume that both draws the same amount of heat $Q_H$ from the high temperature thermal reservoir and irreversible heat engine is more efficient than the reversible one ($\eta_{irrev} > \eta_{rev}$). This will result in more work output of the irreversible heat engine compared to the reversible engine ($W_{irrev} > W_{rev}$). Therefore, the amount of heat rejected by the two will be different. Consider that $Q_L$ is the heat rejected by the reversible heat engine and $\hat{Q}_L$ is the amount of heat rejected by the irreversible heat engine. Based on the energy balance $\hat{Q}_L < Q_L$. Since the processes inside the reversible heat engine can be reversed to achieve a reversible refrigerator which will consume $W_{rev}$ amount of work to transfer $Q_L$ amount of heat from the low temperature thermal reservoir. Let us consider now that the operation of the reversible heat engine is reversed to operate it as a refrigerator and it receives required work from the irreversible heat engine as shown in Fig. 5.24-b. Under this condition the net work output produced by the combined (irreversible heat engine and reversible refrigerator) system will be $W_{net} = W_{irrev} - W_{rev}$. Now if we see the high temperature thermal reservoir, it is supplying $Q_H$ amount of heat to irreversible heat engine and receiving the same amount of heat from the reversible refrigerator. Since net energy change is zero for the high temperature thermal reservoir, therefore, it can be replaced and the heat rejected by the refrigerator can be directly fed to the irreversible heat engine (Fig. 5.24-c). Under this situation, we obtain a system that exchanges heat with a single thermal reservoir and produces a net amount of work output. This is the violation of the Kelvin-Planck statement of the second law of thermodynamics. It mean that our initial assumption of $\eta_{irrev} > \eta_{rev}$ is incorrect. Hence, we can say that $\eta_{rev} > \eta_{irrev}$ for all the heat engines operating between the same temperature limits.

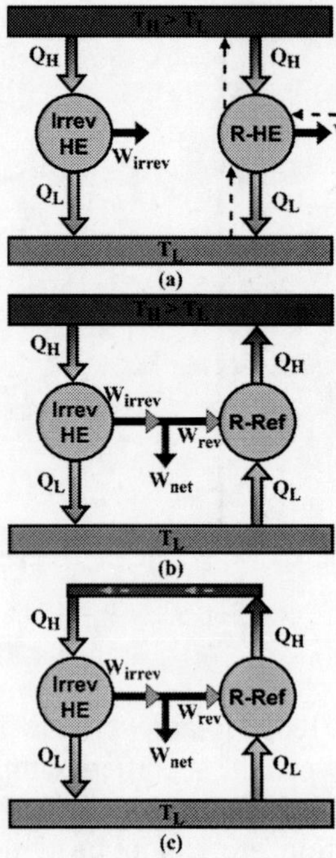


**Fig. 5.24:** Demonstration of first proposition of Carnot.

The second proposition of the Carnot can also be proved in a similar manner. Here, we will consider that we have two reversible heat engines operating between the same temperature limits. Replace the irreversible heat engine in the previous case (Fig. 5.24) with reversible heat engine 1 and consider that it is more efficient. Further proceeding in a similar manner as done in case of Fig. 5.24 will result in a combined system that will violate the second law of thermodynamics. Therefore, it can be concluded that all the reversible heat engines operating between the same temperature limits have the same efficiency irrespective of the type of working fluid used and the kind of arrangement used to achieve the cycles.

## 5.7 Thermodynamic Temperature Scale

We observed in Chapter 1 that the zeroth law of thermodynamics defined the property temperature. Though a number of scales were discussed to measure temperature, these are dependent on the properties of the substances used for temperature measurement. A temperature measurement scale that is independent of the properties of the substances used for measurement of temperature is called an **absolute temperature scale** or a **thermodynamic temperature scale**. The second proposition of Carnot discussed in the previous section forms the basis of the absolute temperature scale. As evident from this statement, the efficiency of the reversible heat engines is solely dependent upon the temperature of the two thermal reservoirs and is independent of the nature working fluid used and the kind of arrangement employed to achieve the cycle. Therefore, $\eta_{rev} = f(T_H, T_L)$. Since $\eta_{rev} = 1 - Q_L/Q_H$, thus $Q_H/Q_L = g(T_H, T_L)$. Here, $T_H$ and $T_L$ are the temperatures of high and low temperature thermal reservoirs, respectively. To further simplify this function relationship, let us consider three reversible heat engines operating between three temperature limits as shown in Fig. 5.25. The reversible heat engines 1 and 2 receives $Q_A$ amount of heat from thermal reservoir at temperature $T_A$. Reversible heat engine 2 rejects $Q_B$ amount of heat at temperature $T_B$ and the same heat is received by the reversible heat engine 3. Reversible heat engine 1 rejects $Q_C$ amount of heat to thermal reservoir at temperature $T_C$. The reversible heat engines 2 and 3 can be replaced with an equivalent heat engine which is identical to heat engine 1 as these are also operating between the same temperature limits. Therefore, reversible heat engine 3 must also reject the same amount of heat $(Q_C)$ to low temperature thermal reservoir as done by heat engine 1. For the three reversible heat engines, the following functional form can be written

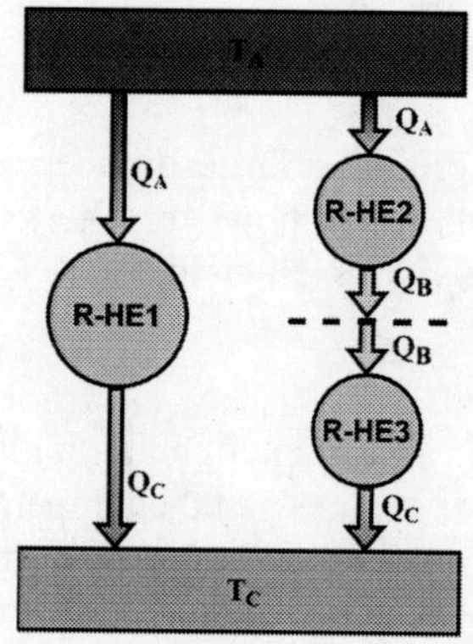


**Fig. 5.25:** A number of reversible heat engines operating between different temperature limits.

$$\frac{Q_A}{Q_C} = g\left(\frac{T_A}{T_C}\right), \frac{Q_A}{Q_B} = g\left(\frac{T_A}{T_B}\right), \frac{Q_B}{Q_C} = g\left(\frac{T_B}{T_C}\right) \quad (5.5)$$

The heat transfer ratios can be correlated as

$$\frac{Q_A}{Q_C} = \frac{Q_A}{Q_B}\frac{Q_B}{Q_C} \quad (5.6)$$

This gives, $g\left(\frac{T_A}{T_C}\right) = g\left(\frac{T_A}{T_B}\right) . g\left(\frac{T_B}{T_C}\right)$ (5.7)

The left hand side of Eqn. (5.7) has only functional dependence on $T_A$ and $T_C$, therefore, the product of two functions on right hand side of the same equation should also be independent of $T_B$. This is only possible if the function $g$ has the following form: $g\left(\frac{T_A}{T_B}\right) = \frac{\theta(T_A)}{\theta(T_B)}$ and $g\left(\frac{T_B}{T_C}\right) = \frac{\theta(T_B)}{\theta(T_C)}$. Substitution of these in Eqn. (5.7) results in cancellation of $\theta(T_B)$. This yields

$$\frac{Q_A}{Q_C} = g\left(\frac{T_A}{T_C}\right) = \frac{\theta(T_A)}{\theta(T_C)} \quad (5.8)$$

Thus for a reversible heat engine the more specific functional form can be written as

$$\frac{Q_H}{Q_L} = \frac{\theta(T_H)}{\theta(T_L)} \tag{5.9}$$

There are a large number of functions $\theta(T)$ possible which can satisfy Eqn. (5.9), however, the simplest form $\theta(T) = T$ is proposed by Lord Kelvin which yields

$$\frac{Q_H}{Q_L} = \frac{T_H}{T_L} \tag{5.10}$$

It is important to mention that Eqn. (5.10) is only applicable for reversible heat engines and the temperature used here is called absolute temperature. Therefore, on an absolute temperature scale, the ratio of temperatures is equal to the ratio of heat transfer between reversible heat engine and the thermal reservoirs and it does not depend upon any other parameter including the properties of the substance. The value of temperature on this scale can vary from zero to infinity. However, it is important to mention that yet we are able to define the ratio of temperatures on this scale and not the value of temperature at a single point.

It is found that Eqn. (5.10) is satisfied upon substituting the absolute temperature with corresponding values in kelvin (K). Also, it is not practically possible to achieve a reversible heat engine for temperature measurement on an absolute scale. Therefore, absolute temperatures can be measured using other methods such as the constant volume ideal gas thermometer by the method of extrapolation as explained in Chapter 1. It is important to mention that the ideal gas temperature scale is identical to the thermodynamic temperature scale. This can be easily demonstrated by using the ideal gas as working medium in a piston cylinder based Carnot heat engine and thermodynamically analyzing the four processes. Such an analysis and its association to thermodynamic temperature scale is demonstrated in Example 5.4. Further, using this knowledge, we can now determine the maximum possible efficiency of a reversible heat engine as

$$\eta_{rev} = 1 - \frac{Q_L}{Q_H} = 1 - \frac{T_L}{T_H} \tag{5.11}$$

## 5.8 Ideal and Real Machines

The ideal machines, which are formed by all the reversible processes in a cycle, give the highest performance. However, the performance of real machines which include the irreversibilities is always less than the ideal ones operating under the similar conditions. Here, we have explicitly mentioned the performance indicators for ideal and real machines and attempted to compare their performance. The performance of reversible and irreversible heat engines, refrigerators and heat pumps can be expressed as

$$\eta = 1 - \frac{Q_L}{Q_H},\ \beta = \frac{Q_L}{Q_H - Q_L},\ \beta' = \frac{Q_H}{Q_H - Q_L} \tag{5.12}$$

These expressions specifically for reversible machines can be expressed as

$$\eta_{rev} = 1 - \frac{Q_L}{Q_H} = 1 - \frac{T_L}{T_H}$$

$$\beta_{rev} = \frac{Q_L}{Q_H - Q_L} = \frac{T_L}{T_H - T_L} \tag{5.13}$$

$$\beta'_{rev} = \frac{Q_H}{Q_H - Q_L} = \frac{T_H}{T_H - T_L}$$

Here, the values of temperature must be substituted in Kelvin (K). The use of temperature in any other scale does not satisfy Eqn. (5.10). The value of efficiency of a heat engine and COP of a refrigerator and a heat pump presented in terms of temperature is only applicable for reversible devices. These represent the theoretically highest possible values that a device can have while operating between the specified temperature limits. Therefore, for actual machines these parameters can be expressed as

$$\eta_{actual} = 1 - \frac{Q_L}{Q_H} < 1 - \frac{T_L}{T_H}$$

$$\beta_{actual} = \frac{Q_L}{Q_H - Q_L} < \frac{T_L}{T_H - T_L} \tag{5.14}$$

$$\beta'_{actual} = \frac{Q_H}{Q_H - Q_L} < \frac{T_H}{T_H - T_L}$$

Let us further discuss the efficiency of a reversible heat engine. It is observed from its expression that the efficiency value increases with a decrease in temperature of the heat sink. Therefore, the net work output increases and the amount of heat rejected decreases. For a limiting case of maximum possible efficiency, the heat rejection will become zero and the temperature of the thermal reservoir corresponding to this is absolute zero. The efficiency formulation of a reversible heat engine here assists in defining the absolute zero temperature on the thermodynamic temperature scale. Conversely, the COP of a refrigerator decreases when the temperature of the refrigerated space is decreased and it requires more work input to produce a refrigerating effect. For the limiting case of absolute zero temperature of the refrigerated space, the refrigerator will require an infinite amount of work to produce a finite amount of refrigeration and its COP will become zero.

Another observation reveals that the efficiency of a reversible heat engine increases upon increasing the temperature of the heat source while the temperature of the heat sink is kept unchanged. Therefore, the fraction of supplied heat that converts into work increases upon increasing the temperature of the heat source for the same temperature of heat sink. It means that a heat input to a reversible heat engine will produce more work if it is supplied from a high temperature source. This clearly explains that energy has quality along with its quantity. The same quantity of energy has high quality if available at higher temperatures. Further, an amount of heat cannot be fully converted to equivalent amount of work. On the other hand, work can be converted fully to equivalent amounts of heat.

## UNIT SUMMARY

The present chapter elucidated the second law of thermodynamics in detail. First, we introduced the essence of studying the second law of thermodynamics which explains that solely the first law of thermodynamics cannot decide the feasibility of a process. It is found that positive heat and work interactions are not possible spontaneously by operating a system in a cyclic manner. A device, named heat engine, is introduced which makes positive heat and work interactions possible by operating in a cyclic manner. Another class of devices, named refrigerator and heat pump, are introduced which makes it possible to transfer heat from low temperature to high temperature by consuming work and operating in a cyclic manner. Two statements of the second law of thermodynamics are introduced which put the restrictions on different cyclic processes to occur in a certain way. It is also demonstrated that both the statements of the second law of thermodynamics are equivalent and violation of one for a system results in violation of the other too. The reversible processes are described as the idealized processes that are mainly used for the purpose of thermodynamic analysis and comparing performance of the actual processes with their ideal counterparts. Different factors that introduce the irreversibilities in actual processes are also illustrated. Internal and external irreversibilities are separately elucidated based on their occurrence within or outside the boundary of the system.

An ideal heat engine formed by the combination of all the reversible processes, called the Carnot heat engine, is explained with associated examples. Two propositions of Carnot based on the reversible heat engine are also illustrated. It is found that the temperature in Kelvin (K) scale satisfies the thermodynamic temperature scale which is independent of the properties of the substance used for measurement of temperature. The thermodynamic temperature scale also assisted in determining the maximum possible efficiency of a reversible heat engine and maximum COP of a reversible refrigerator and a reversible heat pump. All the actual machines have efficiency and COP less than the maximum possible limits specified for their reversible counterparts while operating between the same temperature limits.

## Solved Examples

**Example 5.1:** What will happen if we operate a heat engine without heat rejection?

**Solution:** Analyse a heat engine without heat rejection.

**Assumptions:** Consider a Carnot heat engine operating between the two temperature limits $T_H$ and $T_L$.

**Discussion:** The $P-V$ diagram for the Carnot heat engine under consideration is shown in Fig. E-5.1-a. In this cycle, the process 3-4 is the reversible isothermal heat rejection process. If we omit this process and directly perform the reversible adiabatic compression without heat rejection, we end up with reverse process 3-2 as shown in Fig. E-5.1-b. Therefore, the work output obtained during adiabatic expansion of the gas has been consumed during its compression without heat exchange. Thus, we did not get any work output. Moreover, for the cycle to complete, the process 1-2 is also to be reversed. Such a cycle without heat rejection will not enclose any area on the $P-V$ diagram, hence, results in zero work output. Therefore, to produce a finite work output during a cyclic process heat must be exchanged with two thermal reservoirs.

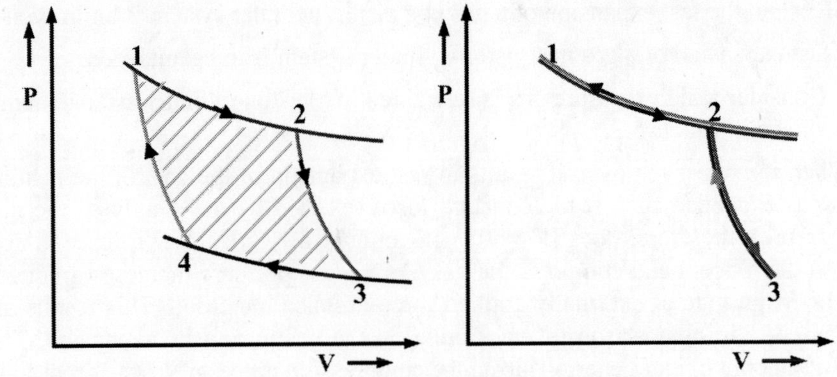


Fig. E-5.1

**Example 5.2:** Explain how the violation of Clausius statement for a system also results in violation of Kelvin-Planck statement for the same system.

**Solution:** The equivalence of two statements of the second law of thermodynamics is to be established.

**Assumptions:** Consider a refrigerator operating between two thermal reservoirs by violating the Calusius statement.

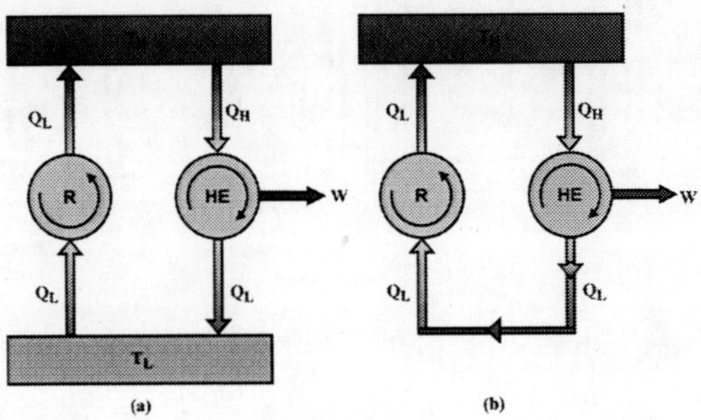


Fig. E-5.2

**Discussion:** A refrigerator transfers $Q_L$ amount of heat from a low temperature reservoir to a high temperature reservoir without any work input as shown in Fig. E-5.2-a. This is the violation of the Clausius statement of the second law of thermodynamics. A heat engine is also operating between the same reservoirs which receives $Q_H$ amount of heat and rejects $Q_L$ amount of heat by producing a work output of magnitude $W = Q_H - Q_L$. In this process, the low temperature thermal reservoir has net zero energy exchange, therefore, it can be removed and heat rejected by the heat engine can be directly fed to the refrigerator (Fig. E-5.2-b). This forms a combined system that exchanges heat with a single thermal reservoir and produces net work output. This will be the violation of the Kelvin-Planck statement of the second law of thermodynamics.

Thus it can be stated that the two statements of the second law of thermodynamics are equivalent to each-other and violation of one for a system results in violation of the other for the same system.

**Example 5.3:** Explain the fast expansion of a gas in a piston-cylinder system is an irreversible process.

**Solution:** The fast expansion of a gas in a piston-cylinder system is to be analyzed.

**Assumptions:** Consider a high pressure gas is restrained in a piston-cylinder arrangement by using an external pin as shown in Fig. E-5.3.

**Discussion:** When the pin is removed it results in fast expansion of the gas till the piston reaches the stops. The gas has expanded against the constant force (exerted by the weight of the piston and the surrounding pressure), therefore, does $W = P_{ext}\Delta V$ amount of work. Now a force is applied on the piston to reverse the process and compress the gas. As the gas pressure increases gradually during its compression, the magnitude of externally applied force has also increased. This results in more work done on the gas ($W'$) in comparison to work done by the gas during its expansion. Therefore, gas receives $W' - W$ amount of extra energy during the compression process. Now in order to bring the gas to initial conditions, heat ($Q$) equivalent to $W' - W$ must be supplied to the surroundings. In this complete process, though the system has reached initial conditions, the surroundings have suffered net loss of work and gain of heat. As surroundings could not be returned to their original state, therefore, the expansion of gas in this case is an irreversible process.

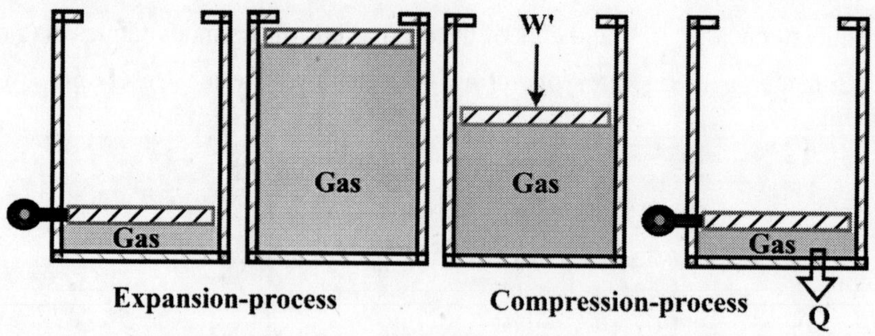


Fig. E-5.3

**Example 5.4:** Perform the thermodynamic analysis of a Carnot heat engine operated using piston-cylinder arrangement and considering an ideal gas as the working medium. Also, determine how this analysis shows resemblance of ideal gas temperature scale to a thermodynamic temperature scale.

**Solution:** The thermodynamic analysis of the Carnot heat engine is to be performed.

**Assumptions:** Consider a Carnot heat engine operates between two temperature limits $T_H$ and $T_L$ using the ideal gas as the working medium. The $P - V$ diagram for such a heat engine is shown in Fig. E-5.4. The processes involved in the cycle are as follows: 1-2: Reversible isothermal heat addition, 2-3: Reversible adiabatic expansion, 3-4: Reversible isothermal heat rejection and 4-1: Reversible adiabatic compression.

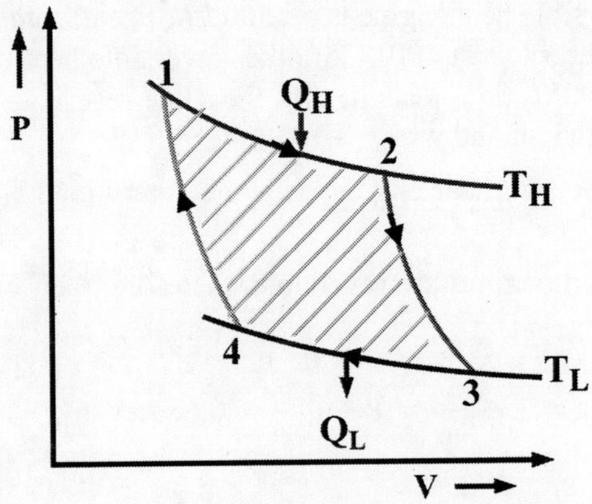


**Fig. E-5.4**

**Analysis:** The present system is a closed system, therefore, the first law of thermodynamics for the same on per unit mass basis is $\delta q - pdv = du$. Since working fluid is an ideal gas, thus $Pv = RT \Rightarrow P = RT/v$ and $du = C_{v0}dT$. Hence, $\delta q = C_{v0}dT + RT/v\, dv$.

For process 1-2: $q_H = RT_H \int_{v_1}^{v_2} \frac{dv}{v} = RT_H ln\frac{v_2}{v_1}$ (5.12)

For process 2-3 (divide equation throughout by *T*): $0 = \int_{T_H}^{T_L} \frac{C_{v0}}{T} dT + R\int_{v_2}^{v_3} \frac{dv}{v}$.

$\Rightarrow \int_{T_H}^{T_L} \frac{C_{v0}}{T} dT = -Rln\frac{v_3}{v_2}$ (5.13)

For process 3-4: $-q_L = RT_L \int_{v_3}^{v_4} \frac{dv}{v} = RT_L ln\frac{v_4}{v_3}$ (5.14)

For process 4-1 (divide equation throughout by *T*): $0 = \int_{T_L}^{T_H} \frac{C_{v0}}{T} dT + R\int_{v_4}^{v_1} \frac{dv}{v}$.

$\Rightarrow \int_{T_L}^{T_H} \frac{C_{v0}}{T} dT = -Rln\frac{v_1}{v_4}$ (5.15)

From Eqns. (5.13) and (5.15), we get $Rln\frac{v_1}{v_4} = -Rln\frac{v_3}{v_2} \Rightarrow ln\frac{v_1}{v_4} = ln\frac{v_2}{v_3}$.

$\Rightarrow \frac{v_1}{v_4} = \frac{v_2}{v_3} \Rightarrow \frac{v_2}{v_1} = \frac{v_3}{v_4}$.

Using Eqns. (5.12) and (5.14), we get $\frac{q_H}{q_L} = \frac{RT_H ln\frac{v_2}{v_1}}{RT_L ln\frac{v_3}{v_4}} = \frac{T_H}{T_L}$.

**Discussion:** The thermodynamic analysis of the Carnot cycle using ideal gas as the working medium shows that the ratio of the associated heat exchanges is equal to the ratio of the temperatures of the ideal gas during the two constant temperature processes. This expression satisfies the thermodynamic temperature scale obtained for a generic system. Hence, it can be said that the thermodynamic temperature scale and the ideal gas temperature scale are identical.

**Example 5.5:** Consider a reversible heat engine takes 500 $kJ$ of heat from a source maintained at $1200K$ and rejects heat to a thermal reservoir at $400K$. Another reversible heat engine takes the same amount of heat from a reservoir at $1600K$ and rejects heat to the same reservoir at $400K$. Out of the two heat engines, which one is more efficient and why?

**Solution:** The efficiency of the two heat engines is to be determined and their performance is to be compared.

**Assumptions:** As both the hean engines are reversible, therefore, their efficiency can be expressed as $\eta = 1 - \frac{T_L}{T_H}$.

**Analysis:** For heat engine 1, $\eta_1 = 1 - \frac{T_L}{T_{H1}} = 1 - \frac{400}{1200} = 0.667$.

Since, $\eta_1 = \frac{W_1}{Q_H} \Rightarrow W_1 = 333.5\ kJ$.

For heat engine 2, $\eta_2 = 1 - \frac{T_L}{T_{H2}} = 1 - \frac{400}{1600} = 0.75$.

Since, $\eta_2 = \frac{W_2}{Q_H} \Rightarrow W_2 = 375\ kJ$.

It is observed from the previous analysis that the same amount of heat at higher temperature has higher potential to produce work. Therefore, heat energy carries quality along with its quantity and for the same quantity of heat its quality improves as its temperature increases. For example, 500 $kJ$ of heat here produced 333.5 $kJ$ of work when it was available at $1200K$. However, the same amount of heat could produce 375 $kJ$ of work when it was available at $1600K$ without any change in sink temperature.

**Example 5.6:** 5 $kg$ water at 10°C and atmospheric pressure is to be converted to ice at -5°C by operating a refrigerator having COP 1.8. Determine the refrigerating effect and the work required to operate the device.

**Solution:** Liquid water is to be converted to ice in a refrigerator. The refrigerating effect and work required to operate the refrigerator need to be calculated.

**Assumptions:** Assume that specific heats of liquid water and ice are constant. $C_{p,water} = 4.18\ kJ/kg - K$ and $C_{p,ice} = 2.06\ kJ/kg - K$ and enthalpy of fusion of water at 0°C is $333.5\ kJ/kg$.

**Analysis:** In order to convert liquid water to ice at atmospheric pressure, first its temperature is to be decreased from 10°C to 0°C and then heat equal to its enthalpy of fusion must be removed. Once the ice is obtained at 0°C, then it is to be further cooled to reach the desired temperature of -5°C.

Thus, the refrigerating effect (or amount of heat to be removed from water) is $Q_{ref.} = m[C_{p,water}(10 - 0) + 333.5 + C_{p,ice}(0 - (-5))] = 1928\ kJ$

COP of refrigerator, $\beta = 1.8 = Q_{ref.}/W$. Thus, $W = 1928/1.8 = 1071.11\ kJ$.

Therefore, 1928 $kJ$ of heat is to be removed from water to get the ice at the desired temperature and it requires 1071.11 $kJ$ of work to be supplied to the given refrigerator.

## EXERCISES

**Multiple Choice Questions**

| | | | | |
|---|---|---|---|---|
| 5.1 | The spontaneous transfer of heat from low temperature body to a high temperature body is the violation of | | | |
| | A. first law of thermodynamics | B. Kelvin-Planck statement of the second law | C. Clausius statement of the second law | D. zeroth law of thermodynamics |
| 5.2 | The violation of the Kelvin-Planck statement of the second law of thermodynamics for a system warrants the violation of__________________ for the same system. | | | |
| | A. zeroth law of thermodynamics | B. first law of thermodynamics | C. Clausius statement of the second law | D. all of the above |
| 5.3 | The occurrence of a process is feasible if it satisfies:<br>i. Zeroth law of thermodynamic<br>ii. First law of thermodynamics<br>iii. Second law of thermodynamics | | | |
| | A. i and ii | B. ii and iii | C. i and iii | D. i, ii and iii |
| 5.4 | An inventor claims a device that produces net heat transfer and work output without any energy input. His invention violates | | | |
| | A. first law of thermodynamics | B. zeroth law of thermodynamics | C. second law of thermodynamics | D. all of the above |
| 5.5 | An invention claims to convert the total amount of heat supplied into equivalent amounts of work by working in a cyclic manner. Such an invention is not possible due to violation of | | | |
| | A. first law of thermodynamics | B. second law of thermodynamics | C. zeroth law of thermodynamics | D. all of the above |
| 5.6 | A cyclic process takes □ amount of heat from its surroundings during the forward process and supplies equivalent amount of work to its surroundings during the reverse process. The combined (forward and reverse) process can be considered | | | |
| | A. reversible | B. irreversible | C. internally reversible | D. none of these |

| 5.7 | The fast expansion of gas in a piston-cylinder system is a/an__________process. | | | |
|---|---|---|---|---|
| | A. internally reversible process | B. internally irreversible process | C. externally reversible process | D. externally irreversible process |
| 5.8 | A household refrigerator operating between 5°C and 45°C has a COP of 7. The device under consideration is | | | |
| | A. reversible | B. irreversible | C. not possible | D. both B and C |
| 5.9 | Water is heated over a gas stove at 100°C in an open pan. The process of heat transfer is<br>i. internally reversible<br>ii. internally irreversible<br>iii. externally reversible<br>iv. externally irreversible | | | |
| | A. i and iv | B. i and iii | C. ii and iii | D. ii and iv |
| 5.10 | The maximum possible COP of a refrigerator operating between 5°C and 55°C will be | | | |
| | A. 0.1 | B. 5.56 | C. 1 | D. ∞ |

**Answers of Multiple Choice Questions**

**5.1 C, 5.2 C, 5.3 B, 5.4 A, 5.5 B, 5.6 B, 5.7 B, 5.8 C, 5.9 A, 5.10 B**

**Short and Long Answer Type Questions**

5.1 In an experiment heating a small amount of water to 180°C is claimed by using a high pressure steam at 160°C without using any refrigerator or heat pump. Access the feasibility of this claim.

5.2 Explain the limitations of the first law of thermodynamics.

5.3 Explain a hypothetical process that satisfies the second law of thermodynamics but violates the first law.

5.4 Name any five thermal reservoirs. Explain with justification why these can be called thermal reservoirs.

5.5 Can the products of combustion used to produce steam in a thermal power plant be treated as a thermal reservoir?

5.6 Can a heat engine have 100% efficiency in absence of all the irreversibilities?

5.7 Two heat engines operating between the same thermal reservoirs receive equal amounts of heat. One heat engine is reversible and the other is irreversible. Which out of the two will reject more heat?

5.8 Explain the validity of a 100% efficient heat engine in terms of the first and the second laws of thermodynamics.

5.9 Can the efficiency of a hydroelectric power plant be governed by the second law of thermodynamics? Explain.

5.10 A heat pump transfers the heat from a colder outdoor to a warmer space. How does the violation of the second law of thermodynamics be avoided in this case?

5.11 A cold water bottle is taken out of a refrigerator which warms up to room temperature. The same bottle is again placed inside the refrigerator to cool it. Is it a reversible process?

5.12 Explain a process that is internally irreversible and externally reversible.

5.13 Water is heated from 25°C to 125°C using the following two methods. i. Water pan is placed on an induction heater which is maintained at 130°C. ii. Water pan is placed on a gas stove whose flame temperature is around 400°C. Which of the methods is most efficient?

5.14 It is proposed to transfer the rejected heat of a heat engine to the source of heat at high temperature using a refrigerator in order to save energy. Is this idea feasible?

5.15 Propose the strategies to reduce the energy consumption of a household refrigerator.

**Numerical Problems**

5.1 A steam power plant of 1000 $MW$ capacity receives its input energy from a coal fired furnace. The coal used in the plant has the calorific value of 26000 $kJ/kg$. Determine the amount of coal required in a day if the efficiency of the plant is 40%.

5.2 A reversible heat engine operating between two temperature limits has 50% efficiency. If the heat engine is reversed to operate as a heat pump, what will be the COP of the heat pump?

5.3 The condenser of a room air-conditioning unit rejects heat to the ambient at the rate of 2.5 $kW$. The compressor of the same unit consumed 1 $kW$ power. Determine the refrigerating effect produced by the air-conditioning unit and its COP.

5.4 A water cooler is desired to cool the water at the rate of 40 $L/h$ from 35°C to 20°C. During the cooling process, a heat leakage at the rate of 80 $kJ/s$ also takes place in the water reservoir. A refrigerating unit having COP of 3 is employed for this purpose. Determine the rate of cooling required and power consumption of the refrigerating unit.

5.5 0.5 $kg$ water at 15°C is to be converted into ice using a refrigerator. Determine the refrigerating effect involved in this process. Also, estimate the amount of work required if a refrigerator having COP of 2.5 is used for this purpose. If the compressor of the refrigerator consumes power at the rate of 1.5 $kW$, find the time required to obtain ice from the water.

5.6 Two reversible heat engines are connected to a thermal reservoir at 850$K$. Both receives a total of 2200 $kJ$ heat from this reservoir and jointly produces a work output of 300 $kJ$. If the first heat engine rejects heat to a heat sink at 750$K$ and second rejects to another heat sink at 650$K$, determine the amount of heat received by the two sinks.

5.7 A solar powered heat pump operates between temperature limits of $220K$ and $320K$. It received heat at the rate of 10 $kW$ from the low temperature reservoir. Determine the rate of work required if the heat pump operates in reversible mode. The solar collector used to receive the energy from solar radiations and supply it to heat pump has energy collection capacity of $0.18\ kW/m^2$. Using this information, find the required area of the collectors to operate the heat pump.

5.8 Two Carnot heat engines are connected in series between the temperature limits of 220°C and 25°C. Determine the intermediate temperature if both produce the same amount of net work output.

5.9 A heat pump is used to maintain a room at a temperature of 25°C when the ambient is at -12°C. The rate of heat leakage from the room is 22 $kW$. Determine the minimum amount of work required to operate the heat pump.

5.10 A coal based thermal power plant is located on a river which has the approximate water flow rate of $5 \times 10^5\ kg/s$ at the temperature of 25°C. The plant has an efficiency of 35%. If the river water is used to cool the condenser such that its temperature should not rise more than 2°C, calculate the maximum possible power produced by the plant.

5.11 A cyclic heat engine receives 625 $kJ$ of energy from a heat source at $1200K$ and rejects 300 $kJ$ of energy to a heat sink at $400K$. It produces a work output of 325 $kJ$. Is this heat engine reversible, irreversible or impossible? Explain.

5.12 A heat pump receives heat from ambient at 45°C and delivers to a boiler which is operated at 280 $kPa$. The boiler receives saturated liquid and supplies saturated vapour at the given pressure. A motor of 50 $kW$ capacity is used to drive the heat pump. The COP of the heat pump is 55% to that of a Carnot heat pump. Determine the mass flow rate at which the boiler can supply the steam.

## PRACTICAL

**Activity 1:** Consider the operation of your home refrigerator equivalent to that of a Carnot refrigerator and estimate its COP in different seasons for different conditions of the maintained refrigerated space. Also, make efforts to find its actual COP and compare it with Carnot COP obtained previously. List all the assumptions made during the analysis.

## REFERENCES AND SUGGESTED READINGS

1. Schneider, E.D. and Kay, J.J., 1994. Life as a manifestation of the second law of thermodynamics. *Mathematical and computer modelling*, *19*(6-8), pp.25-48.

# 6 Entropy

## UNIT SPECIFICS

*Through this unit we have discussed the following aspects:*

- *Clausius Inequality;*
- *Definition of entropy;*
- *Illustration of entropy as property;*
- *Estimation of entropy change for pure substances with state change;*
- *Representation of entropy changes;*
- *Entropy balance for different systems;*
- *Concept of isentropic efficiency of different steady flow processes;*
- *Physical inferences of the concept of entropy.*

*The second law of thermodynamics points towards the limitation on the conversion of heat into work while a thermodynamic system undergoes a cyclic process. In this unit, a new thermodynamic property named entropy is discussed in detail which helps in quantifying the amount of heat retained by the system if temperature history of the system is known during the process. Need for identifying this property is established through the limiting conditions of the Clausius inequality. Mathematical analysis has been done to establish the path independence of entropy so that it can be appreciated as a thermodynamic property. Methods of estimating entropy changes during different thermodynamic processes involving various phases of the substances are also described in this chapter.*

## RATIONALE

*The clear understanding of the concept of entropy and estimation of entropy change during a process along with the principle of entropy increase in the universe is helpful in deciding the feasibility and spontaneity of any process. For developing this understanding, in this chapter the concept of entropy is introduced starting from the Clausius inequality and methods of estimation of entropy change in various processes involving different states of pure substances are analysed.*

*Concept of isentropic processes is established as reversible processes without involvement of heat transfer. These idealised processes concepts are used to define the isentropic efficiencies of various steady flow devices such as turbines, compressors, etc. At the end of this chapter physical inferences of the entropy are established including the description of the microscopic states of the system.*

## PRE-REQUISITES

*Second law of thermodynamics; first law of thermodynamics; basic concept of thermodynamic states*

## UNIT OUTCOMES

*List of outcomes of this unit is as follows:*

*U6-O1: Appreciation of concept of entropy*

*U6-O2: Understanding of entropy as a thermodynamic property*

*U6-O3: Estimation of entropy change in a thermodynamic process*

*U6-O4: Developing skill to estimate entropy change during reversible process*

*U6-O5: Understanding different modes of entropy transfer and generation*

| Unit-6 Outcomes | EXPECTED MAPPING WITH COURSE OUTCOMES (1- Weak Correlation; 2- Medium correlation; 3- Strong Correlation) | | | | | |
|---|---|---|---|---|---|---|
| | CO-1 | CO-2 | CO-3 | CO-4 | CO-5 | CO-6 |
| U6-O1 | 3 | 1 | 2 | 3 | 2 | 2 |
| U6-O2 | 3 | 1 | 2 | 3 | 3 | 3 |
| U6-O3 | 3 | 1 | 2 | 3 | 3 | 3 |
| U6-O4 | 3 | 1 | 2 | 3 | 3 | 3 |
| U6-O5 | 1 | 1 | 2 | 3 | 3 | 3 |

## 6.1 Clausius Inequality

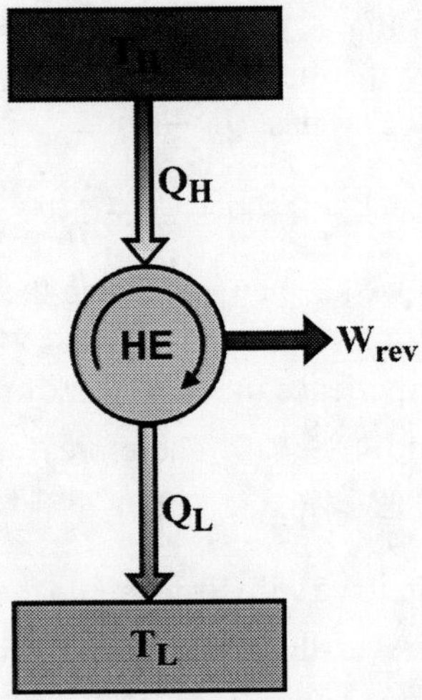

**Fig. 6.1:** A reversible heat engine operating between $T_H$ and $T_L$.

We presented the second law of thermodynamics for a cyclic process in the previous chapter. Now our emphasis will be to discuss its applicability for a process. Like we observed during our discussion on the first law of thermodynamics that for a cyclic process the heat and work balances each-other, however, the need of a property called internal energy arises when it is applied for a process involving a closed system. Similarly the application of the second law of thermodynamics for a process gives birth to an important thermodynamic property, called **entropy**. We now start with the fundamental interpretation of the second law of thermodynamics which forms the basis for defining entropy. Such an important outcome of the second law of thermodynamics is known as the **Clasius inequality** and is expressed as $\oint \frac{\delta Q}{T} \leq 0$. This inequality holds good for all the reversible and irreversible heat engines and refrigerators.

Let us now demonstrate the applicability of Clausius inequality for heat engines. Consider a reversible heat engine operating between two temperature limits as shown in Fig. 6.1. For such a heat engine, using the balance of energy, it is obtained that cyclic integral of heat energy is greater than zero, i.e. $\oint \delta Q = Q_H - Q_L > 0$. The temperature limits $T_H$ and $T_L$ are constant for thermal reservoirs, considered in absolute temperature scale and as the heat engine is reversible, therefore,

$$\oint \frac{\delta Q}{T} = \frac{Q_H}{T_H} - \frac{Q_L}{T_L} = 0 \left(because\ \frac{Q_H}{Q_L} = \frac{T_H}{T_L}\right).$$

For a limiting situation if $T_H$ approaches $T_L$, the $\oint \delta Q = 0$ whereas the $\oint \delta Q/T = 0$ holds good for this condition as well. Therefore, for all reversible heat engines, it can be concluded that $\oint \delta Q \geq 0$ and $\oint \delta Q/T = 0$.

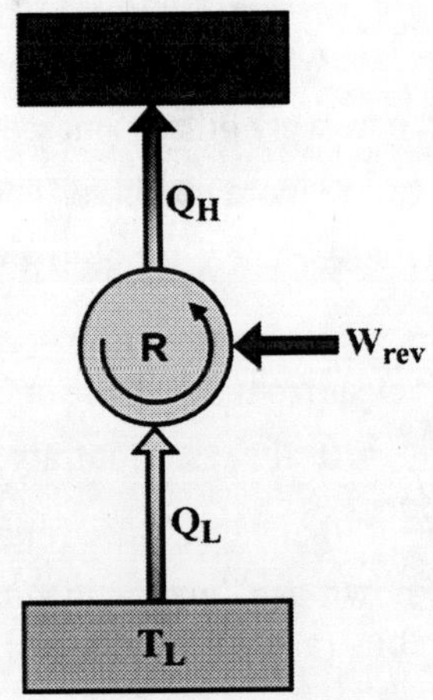

**Fig. 6.2:** A reversible refrigerator operating between $T_H$ and $T_L$.

Now let us assume an irreversible heat engine operates between the same temperature limits as shown in Fig. 6.1 and takes the same supply of heat ($Q_H$) from the heat source. Consider the work produced by this heat engine is $W_{irrev}$. From the second law of thermodynamics, it can be said that $W_{irrev} < W_{rev}$. Thus, $Q_H - Q_{L,irrev} < Q_H - Q_{L,rev}$ and $Q_{L,irrev} > Q_{L,rev}$. This results in

$\oint \delta Q > 0$ and $\oint \frac{\delta Q}{T} = \frac{Q_H}{T_H} - \frac{Q_{L,irrev}}{T_L} < 0 \left( because \ \frac{Q_H}{Q_{L,rev}} = \frac{T_H}{T_L} \Rightarrow \frac{Q_{L,irrev}}{T_L} > \frac{Q_{L,rev}}{T_L} = \frac{Q_H}{T_H} \right)$.

If we keep on increasing the irreversibility of the heat engine by keeping the same values of $T_H$, $T_L$ and $Q_H$, the value of $Q_{L,irrev}$ will keep on increasing. For a limiting situation of irreversibility, we will experience the zero work output of the heat engine. Corresponding to this situation, we get $\oint \delta Q = 0$ and $\oint \frac{\delta Q}{T} < 0$. Therefore, for all irreversible heat engine cycles, it can be concluded that $\oint \delta Q \geq 0$ and $\oint \frac{\delta Q}{T} < 0$.

Now we will demonstrate the Clausius inequality for reversible and irreversible refrigeration cycles. For a reversible refrigeration cycle, as shown in Fig. 6.2, we obtain $\oint \delta Q = -Q_H + Q_L < 0$ and $\oint \frac{\delta Q}{T} = -\frac{Q_H}{T_H} + \frac{Q_L}{T_L} = 0 \left( because \ \frac{Q_H}{Q_L} = \frac{T_H}{T_L} \right)$. The cyclic integral of heat transfer, $\oint \delta Q = 0$ when $T_H$ approaches $T_L$ whereas the $\oint \frac{\delta Q}{T} = 0$ even for this limiting case. Therefore, for all reversible refrigeration cycles, it can be said that $\oint \delta Q \leq 0$ and $\oint \delta Q/T = 0$. Further, let us consider an irreversible refrigerator operating between the same temperature limits and removes same amount of heat ($Q_L$) as done in case of Fig. 6.2. The irreversible refrigerator will consume more amount of work for producing the same amount of refrigeration effect as per the second law of thermodynamics, therefore, $W_{irrev} > W_{rev}$. This implies $Q_{H,irrev} - Q_L > Q_{H,rev} - Q_L$, hence, $Q_{H,irrev} > Q_{H,rev}$. Thus, for an irreversible refrigerating cycle

$\oint \delta Q < 0$ and $\oint \frac{\delta Q}{T} = -\frac{Q_{H,irrev}}{T_H} + \frac{Q_L}{T_L} < 0 \left( because \ \frac{Q_H}{Q_{L,rev}} = \frac{T_H}{T_L} \Rightarrow \frac{Q_{H,irrev}}{T_H} > \frac{Q_{H,rev}}{T_H} = \frac{Q_L}{T_L} \right)$.

If the irreversibility is increased further upon keeping $T_H$, $T_L$ and $Q_L$ constant, the $\oint \delta Q$ and $\oint \frac{\delta Q}{T}$ become more and more negative. Thus for an irreversible refrigeration cycle, the limiting case of $\oint \delta Q = 0$ does not exist. Hence, for all irreversible refrigeration cycles, it can be concluded that $\oint \delta Q < 0$ and $\oint \frac{\delta Q}{T} < 0$.

It is clear from the above discussion that $\oint \delta Q$ can be positive or negative for a reversible cycle, whereas $\oint \frac{\delta Q}{T} = 0$ always. Contrary, for any irreversible cyclic, $\oint \delta Q$ can be positive or negative, whereas $\oint \frac{\delta Q}{T} < 0$ always. Thus, for all the cyclic process, it can be stated that $\oint \delta Q/T \leq 0$. This generic relation is known as Clausius inequality. The applicability of Clausius inequality for a steam power plant is demonstrated in Example 6.1.

## 6.2 Entropy

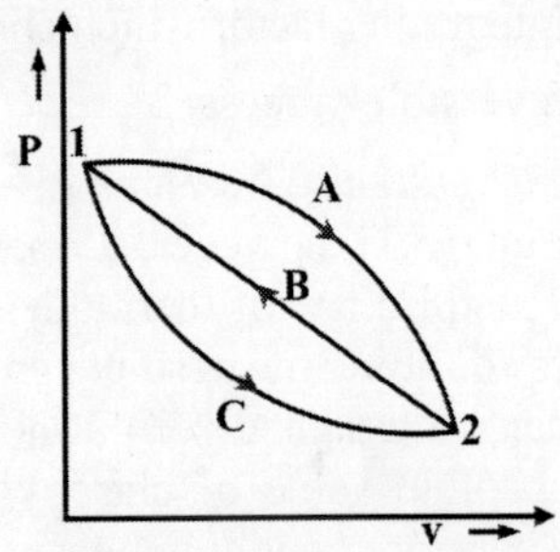


**Fig. 6.3:** Three reversible processes occurring through paths A, B and C between states 1 and 2.

Here, Clausius inequality, the expression driven from the second law of thermodynamics, is used to define an important property called entropy. Let us consider a closed mass that has gone through a reversible cyclic process through paths A and B as shown in Fig. 6.3. Through path A, the system has changed its state from state 1 to state 2, whereas it has moved back from state 2 to state 1 through path B. For such a reversible cycle, the Clausius inequality can be given as

$$\oint \frac{\delta Q}{T} = 0 \Rightarrow \int_1^2 \left(\frac{\delta Q}{T}\right)_A + \int_2^1 \left(\frac{\delta Q}{T}\right)_B = 0 \qquad (6.1)$$

Now consider one more reversible cycle that first progresses through path C and then returns to initial conditions via path B (Fig. 6.3). Clausius inequality for this cycle can be written as

$$\oint \frac{\delta Q}{T} = 0 \Rightarrow \int_1^2 \left(\frac{\delta Q}{T}\right)_C + \int_2^1 \left(\frac{\delta Q}{T}\right)_B = 0 \qquad (6.2)$$

Subtracting Eqn. (6.2) from Eqn. (6.1) gives $\int_1^2 \left(\frac{\delta Q}{T}\right)_A = \int_1^2 \left(\frac{\delta Q}{T}\right)_C$.

This shows that $\int \delta Q/T$ is the same for all reversible paths between states 1 and 2. As this quantity has become independent of the path of the process, therefore, it can be regarded as the property of the system that depends only on the state points. The property introduced here is called entropy. It is designated by symbol $S$ and is defined as

$$dS = \left(\frac{\delta Q}{T}\right)_{rev} \qquad (6.3)$$

Entropy is an extensive property and has units $kJ/K$ in the SI system. Like other properties, entropy on per unit mass basis is called specific entropy and is designated as $s$. The change in entropy during a process can be defined by integrating Eqn. (6.3) between initial and final states as

$$S_2 - S_1 = \int_1^2 \left(\frac{\delta Q}{T}\right)_{rev} \qquad (6.4)$$

It is important to note that using Eqn. (6.3) entropy is defined along a reversible path only. Therefore, the integration of Eqn. (6.4) should be performed only along a reversible path. To perform this integration, a relationship between $Q$ and $T$ is required. Usually this relation is available for a very few processes, therefore, this integration can be performed only for such processes. However, in majority cases, we rely on the property tables for estimation of entropy. It is also utmost important to highlight that though entropy change between two states is to be determined only for a reversible process, yet once it is obtained, it can be used for all irreversible processes between the same two states because entropy is a property. Moreover, the integration of Eqn. (6.4) along irreversible paths will result in different values of entropy change. Thus, it must be emphasized that the use of integration of Eqn. (6.4) along an irreversible process is inappropriate.

We established in earlier chapters that heat transfer is a path function, therefore, $\delta Q$ is an inexact differential. However, the term $\frac{\delta Q}{T}$ for a reversible process in Eqn. (6.3) becomes a property of the system, therefore, it is a state function. Thus, we can mathematically say that $1/T$ is acting as an integrating factor which converts an inexact differential $\delta Q$ to an exact differential $\frac{\delta Q}{T}$ specifically for a reversible process.

It should also be emphasized that we can only estimate the change in entropy during a process using Eqn. (6.3), however, it does not provide any idea about the absolute value of entropy at a given state. The third law of thermodynamics, which is discussed in a later section of this chapter, forms the basis for defining the absolute entropy. It basically specifies that for all pure substances, the value of entropy can be assigned zero at absolute zero temperature. As an engineer, we are not concerned about the absolute value of entropy, rather our interest is to estimate the entropy change during a process. Therefore, any commensurate reference state can be selected for a substance at which its entropy is assigned zero and then Eqn. (6.4) can be used to obtain entropy at the desired state. When the entropy of a substance at different states is estimated with respect to a single reference state, it becomes quite convenient to estimate entropy change between any two states as reference value gets cancelled. Usually the same procedure is used for construction of property tables for entropy as demonstrated for other properties with the help of steam tables in previous chapters.

## 6.3 Entropy of a Pure Substance

Entropy is a property thus it has a unique value at the given state of the system. As we know that for a simple compressible substance, two independent intensive properties are required to specify its state and estimate the value of all other properties, including its entropy. Thus we have property tables available for entropy as well like we demonstrated for other properties in Chapter 3. For constructing a property table, a reference state is required. Therefore, in case of steam tables (Table B.1), the value of entropy of a saturated liquid $(s_f)$ is assigned zero at 0.01°C. Usually for refrigerant R-134a, it is specified zero at -40°C. Likewise different reference states can be selected for different substances. The values of entropy at temperatures less than the reference temperature are negative.

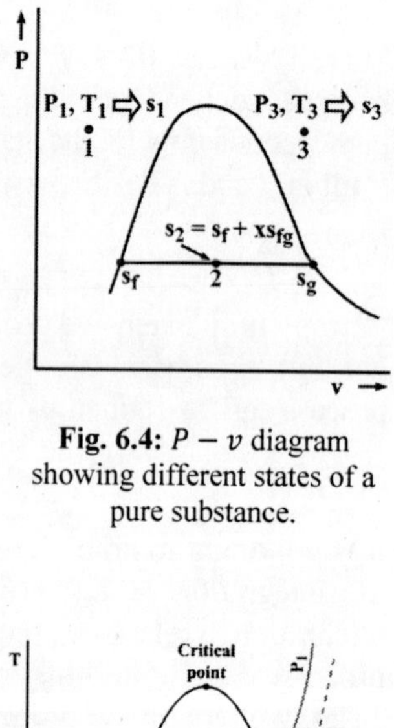


**Fig. 6.4:** $P-v$ diagram showing different states of a pure substance.

The entropy for different states of a substance can be obtained using the same procedure as demonstrated for other properties using steam tables in Chapter 3. Therefore, tables B.3 and B.4 can be used to specify entropy of superheated and compressed states, respectively for water. These tables can be used once the two independent intensive properties are known at these states. In the saturated liquid-vapour mixture region, the entropy estimation requires knowledge of quality in addition to saturated tables B.1 and B.2. The entropy for saturated mixture is determined as $s = s_f + xs_{fg}$. Here $x$ is quality, $s_f$ is entropy of saturated liquid and $s_{fg} = s_g - s_f$, where $s_g$ is entropy of saturated vapour. The values of $s_f$ and $s_g$ are available in saturated steam tables B.1 and B.2. A graphical depiction of entropy estimation for different states of a substance is shown on $P-v$ diagram in Fig. 6.4. If the compressed state property tables are not available, then entropy of a compressed liquid can be considered

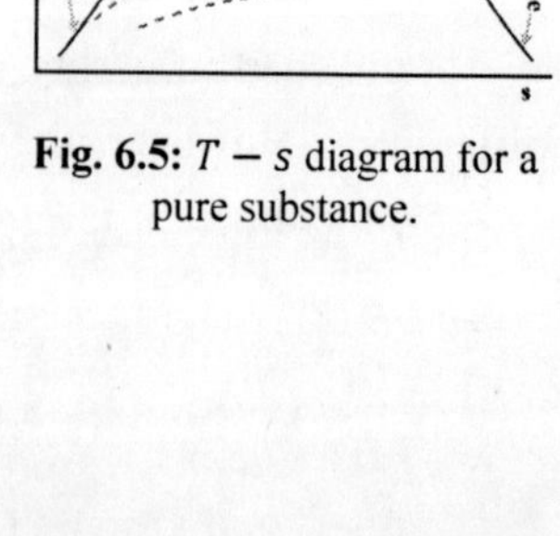


**Fig. 6.5:** $T-s$ diagram for a pure substance.

equivalent to saturated liquid entropy at the given temperature, i.e. $s_{comp} \simeq s_{f@T}$ (as shown in Fig. 6.5).

When studying the second law of thermodynamics for a process, it is usual practice to use entropy as one of the coordinates on property diagrams such as $T - s$ and $h - s$ plots. The $h - s$ diagram is famously known as the Mollier diagram and is an important plot which is used often in place of steam tables to conveniently locate the states and determine associated properties. A $T - s$ diagram for water is shown in Fig. 6.5. It shows that constant specific volume lines have positive slope and the slope is smaller for larger values of specific volumes. The constant pressure and constant temperature lines are horizontal within the saturated liquid-vapour region. The constant pressure lines in the subcooled/compressed region almost coincide with the saturated liquid curve. Similarly a $h - s$ diagram for water is shown in Fig. 6.6.

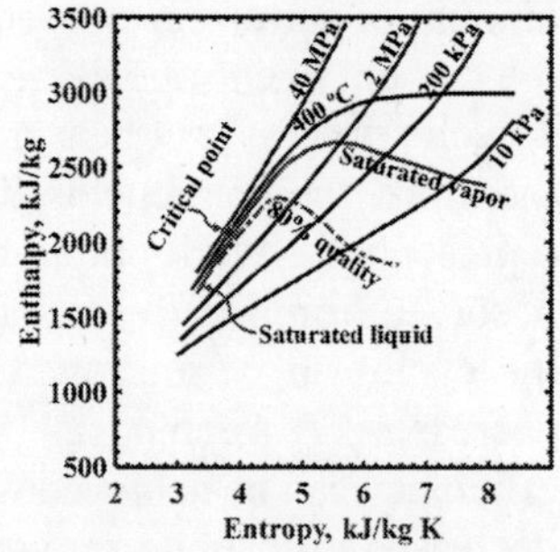


**Fig. 6.6:** $h - s$ diagram for a pure substance.

## 6.4 Entropy Change for Internally Reversible Processes

After defining entropy, our aim is to establish its significance for different processes. In this attempt, let us consider the working fluid of a reversible Carnot cycle is our system. The first process of this cycle is an isothermal heat addition to the working fluid from a high temperature thermal reservoir. Such a process 1-2 is shown on $T - S$ diagram in Fig. 6.7. For this process, we can write $S_2 - S_1 = \int_1^2 \left(\frac{\delta Q}{T}\right)_{rev}$. Since temperature of working fluid is constant during this process, therefore, $S_2 - S_1 = \frac{1}{T_H}\int_1^2 \delta Q = \frac{Q_{12}}{T_H} \Rightarrow Q_{12} = T_H(S_2 - S_1)$. Thus, the area under the line 1-2 on $T - S$ diagram represents the heat added to the working fluid.

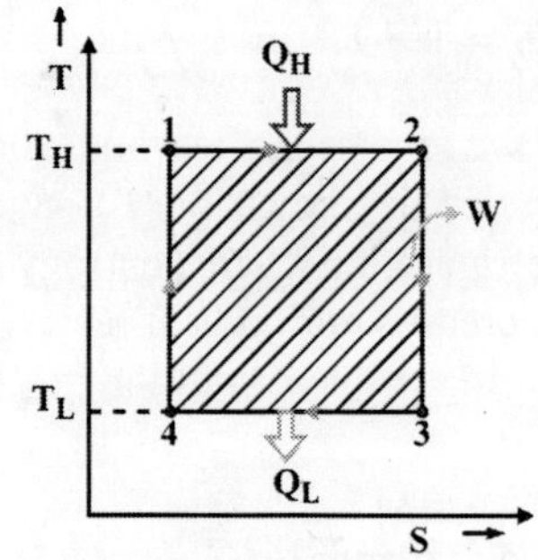


**Fig. 6.7:** $T - S$ diagram for a Carnot heat engine.

The second process 2-3 is a reversible adiabatic process. For this process, we get $S_3 - S_2 = \int_2^3 \left(\frac{\delta Q}{T}\right)_{rev} = 0 \Rightarrow S_3 = S_2$. Hence, entropy remains constant during a reversible adiabatic process and it is called an **isentropic process**. Such a process will be represented by vertical line on the $T - S$ diagram and this line extends till the temperature of low temperature thermal reservoir is achieved. The third process is isothermal heat rejection and entropy change equation for this can be expressed as $S_4 - S_3 = \int_3^4 \left(\frac{\delta Q}{T}\right)_{rev}$. Thus $S_4 - S_3 = \frac{1}{T_L}\int_3^4 \delta Q = \frac{Q_{34}}{T_L} \Rightarrow Q_{34} = T_L(S_4 - S_3)$. Since the working fluid loses heat during this process, therefore, entropy change will be negative which signifies that entropy of the working fluid decreases during heat rejection. Moreover, the area under the line 3-4 in Fig. 6.7 represents the amount of heat rejected. The last process, that completes the cycle, is reversible adiabatic compression, hence isentropic. Therefore, it will be represented by a vertical line 4-1 on $T - S$ diagram. Furthermore, it draws an important conclusion that the

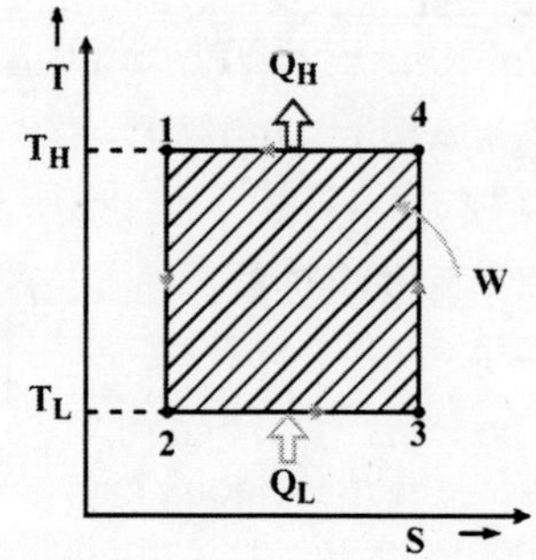


**Fig. 6.8:** $T - S$ diagram for a Carnot refrigerator.

amount of entropy increased during the heat addition process is equal to the amount of entropy decreased during the heat rejection process. As for a cyclic process, net heat transfer is equal to net amount of work done, therefore, the area enclosed by the region 1-2-3-4-1 on $T - S$ diagram represents the work output of the Carnot heat engine. Further, it can be observed from the $T - S$ diagram shown in Fig. 6.8 that the efficiency of the cycle can be improved by increasing $T_H$ keeping $T_L$ fixed and by decreasing $T_L$ keeping $T_H$ fixed. Also, the limiting condition of 100% efficiency can be achieved when either $T_H$ is increased to infinity or $T_L$ is decreased to absolute zero temperature.

If instead of a totally reversible cycle, the processes shown in the above cycle (Fig. 6.7) are only internally reversible, then also the temperature of working fluid will remain $T_H$ during heat addition process and $T_L$ during the heat rejection process. Thus such an internally reversible cycle can also be represented identically on $T - S$ diagram and will have equal amounts of heat and work interactions. Therefore, the entropy analysis presented above is applicable for a thermodynamic cycle that comprises of internally reversible processes. Thus, the entropy change Eqn. (6.3) is applicable for both reversible and internally reversible processes. Hence, we can write its generic form as

$$dS = \left(\frac{\delta Q}{T}\right)_{rev} = \left(\frac{\delta Q}{T}\right)_{int,rev} \tag{6.5}$$

If we reverse the heat and work interactions of reversible Carnot cycle, we get the reversible refrigeration cycles as shown on $T - S$ diagram in Fig. 6.8. It is important to note that for such an reversible refrigeration cycle, entropy of working fluid increases during heat addition at temperature $T_L$ and decreases during heat rejection at temperature $T_H$. This concludes that entropy of a substance increases during the heat addition process and it decreases during the heat rejection process.

## 6.5 Thermodynamic Property Relations

In this section, we will attempt to formulate the equations which solely include thermodynamic properties. Consider a closed system containing a simple compressible substance undergoing a reversible process. First law of thermodynamics for such a situation on per unit mass basis is given as $\delta q - \delta w = du$. For a reversible process, $\delta q = Tds$ and $\delta w = Pdv$. Thus

$$Tds = du + Pdv \tag{6.6}$$

Now we know $h = u + Pv \Rightarrow dh = du + Pdv + vdP$. Using Eqn. (6.6), we get $dh = Tds + vdP$ which becomes

$$Tds = dh - vdP \tag{6.7}$$

The Eqns. (6.6) and (6.7) are the thermodynamic property relations known as $Tds$ or **Gibbs equations.** Note that these equations involve all the property terms only and as the properties are the state functions, therefore, these are applicable for both reversible and irreversible processes. However, it is emphasized that integration of these equations must be performed only along a reversible process because the terms involving multiple properties form heat and work which are path functions.

## 6.6 Entropy Change for Solids and Liquids

Thermodynamic property relations, Eqns. (6.6) and 6.7 can be expressed as $ds = \frac{du}{T} + \frac{Pdv}{T}$ and $ds = \frac{dh}{T} - \frac{vdP}{T}$. For solids and liquids due to their incompressible nature $dv \simeq 0$ and as their density is large, thus $v \simeq 0$. Also, as discussed in Chapter 2 for solids and liquids that $C_p \approx C_v = C \Rightarrow dh = du = CdT$, therefore, $ds = \frac{CdT}{T}$. Then entropy change for a process can be obtained by integration of this equation as

$$s_2 - s_1 = \int_{T_1}^{T_2} \frac{C(T)dT}{T} = C_{avg} ln\frac{T_2}{T_1} \quad (6.8)$$

Here, $C_{avg}$ is the average specific heat for the specified temperature range. It is also to be noted that entropy of a purely incompressible substance is the sole function of temperature and does not depend upon other properties. This equation will not give accurate results for liquids that significantly expands with temperature. Such situations mainly arise when the operating temperature range is large. If we subject Eqn. (6.8) to an isentropic process, it gives $T_1 = T_2$. It means that an isentropic process involving perfectly incompressible solids and liquids is isothermal as well. This is the reason that the temperature of common liquids such as water does not change during their pumping.

## 6.7 Entropy Change for an Ideal Gas

The entropy change equation for ideal gases can be obtained by using Eqns. (6.6) and (6.7) together with the ideal gas equation $Pv = RT$. Using Eqns. (6.6) and (6.7), we get $ds = \frac{du}{T} + \frac{Pdv}{T}$ and $ds = \frac{dh}{T} - \frac{vdP}{T}$. By invoking ideal gas equation in second term of each entropy change equation, we get $ds = \frac{du}{T} + \frac{Rdv}{v}$ and $ds = \frac{dh}{T} - \frac{RdP}{P}$. For ideal gases, $du = C_{v0}dT$ and $dh = C_{p0}dT$. Thus first entropy change equation becomes $ds = \frac{C_{v0}dT}{T} + \frac{Rdv}{v}$. Upon integration

$$s_2 - s_1 = \int_1^2 \frac{C_{v0}dT}{T} + Rln\frac{v_2}{v_1} \quad (6.9)$$

Similarly, the second entropy change equation becomes $ds = \frac{C_{p0}dT}{T} - \frac{RdP}{P}$. By integrating this equation, we get

$$s_2 - s_1 = \int_1^2 \frac{C_{p0}dT}{T} - Rln\frac{P_2}{P_1} \quad (6.10)$$

Equations (6.9) and (6.10) give the entropy change for a process involving ideal gas as the working fluid. The first term of both equations requires the temperature dependence on $C_{v0}$ and $C_{p0}$. This integration can be performed in two ways. In first case, if the temperature range is not large then average values of $C_{v0}$ and $C_{p0}$ can be used for given temperature range and we get

$$s_2 - s_1 = C_{v0,avg} ln\frac{T_2}{T_1} + Rln\frac{v_2}{v_1} \quad (6.11)$$

$$s_2 - s_1 = C_{p0,avg} ln\frac{T_2}{T_1} - Rln\frac{P_2}{P_1} \quad (6.12)$$

The second approach requires a polynomial expression between specific heats and temperature which can be then substituted in Eqns. (6.9) and (6.10) to obtain the entropy change. However, performing this integration every time is cumbersome, therefore, a common approach is to obtain the first term of entropy change equations for different temperatures using some reference temperature and tabulate this data. The tabulated data then can be used as and when required. Such a process is demonstrated here for Eqn. (6.10), considering the absolute zero temperature as the reference state. The parameter $s_T^0$ for this purpose is defined as

$$s_T^0 = \int_0^T \frac{C_{p0}(T)dT}{T} \tag{6.13}$$

The term $s_T^0$ is called **standard entropy**, it is a sole function of temperature and its value is zero at absolute zero temperature. This value can be obtained for different temperatures using Eqn. (6.13) and data can be tabulated as the function of temperature. The integration of the first term of Eqn. (6.10) gives

$$\int_1^2 \frac{C_{p0}(T)dT}{T} = s_{T_2}^0 - s_{T_1}^0 \tag{6.14}$$

Here, $s_{T_2}^0$ and $s_{T_1}^0$ are values of standard entropy at temperatures $T_2$ and $T_1$, respectively. Using Eqn. (6.14), Eqn. (6.12) can be expressed as

$$s_2 - s_1 = s_{T_2}^0 - s_{T_1}^0 - Rln\frac{P_2}{P_1} \tag{6.15}$$

To calculate the entropy change using Eqn. (6.15), standard entropy can be obtained from entropy tables at different temperature conditions. It must be emphasized that entropy change of an ideal gas is not a sole function of temperature like its internal energy and enthalpy. Therefore, usually the property tables for ideal gas include the standard entropy as function of temperature along with its internal energy and enthalpy whereas the entropy change should be calculated using Eqn. (6.15) which shows dependence on other properties as well.

Now let us analyze the entropy change relations, Eqns. (6.11) and (6.12) for a special case of isentropic process. Using Eqn. (6.11), we get

$$0 = C_{v0,avg}ln\frac{T_2}{T_1} + Rln\frac{v_2}{v_1} \Rightarrow ln\frac{T_2}{T_1} = -\frac{R}{C_{v0,avg}}ln\frac{v_2}{v_1} \Rightarrow ln\frac{T_2}{T_1} = ln\left(\frac{v_1}{v_2}\right)^{\frac{R}{C_{v0,avg}}}.$$

Since $R = C_{p0,avg} - C_{v0,avg} \Rightarrow \frac{R}{C_{v0,avg}} = k - 1$ where $k = \frac{C_{p0,avg}}{C_{v0,avg}}$. Therefore,

$$\frac{T_2}{T_1} = \left(\frac{v_1}{v_2}\right)^{k-1} \text{ (isentropic process for an ideal gas)} \tag{6.16}$$

Similarly using Eqn. (6.12), we get

$$0 = C_{p0,avg}ln\frac{T_2}{T_1} - Rln\frac{P_2}{P_1} \Rightarrow ln\frac{T_2}{T_1} = \frac{R}{C_{p0,avg}}ln\frac{P_2}{P_1} = ln\left(\frac{P_2}{P_1}\right)^{\frac{R}{C_{p0,avg}}}$$

Since $\frac{R}{C_{p0,avg}} = \frac{C_{p0,avg} - C_{v0,avg}}{C_{p0,avg}} 1 - \frac{1}{k} = \frac{k-1}{k}$. Hence,

$$\frac{T_2}{T_1} = \left(\frac{P_2}{P_1}\right)^{\frac{k-1}{k}} \text{ (isentropic process for an ideal gas)} \tag{6.17}$$

Equating the right hand side of Eqns. (6.16) and (6.17), we get

$$\frac{P_2}{P_1} = \left(\frac{v_1}{v_2}\right)^{k} \text{ (isentropic process for an ideal gas)} \tag{6.18}$$

Writing Eqns. (6.16)-(6.18) in compact form, we obtain $Tv^{k-1} = const.$, $TP^{\frac{1-k}{k}} = const.$ and $Pv^{k} = const.$ Therefore, it can be concluded that isentropic process using ideal gas is a special case of polytropic process where polytropic index $(n)$ is equal to ratio of specific heats $(k)$. However, these equations are only valid when the operating temperature range is small and the assumption of average specific heats is appropriate.

## 6.8 Property Diagrams Involving Entropy

As explained in Section 6.3 that it is convenient for the second law analysis of a thermodynamic process to present it on a property diagram involving entropy as one of the coordinates, we attempted here to present different processes on a $T - s$ diagram. It is well established in Chapter 2 that many processes involving simple compressible substance can be easily represented with a simple mathematical expression $PV^{n} = C$, known as a polytropic process. Here, $n$ is polytropic index and $C$ is a constant. Also, values of $n$ for some familiar processes is given as

$n = 0, P = C$: Isobaric process,

$n = 1, T = C$: Isothermal process,

$n = k, s = C$: Isentropic process,

$n = \infty, v = C$: Isochoric process.

Now we present the $P - v$ diagram in Fig. 6.9-a to represent the above process and then we further plot the same processes on $T - s$ diagram in Fig. 6.9-b. A horizontal line on $T - s$ diagram represents the entropy change (either increase or decrease depending on the direction of the process) during an isothermal process and a vertical line on the same diagram depicts no change of entropy, hence an isentropic process. It is also evident that slope of an isochoric process is more than the slope of an isobaric process on the $T - s$ diagram. Note that Fig. 6.9 is just a schematic representation for different polytropic processes involving a simple compressible substance.

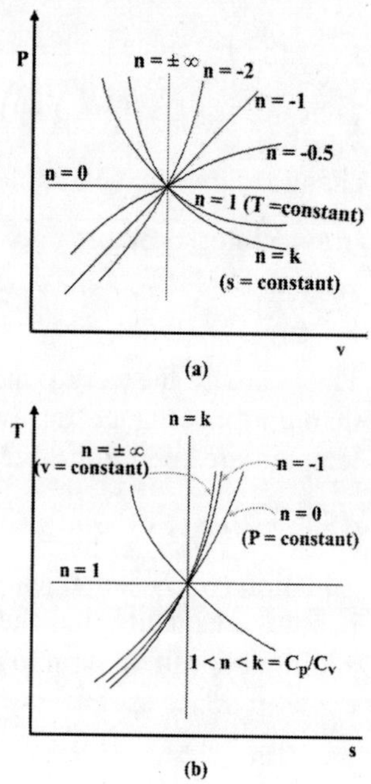


**Fig. 6.9:** Representation of different polytropic processes on: (a) $P - v$ diagram and (b) $T - s$ diagram.

## 6.9 Entropy Change for an Irreversible Process and Entropy Equation

In the earlier discussion, we have mainly concentrated on entropy analysis for reversible processes. Here, we will utilize the Clausius inequality to define entropy change expression for an irreversible process. Consider three different paths A, B and C for the processes occurring between states 1 and 2 as shown in Fig. 6.10. The cyclic process occurring through paths A and B is reversible and for this Clausius inequality can be expressed as

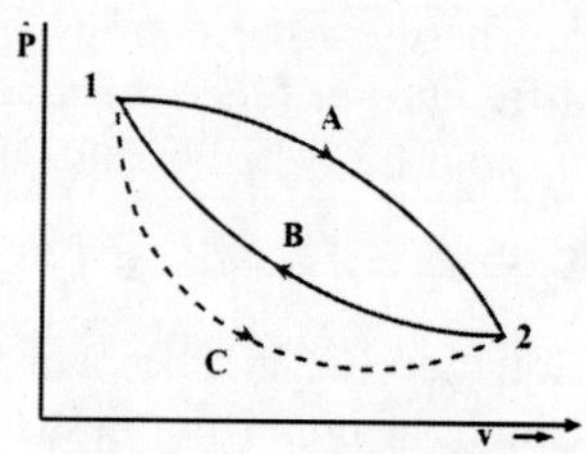


**Fig. 6.10:** Three different processes through paths A, B and C on $P - v$ diagram between states 1 and 2.

$$\oint \frac{\delta Q}{T} = \int_1^2 \left(\frac{\delta Q}{T}\right)_A + \int_2^1 \left(\frac{\delta Q}{T}\right)_B = 0.$$

Now consider a cycle consisting of an irreversible process C and reversible process B. This is an irreversible cycle, therefore, Clausius inequality for this results

$$\oint \frac{\delta Q}{T} = \int_1^2 \left(\frac{\delta Q}{T}\right)_C + \int_2^1 \left(\frac{\delta Q}{T}\right)_B < 0.$$

Upon comparing the two equations, we can say that $\int_1^2 \left(\frac{\delta Q}{T}\right)_A > \int_1^2 \left(\frac{\delta Q}{T}\right)_C$. Since path A is reversible, therefore, $\int_1^2 \left(\frac{\delta Q}{T}\right)_A = S_2 - S_1$. Thus, $S_2 - S_1 > \int_1^2 \left(\frac{\delta Q}{T}\right)_C$. It can be concluded from this that entropy change between two states is greater than integral of $\frac{\delta Q}{T}$ along an irreversible process, i.e. $\int_1^2 dS > \int_1^2 \left(\frac{\delta Q}{T}\right)_C$. It means $dS > \left(\frac{\delta Q}{T}\right)_{irrev}$.

This leads to the very important conclusion that entropy change for an irreversible process is greater than entropy change for a reversible process involving the same amount of heat transfer at the same temperature conditions. Further, we can write the more generic expression for entropy change as

$$dS \geq \frac{\delta Q}{T} \tag{6.19}$$

Equation (6.19) is applicable for both reversible and irreversible processes. For irreversible processes, it holds inequality and the left side of Eqn. (6.19) is greater than the right side, thus we need one additional positive term to make this equality. The Eqn. (6.19) takes the following form after including the term $\delta S_{gen}$ as

$$dS = \frac{\delta Q}{T} + \delta S_{gen} \tag{6.20}$$

Equation (6.20) is the most generic expression of entropy change and it is applicable for both reversible and irreversible processes. For reversible process $\delta S_{gen} = 0$, whereas it is positive for all irreversible processes. Thus $\delta S_{gen} \geq 0$. It must be emphasized that $\delta S_{gen}$ is called **entropy generation** and it is generated within a system due to the irreversibilities associated with a process as elucidated in Chapter 5. There can be entropy generation outside the system due to transfer of heat at finite temperature difference or other causes of irreversibility. The integration of Eqn. (6.20) for a process 1-2 occurring in a closed system results in

$$S_2 - S_1 = \int_1^2 dS = \int_1^2 \frac{\delta Q}{T} + S_{12,gen} \tag{6.21}$$

Equation (6.21) is known as the **entropy balance** equation for a closed system. Unlike energy which cannot be created or destroyed, entropy can be generated due to the irreversibilities, however, it cannot be destroyed. It is evident from Eqn. (6.21) that there are two possible ways for entropy change of a closed system. Firstly it increases with heat addition and secondly it can be generated due to irreversibilities in a system. Further as entropy generation is always either positive or zero, therefore, entropy can decrease only if heat transfers out of the system. It must also be pointed out that entropy change due to heat transfer takes place across the boundary of the system, i.e. entropy gain for a system due to heat transfer represents the loss of equivalent entropy to its surroundings. Contrary, the entropy generation within a system is due to irreversibilities in a process. It should be further emphasized that for an adiabatic process, increase of entropy is only due to irreversibilities in a system.

It is also important to understand how the irreversibilities associated with a process affect its work output. To demonstrate this, consider a closed system undergoes a reversible process for which entropy change is $dS$. For such a process $\delta Q = TdS$ and $\delta W = PdV$. If the process takes place through an irreversible path between the same end states, hence same $dS$, then using Eqn. (620), $dS = \frac{\delta Q_{irrev}}{T} + \delta S_{gen}$. Thus, $\delta Q_{irrev} = TdS - T\delta S_{gen}$. Also, form the first thermodynamic property relation, we get $TdS = dU + PdV$. Hence, $\delta Q_{irrev} = dU + PdV - T\delta S_{gen}$. Using the first law of thermodynamics for irreversible process, we get $\delta Q_{irrev} = dU + \delta W_{irrev}$. Substituting $\delta Q_{irrev}$ here, we get $dU + PdV - T\delta S_{gen} = dU + \delta W_{irrev}$. Therefore,

$$\delta W_{irrev} = PdV - T\delta S_{gen} = \delta W - T\delta S_{gen} \qquad (6.22)$$

Where $\delta W$ is work done in a reversible process between the same states. The term $T\delta S_{gen}$ is often called **lost work**. It must be noted that $T\delta S_{gen}$ is not any work or energy quantity lost in reality, however, it represents a lost potential to obtain work during a process. This concludes that irreversibilities cause work to be less than the reversible work for processes operating between the same end states. It means that an irreversible expansion process will produce less work than a reversible expansion. On the other hand, a system will consume more work during an irreversible compression in comparison to reversible compression.

Finally, it must be stated heat transfer and work for an irreversible process are not equal to $\int \quad TdS$ and $\int \quad PdV$, respectively. Therefore, the areas under the $T - S$ and $P - V$ diagrams for irreversible processes cannot be considered equal to heat transfer and work, respectively. This is the reason for irreversible processes to be represented by dashed lines on property diagrams in order to differentiate from reversible processes.

## 6.10 Principle of Increase of Entropy

In the previous section, we found Eqn. (6.20) which gives $dS = \frac{\delta Q}{T} + \delta S_{gen}$. For an isolated system there is no heat, work and mass transfer across its boundaries. Therefore, for such a system $dS = \delta S_{gen} \geq 0$. Now we will use this concept to demonstrate the principle of increase of entropy. Let us consider a system and its surroundings as shown in Fig. 6.11. The system can have heat, work and mass interactions with its surroundings. The region of the surroundings in the immediate vicinity of the system will be influenced by the system, whereas away from this region, the surroundings will remain unaffected. If we put together the system and its immediate surroundings as shown in Fig. 6.11. This can be considered as an thermodynamic universe of finite size which does not involve any energy and mass interactions. So thermodynamically, we can consider our thermodynamic universe as an isolated system. The total

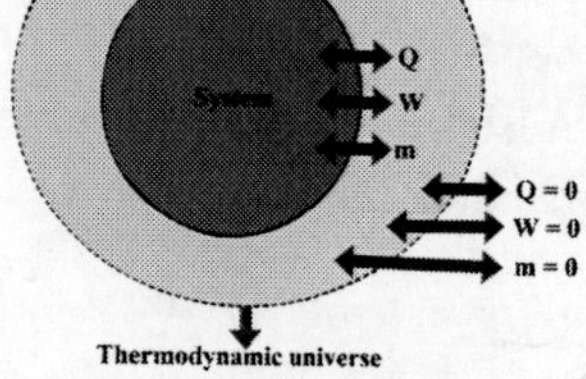


**Fig. 6.11:** Representation of a system and its immediate surroundings as an thermodynamic universe that can be considered an isolated system.

change in entropy for such an isolated system will be $\Delta S_{total} = S_{gen} \geq 0$. As the isolated system comprises of system and its immediate surroundings, hence, $\Delta S_{system} + \Delta S_{immediate-surroundings} = S_{gen} \geq 0$. This explains that the entropy change of the thermodynamic universe can be zero if all the processes occurring within the universe are reversible and it increases if irreversibilities are associated in the universe. This is known as the **principle of increase of entropy**. This principle is further demonstrated for an universe involving control volumes in Example 6.3. We can consider our actual universe also as an isolated system that comprises of multiple systems and their surroundings as part of it. Then, we can say that the entropy of the universe will never decrease and it keeps on increasing due to the irreversible processes occurring within it.

## 6.11 Entropy Rate Equation for a Closed System

The incremental change in entropy of a closed system per unit time $\delta t$ can be expressed as $\frac{dS}{\delta t} = \frac{1}{T}\frac{\delta Q}{\delta t} + \frac{\delta S_{gen}}{\delta t}$. For a limiting case of $\delta t \to 0$, this equation takes the form

$$\frac{dS}{dt} = \frac{1}{T}\frac{\delta Q}{dt} + \frac{\delta S_{gen}}{dt} \Rightarrow \dot{S} = \frac{\dot{Q}}{T} + \dot{S}_{gen} \tag{6.23}$$

Equation (6.23) is known as entropy rate equations for a closed system and it represents the rate of change of entropy of a closed system is equals to flux of entropy across its boundary due to heat transfer and rate of heat generation within the system due to irreversibilities. For a system, there can be multiple heat interactions across its different boundary faces, occurring at different temperatures. For such a situation, their effect is to be introduced individually and finally can be superimposed as given in Eqn. (6.24).

$$\dot{S} = \sum \frac{\dot{Q}}{T} + \dot{S}_{gen} \tag{6.24}$$

These entropy rate equations are useful for the analysis of heat engines and refrigerators which continuously operate with time.

Let us demonstrate the significance of Eqn. (6.24) for a heat engine. Consider a heat engine operating between temperatures $T_H$ and $T_L$ and exchanges heat at the rate of $\dot{Q}_H$ and $\dot{Q}_L$ with the thermal reservoirs. Assume that the heat engine is operating under steady state condition $(\dot{S} = 0)$. The energy and entropy balance for such a device is expressed as

$\dot{W}_{HE} = \dot{Q}_H - \dot{Q}_L$ and $0 = \frac{\dot{Q}_H}{T_H} - \frac{\dot{Q}_L}{T_L} + \dot{S}_{gen}$. From entropy balance equation, we get $\dot{Q}_L = \frac{T_L}{T_H}\dot{Q}_H + T_L\dot{S}_{gen}$. Substituting $\dot{Q}_L$ in energy balance gives $\dot{W}_{HE} = \dot{Q}_H - \left(\frac{T_L}{T_H}\dot{Q}_H + T_L\dot{S}_{gen}\right)$.

$$\Rightarrow \dot{W}_{HE} = \dot{Q}_H\left(1 - \frac{T_L}{T_H}\right) - T_L\dot{S}_{gen} \tag{6.25}$$

Further, $\dot{W}_{HE} = \dot{Q}_H\eta_{Carnot} - T_L\dot{S}_{gen} = \dot{Q}_H\eta_{actual}$ (6.26)

Let us further show a similar analysis for a refrigerator operating under the same conditions as used in the above heat engine but in reverse mode. The operation of the refrigerator is also steady and its power input is $\dot{W}_R$. For such a device, the energy and entropy balance can be given as

$-\dot{W}_R = -\dot{Q}_H + \dot{Q}_L$ and $0 = -\frac{\dot{Q}_H}{T_H} + \frac{\dot{Q}_L}{T_L} + \dot{S}_{gen}$. From entropy balance equation, we get $\dot{Q}_H = \frac{T_H}{T_L}\dot{Q}_L + T_H\dot{S}_{gen}$. Substituting $\dot{Q}_H$ in energy balance gives $\dot{W}_R = -\dot{Q}_L + \left(\frac{T_H}{T_L}\dot{Q}_L + T_H\dot{S}_{gen}\right) \Rightarrow \dot{W}_R = \dot{Q}_L\left(\frac{T_H}{T_L} - 1\right) + T_H\dot{S}_{gen}$. Further $\dot{Q}_L = \left(\frac{T_L}{T_H - T_L}\right)\dot{W}_R - \left(\frac{T_H T_L}{T_H - T_L}\right)\dot{S}_{gen}$ (6.27)

Thus, $\dot{Q}_L = \beta_{Carnot}\dot{W}_R - \left(\frac{T_H T_L}{T_H - T_L}\right)\dot{S}_{gen} = \beta_{actual}\dot{W}_R$ (6.28)

It can be inferred from Eqns. (6.26) and (6.28) that maximum benefits are obtained for both heat engine and refrigerator in the reversible mode. The efficiency of a heat engine becomes equal to Carnot efficiency and COP of the refrigerator becomes Carnot COP in reversible mode. The rate of loss of work in a heat engine is proportional to the rate of heat generation. Similarly, a refrigerator produces less refrigeration effect for a given work input in irreversible mode and the rate of loss is proportional to rate of entropy generation.

## 6.12 Entropy Rate Equation for a Control Volume

We discussed the heat transfer and entropy generation as the only possible modes of entropy change for closed systems in the previous sections. However, for a control volume there will be inflow and outflow of mass across its boundaries as shown in Fig. 6.12. As this mass possesses some associated entropy due to its thermodynamic state. Therefore, it also carries entropy equivalent to $S = m \times s$ where $m$ is the amount of mass and $s$ is its specific entropy corresponding to thermodynamic state. Using this information and considering a single heat interaction and one inflow and one outflow of mass as shown in Fig. 6.12 for a control volume, we obtain the entropy balance equation as

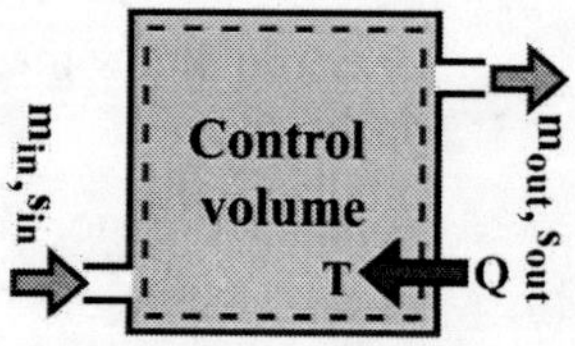


**Fig. 6.12:** Representation of a single inlet, single outlet control volume.

$$(S_2 - S_1)_{cv} = \frac{Q}{T} + m_{in}s_{in} - m_{out}s_{out} + S_{gen} \quad (6.29)$$

In case of multiple heat and mass interactions, Eqn. (6.29) becomes

$$(S_2 - S_1)_{cv} = \sum \frac{Q}{T} + \sum m_{in}s_{in} - \sum m_{out}s_{out} + S_{gen}$$
(6.30)

The entropy balance equation for a control volume in rate form is given as

$$\frac{dS_{cv}}{dt} = \sum \frac{\dot{Q}}{T} + \sum \dot{m}_{in}s_{in} - \sum \dot{m}_{out}s_{out} + \dot{S}_{gen} \quad (6.31)$$

The entropy balance equation for a control volume states that change in entropy of the control volume is the addition of total entropy change associated with heat transfer across its boundary, net exchange of entropy associated with mass transfer across control surface and the total generation of entropy within a control volume due to various irreversibilities.

For a steady flow process rate of change of property of the control volume with time will be zero, i.e. $\frac{dS_{cv}}{dt} = 0$. Therefore, Eqn. (6.31) becomes

$$\sum \frac{\dot{Q}}{T} + \sum \dot{m}_{in}s_{in} - \sum \dot{m}_{out}s_{out} + \dot{S}_{gen} = 0 \qquad (6.32)$$

It must be emphasized that all the heat transfer rates, mass flow rates and the thermodynamic states of the inflowing and outflowing steams should remain constant with time for a steady flow process and hence, for the application of Eqn. (6.32). A steady flow process that involves a single inflowing stream and a single outflowing stream ($\dot{m}_{in} = \dot{m}_{out} = \dot{m}$), Eqn. (6.32) becomes

$$\sum \frac{\dot{Q}}{T} + \dot{m}(s_{in} - s_{out}) + \dot{S}_{gen} = 0 \qquad (6.33)$$

Dividing Eqn. (6.33) by $\dot{m}$ gives

$$\sum \frac{q}{T} + (s_{in} - s_{out}) + s_{gen} = 0 \qquad (6.34)$$

Here, $q = \frac{\dot{Q}}{\dot{m}}$. Using Eqn. (6.34), we get $s_{out} = s_{in} + \sum \frac{q}{T} + s_{gen}$. For an adiabatic process, $s_{out} = s_{in} + s_{gen}$. Since $s_{gen} \geq 0$, therefore, for a adiabatic process $s_{out} \geq s_{in}$. The equality sign will hold when the adiabatic process is reversible as well. Thus, it can be mentioned that a reversible adiabatic process is an isentropic process.

## 6.13 Shaft Work for Steady Flow Devices

We know that reversible boundary work for a closed system is given as ∫ □□□. Now we have attempted to express a similar expression for the shaft work associated with a steady flow control volume. Consider such a control volume has only one inlet and one outlet and □ is the specific work output from the control volume. For such a control volume, the energy equation on per unit mass flow basis is given as

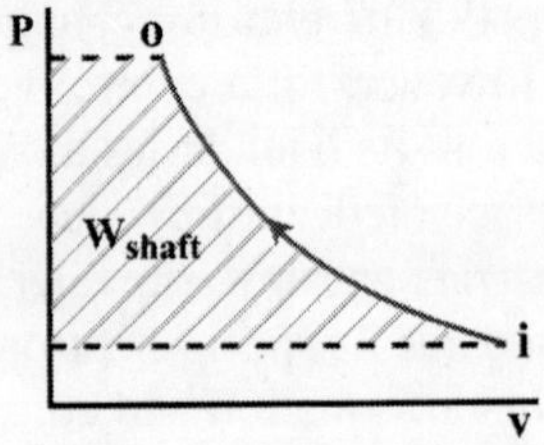


**Fig. 6.13:** Depiction of reversible work in a steady flow process on $P - v$ diagram.

$q + h_i + \frac{1}{2}V_i^2 + gz_i = w + h_o + \frac{1}{2}V_o^2 + gz_o$. The entropy equation for such a process is given as $s_o = s_i + \frac{q}{T} + s_{gen}$. If $s_o - s_i = \int_i^o ds$, then entropy equation becomes $\int_i^o ds = \int_i^o \frac{\delta q}{T} + \int_i^o \delta s_{gen} \Rightarrow q = \int_i^o \delta q = \int_i^o Tds - \int_i^o T\delta s_{gen}$.

From the second thermodynamic property relation $Tds = dh - vdp$. Thus,

$q = \int_i^o dh - \int_i^o vdP - \int_i^o T\delta s_{gen} = h_o - h_i - \int_i^o vdP - \int_i^o T\delta s_{gen}$. Substituting this value of $q$ in energy equation results in

$h_o - h_i - \int_i^o vdP - \int_i^o T\delta s_{gen} + h_i + \frac{1}{2}V_i^2 + gz_i = w + h_o + \frac{1}{2}V_o^2 + gz_o$. Now $w$ becomes

$$w = -\int_i^o vdP + \frac{1}{2}(V_i^2 - V_o^2) + g(z_i - z_o) - \int_i^o T\delta s_{gen} \qquad (6.35)$$

It should be noted from Eqn. (6.35) that last term $\int_i^o T\delta s_{gen}$ is either zero or always positive as $T > 0$ and $\delta s_{gen} \geq 0$. Therefore, it represents that it is always subtracted from the other terms of the equation and decreases the work potential of the control volume. For a reversible process, work is contributed due to changes in pressure, kinetic energy and potential energy. The increase of pressure during a process (like in a pump or compressor) makes the work negative which signifies work is supplied to the control volume. Whereas the decrease of pressure (for example, in a turbine) makes the work positive that specifies it is being produced by the control volume. The specific volume does not change the sign of the work, however, it controls its magnitude. A large specific volume (gas) signifies the higher amount of work whereas a less specific volume (liquid) results in smaller magnitude of work. In the absence of work for a reversible process (flow through a pipe), Eqn. (6.35) takes the form $-\int_i^o vdP + \frac{1}{2}(V_i^2 - V_o^2) + g(z_i - z_o) = 0$. If we consider the fluid involved is incompressible, then we get

$$(P_i - P_o)v + \frac{1}{2}(V_i^2 - V_o^2) + g(z_i - z_o) = 0 \tag{6.36}$$

Equation (6.36) is known as **Bernoulli equation** and it is applicable for flow of incompressible fluids with no irreversibilities such as friction, etc. It is often used in fluid mechanics and can be modified to include irreversibilities and other effects.

Equation (6.35) is essential to determine the work associated with steady flow devices such as compressor, pump and turbine. For these devices, the changes in kinetic and potential energy of the working fluid are negligible. Therefore, shaft work for a reversible process can be written as $w = -\int_i^o vdP$. From this integration, the shaft work associated with the process can be represented by the shaded portion of $P - v$ diagram as shown in Fig. 613. It must be emphasized that this equation is applicable for a steady flow device and should not be confused with expression of boundary work for a closed system. Further, it can also be explained from this equation that the pump of a steam thermal power plant that increases the pressure of liquid from condenser pressure to boiler pressure consumes very less power because specific volume of water is low. Contrary, the turbine operating between the same pressure limits produces large work because associated working fluid in the turbine is in a vapour state which has very high specific volume. This is the reason for large net work output from a steam thermal power plant.

Further the applicability of the first term of Eqn. (6.35) can be discussed for a polytropic process involving ideal gas as the working fluid. As we know that for a polytropic process, $Pv^n = C \Rightarrow v = C^{1/n}/P^{1/n}$. This results in $w = -\int_i^o vdP = -\int_i^o C^{1/n}/P^{1/n}dP$. Further, $w = -C^{1/n}\frac{n}{n-1}\left(P_o^{1-\frac{1}{n}} - P_i^{1-\frac{1}{n}}\right) = -\frac{n}{n-1}(P_o v_o - P_i v_i)$. Similarly, the work during an isothermal process $(Pv = C)$ involving ideal gas can be given as $w = -\int_i^o vdP = -\int_i^o C/P^{1/n}dP = -P_i v_i ln\frac{P_o}{P_i}$. The expression of work for a polytropic process becomes work associated with an isentropic process when $n$ is replaced with $k$. Three different compression processes, isentropic, polytropic and isothermal, are shown on $P - v$ diagram in Fig. 6.14 for the compression of an ideal gas between the same pressure limits. The area on the left of each process on $P - v$ diagram depicts the work of compression during the reversible process. It is evident that isentropic compression requires the maximum work of compression and it is least for an isothermal process. The work of compression lies between the two, for a polytropic process and it decreases as the polytropic index is decreased and becomes minimum for $n = 1$. This shows that removal of heat during the compression process requires less work input.

Therefore, most compressors are provided with either fins or cooling water jackets to remove the heat during the work of compression. Intercooling of working fluid during multistage compression is another popular technique used for decreasing compressor work when the pressure ratio involved is large. Such a method is demonstrated in Example 6.4.

## 6.14 Isentropic Efficiency of Different Steady Flow Devices

In the previous chapter, we have defined the performance of devices involving thermodynamic cycle. It was elucidated that the irreversibilities in the cycle decreases its performance. Also, the actual cycles were compared against the reversible cycles. Now we will attempt to define the performance of discrete processes involving steady flow devices such as turbine, pump or compressor and nozzle. It is established for discrete processes as well in the previous sections that irreversibilities decrease the work output of a turbine and increase the work requirement of a pump. Therefore, it is also essential for actual discrete processes to compare their performance against the idealized situations which are free from irreversibilities.

With this attempt, let us consider a turbine that operates between pressure limits of $P_i$ and $P_o$. The state of the working fluid at the inlet of the turbine will always be known. It is discussed in Chapter 4 that the idealized process for the turbine is reversible adiabatic, hence, isentropic. Therefore, the turbine will produce maximum work output during an isentropic process. Such an isentropic process for turbine is shown on $h-s$ diagram in Fig. 6.15. The actual process within the turbine will involve irreversibilities and hence the entropy generation. Therefore, the entropy of the working fluid at the outlet of a turbine will be greater than its entropy at inlet because $s_o = s_i + s_{gen}$. This suggests that the actual outlet state of working fluid from the turbine will be towards the right of the isentropic exit from the same device. Such a process is presented with a dashed line in Fig. 6.15. Consider the actual work done by the turbine is $w_{actual}$ and $w_{ideal}$ is isentropic work. Thus, the **isentropic efficiency** of the turbine is defined as

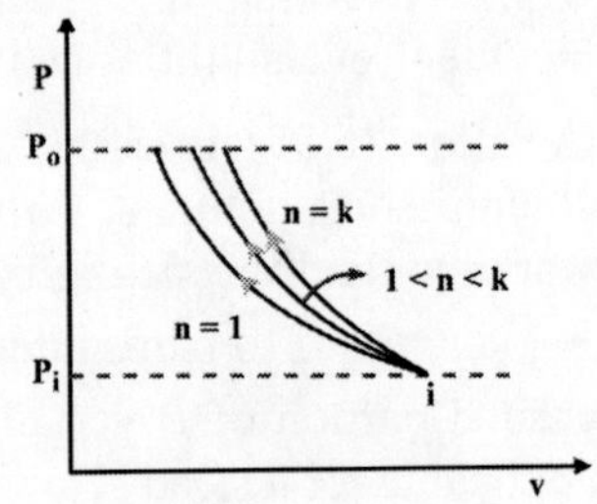


**Fig. 6.14:** Representation of different polytropic processes on $P-v$ diagram.

$$\eta_{turbine} = \frac{w_{actual}}{w_{ideal}} = \frac{h_i - h_o}{h_i - h_{os}} \tag{6.37}$$

The isentropic efficiency of a turbine ranges from 70% to 90%. It is more for the larger turbines and less for the smaller ones.

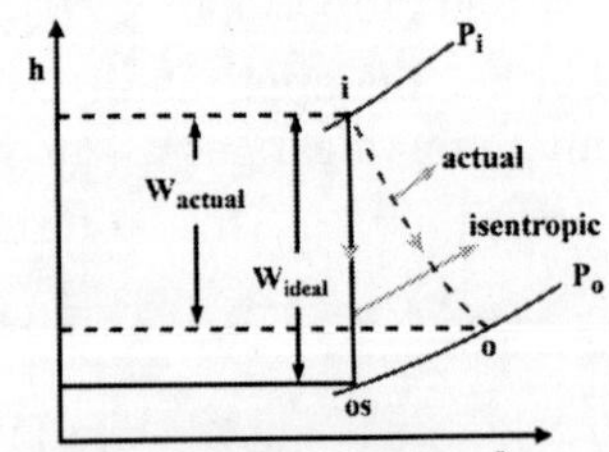


**Fig. 6.15:** Representation of expansion in a turbine on $h-s$ diagram.

It is discussed in Chapter 4 that the processes involved in pump and compressor are adiabatic. The working fluid enters these devices at pressure $P_i$ and discharged at pressure $P_o > P_i$. Ideally the process inside pump or compressor is isentropic which is presented on $h-s$ diagram in Fig. 6.16 with $os$ as the exit state. The irreversibilities in actual process will increase the entropy at the exist, causing actual exit at state $o$. The actual process is represented with a dashed line in Fig. 6.16. The work consumed by the pump or compressor during actual process ($w_{actual}$) is larger than the isentropic work ($w_{ideal}$). Therefore, the isentropic efficiency of the compressor is defined as

$$\eta_{pump/compressor} = \frac{w_{ideal}}{w_{actual}} = \frac{h_{os}-h_i}{h_o-h_i} \quad (6.38)$$

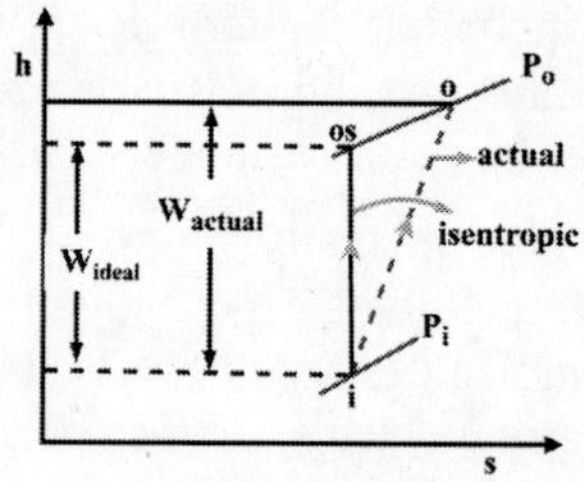


**Fig. 6.16:** Representation of compression in a pump/compressor on $h-s$ diagram.

The typical range of isentropic efficiency for a pump or compressor is 70-90%. The larger devices have more efficiency than the smaller ones.

The compressors are also provided with fins or water jackets to cool the working fluid during the compression process. Such devices cannot be considered adiabatic because working fluid transfers a significant amount of heat to the surroundings. Usually such processes are idealised as reversible isothermal processes. Consider the work input to the compressor during isothermal compression is $w_{isothermal}$. The efficiency of a compressor that is idealized by the reversible isothermal process is referred to as the isothermal efficiency and is defined as

$$\eta_{isothermal,compressor} = \frac{w_{isothermal}}{w_{actual}} \quad (6.39)$$

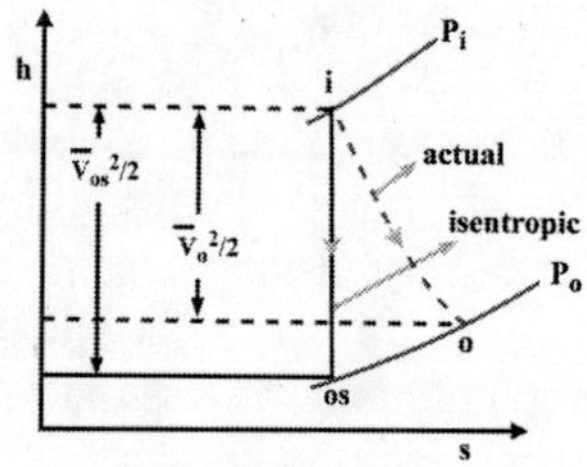


**Fig. 6.17:** Representation of expansion in a nozzle on $h-s$ diagram.

Nozzle is another steady flow device in which expansion of the fluid takes place to increase its kinetic energy at the exit. The process inside the nozzle is also reversible adiabatic. Consider a nozzle operating between pressures $P_i$ and $P_o$. The velocity obtained at the exit of nozzle during an isentropic expansion (shown in Fig. 6.17) is $\underline{V}_{os}$. The actual velocity at the exit of the nozzle is $\underline{V}_o$ as the entropy at the exits increases during the actual process (Fig. 6.17). The isentropic efficiency of the nozzle is defined in terms of the actual and ideal kinetic energies of the fluid stream at its outlet. Thus, it is defined as

$$\eta_{nozzle} = \frac{\underline{V}_o^2/2}{\underline{V}_{os}^2/2} = \frac{h_i-h_o}{h_i-h_{os}} \quad (6.40)$$

As a nozzle does not involve any moving components, therefore, its efficiency is high and ranges from 90-98%.

Scan Me

For more details on the significance of entropy.

## 6.15 Physical Inferences of Entropy

We understood from the above discussion that entropy is an important property that helps in quantifying the lost potential during a process, caused due to irreversibilities. Mathematically the expressions of entropy change for different systems and principle of increasing entropy were elucidated. However, a question, what is entropy?, is still not answered or perhaps it cannot be answered fully. It must be emphasized that like entropy, the physical meaning of energy may not be very clear to us, but we appreciate the term energy due to its common usage and understanding of processes that involve its transformation from one form to another. Here, we will attempt to relate the physical significance of entropy and its understanding from microscopic perspectives.

Fundamentally, entropy can be measured with an amount of molecular disorder or molecular randomness a system can have. Upon increase in disorderness of the system, the predictability of the molecules becomes less and hence, its entropy increases. In the case of solids, the molecules have certain arrangements and they oscillate about their equilibrium positions. Contrary, in gases, the molecules have random motions, they collide with each-other and frequently change their directions. Thus the molecules are less predictable in gases than in solids. Therefore, gases have high entropy whereas solids have less entropy and the entropy of liquids lies between the two.

The microscopic analysis of a system from the concepts of statistical thermodynamics states that for each macroscopic state of the system, there are a large number of corresponding molecular arrangements (microscopic states). This important statement can be understood with the help of a simple example. Consider a perfectly sealed room filled with air at atmospheric pressure. If we isolate the room from its surroundings, its pressure and other macroscopic thermodynamic properties remain the same. However, the air molecules inside the room will be in continuous motion and will change their positions at each and every instant. This means that microscopically the molecules have a large number of arrangements which corresponds to the same pressure (atmospheric) of the room, hence, the macroscopic state. The total number of possible microscopic states of a system corresponding to a unique macroscopic state of that system, is known as **thermodynamic probability** ($p$). Thermodynamic probability is related to the absolute entropy of the system by the **Boltzmann equation** which is given as

$$S = \kappa\, ln(p) \tag{6.41}$$

Here, $\kappa$ is the **Boltzmann constant** and its value is $1.3806 \times 10^{-23}\ J/K$. Equation (6.41) connects the absolute entropy to uncertainty and molecular disorder. The more is the thermodynamic probability (molecular randomness) the more is the value of entropy. This also means that the molecular randomness of an isolated system increases every time it undergoes a process as irreversibilities cannot be avoided in reality.

As stated earlier, molecules in solids have a particular arrangement and they oscillate about their equilibrium positions. Such oscillations of the molecules decrease with decrease of temperature and finally cease when temperature reaches absolute zero condition. The **third law of thermodynamics** states that entropy of a pure crystalline solid is zero at absolute zero temperature. There is no molecular disorder or uncertainty at that condition for pure crystalline solids, therefore, Eqn. (6.41) results in zero value. This condition is termed as the absolute reference state for entropy estimation and its value measured with reference to this point is called **absolute entropy**. Absolute entropy is very useful for the thermodynamic analysis of chemical reactions. It should be noted that entropy of solids other than crystalline solids cannot be assigned zero at absolute zero temperature as these carry certain molecular uncertainty.

As heat transfers from a hot body to a cold body, it is a carrier of entropy as well. Therefore, entropy of the hot body decreases and the entropy of the cold body increases during a heat transfer process. Combining both hot and cold objects, one can say that net entropy change is balanced, however, this is not the case because irreversibilities during the heat transfer process generates entropy. Therefore, the entropy of a combined system including hot and cold objects increases. This concludes that processes always occur in the direction of increased universal entropy. Thus the entropy of the universe, hence the molecular randomness, is getting increased every instant.

## UNIT SUMMARY

This chapter introduces the Clausius inequality and shows its applicability for different cyclic devices. Further, it is used to define the entropy change equation for a reversible and internally reversible process. Entropy change of a pure substance is elucidated for the different regions such as compressible liquid, superheated vapour and saturated liquid-vapour mixture on property diagrams. Then the entropy change for a reversible process is used to derive the thermodynamic property relations. These property relations are further used to define the entropy change equations for solids, liquids and ideal gases. The property diagrams involving entropy as one of the coordinates are also shown for different polytropic processes. It was found that entropy change is contributed due to heat transfer and entropy generation for irreversible processes, whereas entropy generation term is zero for reversible processes.

Different modes of entropy transfer and generation are then used to define the entropy balance equations for closed system and open system. It was observed that in the case of an open system, the entropy exchange across the boundary of the control volume also takes place with inflow and outflow of mass along with heat exchange. For adiabatic steady flow devices, the entropy of the outflowing stream will always be greater than the entropy of the inflowing stream whereas it will be equal for both streams if the process is reversible, hence, isentropic. The shaft work in case of reversible steady flow process becomes $-\int vdP$ when the changes in kinetic and potential energies of the working fluid are neglected. The reversible adiabatic (isentropic) process is used as reference for comparing the performance of different steady flow devices such as turbine, compressor and nozzle. The deviation between the actual and the ideal process for such devices is defined in terms of isentropic efficiency of the device. Finally, the physical understanding of entropy is presented in terms of molecular randomness possessed by a system.

## Solved Examples

**Example 6.1:** Consider a steam power plant operates between pressure limits of 5 $bar$ and 20 $kPa$. The inlet state of the working fluid at the boiler is saturated liquid and its outlet is saturated vapour. The working fluid leaves the turbine at 95% quality and enters the pump with 10% quality. Demonstrate the Clausius inequality for this cycle.

**Solution:** Determine the heat added to the working fluid in the boiler and heat rejected from the working fluid in the condenser. Use this information to calculate $\oint \frac{\delta Q}{T}$.

**Assumptions:** The state of the working fluid at the outlet of one steady flow device is identical to its state at the inlet of consecutive steady flow device that forms the cycle. Let us consider the state of working fluid changes from 1 to 2 during the constant temperature heat addition process inside the boiler. Similarly, working fluid changes from state 3 (turbine outlet) to state 4 (pump inlet) in the condenser.

**Analysis:** Using saturated pressure steam table B.1, we get

$T_1 = T_2 = T_H = T_{sat@500\ kPa} = 151.83°C$,

$h_1 = h_{f@500\ kPa} = 640.09\ kJ/kg,\ h_2 = h_{g@500\ kPa} = 2748.1\ kJ/kg.$

At 20 $kPa$, $T_3 = T_4 = T_L = T_{sat@20\ kPa} = 60.06°C$,

$h_{f@20\ kPa} = 251.42\ kJ/kg,\ h_{fg@20\ kPa} = 2357.5\ kJ/kg,\ x_3 = 0.1$ and $x_4 = 0.95$.

$\Rightarrow h_3 = h_f + x_3 h_{fg} = 487.17\ kJ/kg$ and $h_4 = h_f + x_4 h_{fg} = 2491.045\ kJ/kg$.

The heat added to working fluid on per unit mass basis is

$q_H = h_2 - h_1 = 2108.01\ kJ/kg$.

The heat loss from working fluid on per unit mass basis is

$q_L = h_4 - h_3 = 2003.875\ kJ/kg$.

The Clausius inequality for the cycle results in $\oint \frac{\delta Q}{T} = \frac{q_H}{T_H} - \frac{q_L}{T_L} = \frac{2108.01}{424.98} - \frac{2003.875}{333.21}\ kJ/kg\ K$.

$\Rightarrow \oint \frac{\delta Q}{T} = -1.054\ kJ/kg - K$. This shows that the given steam power cycle does not violate the second law of thermodynamics and is of irreversible nature.

**Example 6.2:** A constant pressure piston-cylinder system contains saturated water at 200 $kPa$. Heat is now added to the water till it reaches saturated vapour state. Determine the change in entropy per unit mass of water during this process.

**Solution:** The change in entropy of water is to be determined during a constant pressure heat addition process.

**Assumptions:** Consider the initial state of the system is 1 and its final state is 2.

**Analysis:** Using steam table B.2, we get $s_1 = s_{f@200\ kPa} = 1.5302\ kJ/kg - K$,

$s_2 = s_{g@200\ kPa} = 7.1270\ kJ/kg - K$. Thus entropy change of water during heat addition process is $\Delta s = s_2 - s_1 = 5.5968\ kJ/kg - K$.

**Alternative approach:** Let us find first heat added during the process, i.e. $q = h_2 - h_1$. Using table B.2, we get $h_1 = h_{f@200\ kPa} = 504.71\ kJ/kg$,

$h_2 = h_{g@200\ kPa} = 2706.3\ kJ/kg \Rightarrow q = h_2 - h_1 = h_{fg@200\ kPa} = 2201.6\ kJ/kg$.

As the state of substance changes from saturated liquid to saturated vapour, therefore, temperature is constant during the heat addition process.

Thus $T = T_{sat@200\ kPa} = 120.21°C$. Therefore, entropy change for this internally reversible process is given as $s_2 - s_1 = \frac{q}{T} = \frac{h_{fg}}{T} = \frac{2201.6}{393.36} = 5.5969\ kJ/kg - K$.

**Discussion:** It is to be noted that entropy calculated using alternate approach matches with the direct estimation of entropy obtained from the property tables. This it is important to highlight that the relationship $s_{fg} = \frac{h_{fg}}{T}$ is used to construct the entropy tables in the saturated liquid-vapour regime as it can be measured directly using experiments.

**Example 6.3:** Illustrate the principle of increase of entropy using control volume approach.

**Solution:** Principle of increase of entropy is to be demonstrated for control volumes.

**Assumptions:** Consider the universe is divided into two control volumes as shown in Fig. E-6.3.

**Discussion:** Control volume CV1 involves mass, heat and work interactions with its surroundings CV2. The $\dot{m}_i$ is the mass entering CV1 from CV2 and $\dot{m}_o$ is the mass leaving CV1 and entering CV2. $\dot{W}$ is the work done by the control volume CV1 and $\dot{Q}$ is the rate of heat added to CV1. This heat is added at temperature $T_s$. The temperature of surroundings (CV2) can be different than $T_s$ away from the location of heat transfer. The rate of entropy change for control volume CV1 is $\frac{dS_{CV1}}{dt} = \frac{\dot{Q}}{T_s} + \dot{m}_i s_i - \dot{m}_o s_o + \dot{S}_{gen,CV1}$. Similarly, rate of entropy change for control volume CV2 is $\frac{dS_{CV2}}{dt} = -\frac{\dot{Q}}{T_s} - \dot{m}_i s_i + \dot{m}_o s_o + \dot{S}_{gen,CV2}$. Thus rate change of total entropy of universe is $\frac{dS_{universe}}{dt} = \frac{dS_{CV1}}{dt} + \frac{dS_{CV2}}{dt} = \dot{S}_{gen,CV1} + \dot{S}_{gen,CV2} \geq 0$.

Therefore, the rate of change of entropy of the universe is equal to the rate of entropy generation due to irreversibilities that will be always positive except for a special case of reversible processes for which it will be zero.

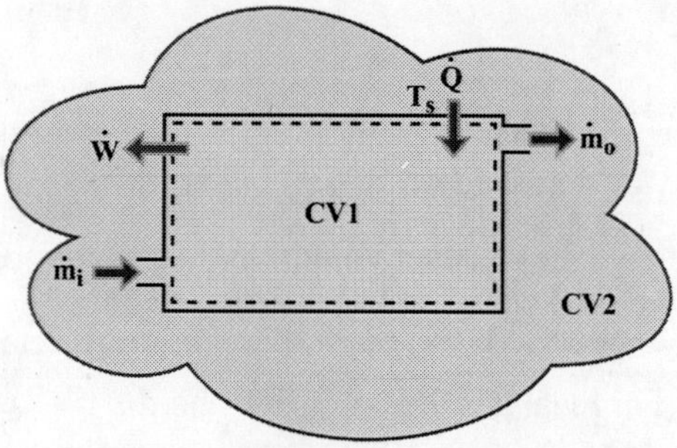


**Fig. E-6.3**

**Example 6.4:** Show that multistage polytropic compression with intercooling takes less work input than a single stage polytropic compression process between the same pressure limits.

**Solution:** We need to compare work of compression between single stage polytropic compression and multistage polytropic compression with intercooling for the same pressure limits.

**Assumptions:** Consider a fluid to be compressed from pressure $P_1$ to $P_2 > P_1$ in a polytropic process.

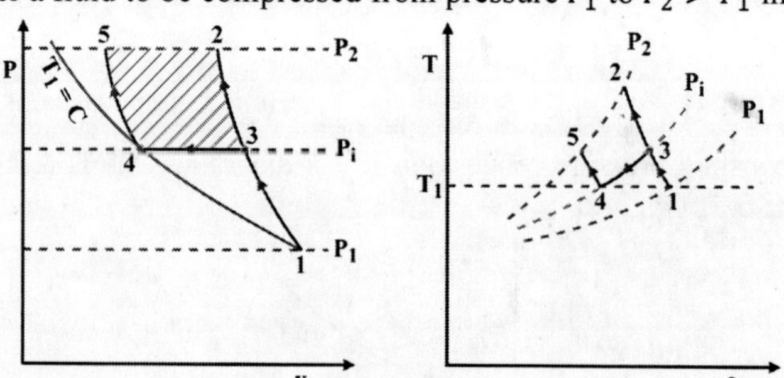


**Fig. E-6.4**

**Discussion:** In the first case, the compression from pressure $P_1$ to $P_2$ is achieved in a single polytropic process and is represented by process 1-2 on $P - v$ and $T - s$ diagram in Fig. E-6.4. In the second situation, a first stage compressor is used to increase pressure from $P_1$ to $P_i$ where $P_1 < P_i < P_2$ as represented by process 1-3 on $P - v$ and $T - s$ diagram in Fig. E-6.4. Then at pressure $P_i$ fluid is passed through a heat exchanger to cool it. The heat exchanger used is known as an intercooler. In an ideal intercooler, the fluid is cooled to its initial temperature $T_1$ as shown by process 3-4 in Fig. E-6.4. Further, the fluid obtained at the exit of intercooler is passed through second stage compressor to increase its pressure from $P_i$ to desired pressure $P_2$. The process 4-5 represents second stage compression in Fig. E-6.4. Therefore, in case of multistage compression with intercooling the desired pressure increase is achieved through processes 1-3-4-5. As the reversible work for a steady flow device is $-\int vdP$, therefore, single stage compression process 1-2 has the highest work requirement as it encloses maximum area on its left on $P-v$ diagram. The area enclosed on $P-v$ diagram by process 1-3-4-5 on its left is less. Moreover, the shaded area 2-3-4-5-2 on $P-v$ diagram shows the amount of work saved upon using the multistage compression with intercooling.

It must be emphasized that multistage compression with intercooling is an effective method to reduce the work requirement of the compressor particularly for high pressure ratios. The use of this technique at low pressure ratios may not justify the cost involved in installation and operation of the additional components involved.

**Example 6.5:** Prove that slope of constant volume process is greater than slope of constant pressure process involving ideal gas as working fluid on $T-s$ diagram.

**Solution:** We need to determine $\frac{dT}{ds}$ for constant volume and constant pressure processes.

**Assumptions:** We will use the thermodynamic property relations to determine the slope $\frac{dT}{ds}$ considering ideal gas as the working fluid. For ideal gas, $du = C_{v0}dT$ and $dh = C_{p0}dT$.

**Discussion:** From first thermodynamic property relation, we get

$Tds = du + Pdv \Rightarrow Tds = C_{v0}dT + Pdv$. For constant volume process, $dv = 0$. Thus, $\left(\frac{dT}{ds}\right)_{v=const.} = \frac{T}{C_{v0}}$.

Similarly, from second thermodynamic property relation, we get

$Tds = dh - vdp \Rightarrow Tds = C_{p0}dT - vdp$. For constant pressure process, $dp = 0$. Thus, $\left(\frac{dT}{ds}\right)_{p=const.} = \frac{T}{C_{p0}}$.

Since, $C_{p0} > C_{v0}$, therefore, $\left(\frac{dT}{ds}\right)_{v=const.} > \left(\frac{dT}{ds}\right)_{p=const.}$. Hence, slope of constant volume process is greater than slope of constant pressure process on $T-s$ diagram. This is represented graphically on $T-s$ diagram in Fig. E-6.5.

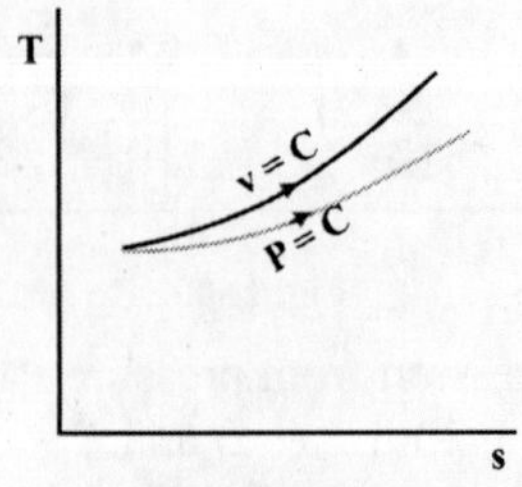


**Fig. E-6.5**

**Example 6.6:** A thermal reservoir at 1300$K$ supplies 3500 $kJ$ of heat to another thermal reservoir at 800$K$. Determine the entropy generation during this heat transfer process. If the temperature of heat sink is increased from 800$K$ to 1100$K$, what happens to irreversibilities associated with the process?

**Solution:** The entropy generated due to the heat transfer process is to be estimated as it quantifies the amount of irreversibilities.

**Assumptions:** The two thermal reservoirs cannot have direct contact between them as it will result in temperature discontinuity at the contact surface. Therefore, for this heat transfer to take place between two thermal reservoirs, an imaginary partition can be assumed as shown in Fig. E-6.6. This partition acts only as a communicator of heat between two reservoirs and does not go through any change in its energy. Thus entropy change for this can be assumed to zero. Also, the heat transfer across a thermal reservoir does not change its temperature, therefore, it can be considered an internally reversible process for the thermal reservoir.

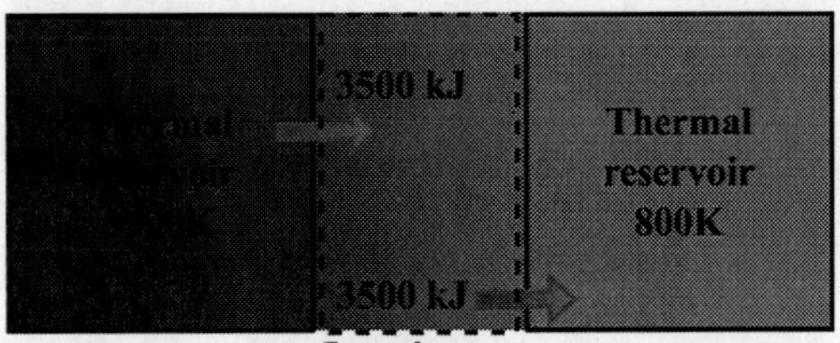


**Imaginary partition**

**Fig. E-6.6**

**Analysis:** Using above assumptions the entropy change for thermal reservoir at 1300$K$ is $\Delta S_{heat\ source} = \frac{Q}{T_{heat\ source}} = \frac{-3500}{1300} = -2.69\ kJ/K$.

Similarly, the entropy change for heat sink at 800$K$ can be given as

$$\Delta S_{heat\ sink} = \frac{Q}{T_{heat\ sink}} = \frac{3500}{800} = 4.375\ kJ/K.$$

When the two reservoirs are combined to form an universe, the entropy generation is expressed as $S_{gen} = \Delta S_{total} = \Delta S_{heat\ source} + \Delta S_{heat\ sink} = 1.685\ kJ/K$.

This entropy is generated due to irreversibilities caused by the heat transfer associated with finite temperature difference.

If the heat sink at 800$K$ is replaced with another heat sink at 1100$K$, then

$$\Delta S_{heat\ sink} = \frac{Q}{T_{heat\ sink}} = \frac{3500}{1100} = 3.182\ kJ/K.$$

Thus, $S_{gen} = \Delta S_{total} = \Delta S_{heat\ source} + \Delta S_{heat\ sink} = 0.492\ kJ/K$.

Upon decreasing the temperature difference between two thermal reservoirs during the heat transfer, the amount of entropy generation has decreased. This signifies the presence of less irreversibilities for a heat transfer occurring at a smaller temperature difference between two objects.

## EXERCISES

### Multiple Choice Questions

| | | | | |
|---|---|---|---|---|
| 6.1 | Clausius inequality states that | | | |
| | A. $\oint \frac{\delta Q}{T} > 0$ | B. $\oint \frac{\delta Q}{T} \geq 0$ | C. $\oint \frac{\delta Q}{T} \leq 0$ | D. $\oint \frac{\delta Q}{T} < 0$ |
| 6.2 | For a heat engine that undergoes a cyclic process whether reversible or irreversible | | | |
| | A. $\oint \delta Q > 0$ | B. $\oint \delta Q \geq 0$ | C. $\oint \delta Q < 0$ | D. $\oint \delta Q \leq 0$ |
| 6.3 | For a reversible refrigerator | | | |
| | A. $\oint \delta Q > 0$ | B. $\oint \delta Q \geq 0$ | C. $\oint \delta Q < 0$ | D. $\oint \delta Q \leq 0$ |
| 6.4 | For an irreversible refrigerator | | | |
| | A. $\oint \delta Q > 0$ | B. $\oint \delta Q \geq 0$ | C. $\oint \delta Q < 0$ | D. $\oint \delta Q \leq 0$ |
| 6.5 | Choose the correct expression for the entropy change of a process. | | | |
| | A. $dS = \frac{\delta Q}{T}$ | B. $dS < \frac{\delta Q}{T}$ | C. $dS \geq \frac{\delta Q}{T}$ | D. $dS > \frac{\delta Q}{T}$ |
| 6.6 | The property relation $Tds = dh - vdP$ is not applicable for a/an | | | |
| | A. reversible process | B. irreversible process | C. internally reversible process | D. none of these |
| 6.7 | The integration of property relation $Tds = dh - vdP$ is not possible along a/an | | | |
| | A. reversible process | B. irreversible process | C. internally reversible process | D. externally irreversible process |

| | | | | |
|---|---|---|---|---|
| 6.8 | The entropy generation for a process cannot be | | | |
| | A. zero | B. positive | C. negative | D. both A and C |
| 6.9 | The isentropic pumping of an incompressible liquid is also an | | | |
| | A. isobaric process | B. isothermal process | C. isochoric process | D. isenthalpic process |
| 6.10 | The pump of a steam thermal power plant consumes very less work due to the low__________________of working fluid. | | | |
| | A. specific volume | B. specific internal energy | C. specific enthalpy | D. specific entropy |
| 6.11 | A process involving ideal gas as the working fluid is reversible adiabatic in nature. The polytropic index for such a process will be | | | |
| | A. $n = 0$ | B. $n = 1$ | C. $n = k$ | D. $n = \infty$ |
| 6.12 | The work loss $(TS_{gen})$ caused due to irreversibilities in a process is always | | | |
| | A. zero | B. positive | C. negative | D. both A and B |

**Answers of Multiple Choice Questions**

| |
|---|
| **6.1 C, 6.2 B, 6.3 D, 6.4 C, 6.5 C, 6.6 D, 6.7 B, 6.8 C, 6.9 B, 6.10 A, 6.11 C, 6.12 D** |

**Short and Long Answer Type Questions**

6.1 Can $\oint \delta Q > 0$ exist for a cycle without violating the Clausius inequality? Explain.

6.2 Is $\oint \delta Q = 0$ a necessary condition for a cyclic process to occur? Elaborate.

6.3 Two processes occur between the same end states, one in reversible manner and other in irreversible manner. Which of the two processes will have larger entropy change? Explain.

6.4 Comment on the value of $\int_1^2 \frac{\delta Q}{T}$ for all the reversible and irreversible processes that occur between states 1 and 2. Elucidate with necessary examples.

6.5 Can the entropy change for a process be negative? Explain with examples.

6.6 What are the modes of entropy transfer across the boundary of a closed system?

6.7 What are the modes of entropy transfer across the boundary of a control volume?

6.8 Can the entropy of a closed system and its surroundings increase simultaneously? Elucidate.

6.9 How the entropy produced in a single inlet and single outlet steady flow process is reflected? Explain.

6.10 Can entropy be destroyed during a process? Elucidate.

6.11 Do all reversible adiabatic processes are isentropic? Explain.

6.12 Do all isentropic processes are reversible adiabatic? Explain.

6.13 How do fins and water jackets are used for a compressor? Elaborate.

6.14 What happens to entropy of an isolated system during a process?

6.15 Prove that an irreversible heat pump consumes more work to produce the same amount of heating in comparison to a reversible heat pump operating between the same temperature limits.

**Numerical Problems**

6.1 A finned air compressor is used to increase the pressure of air at constant temperature conditions of 35°C which is maintained by transferring heat to the surroundings at 15°C. The compressor has rated power of 22 $kW$. Determine the rate of change in entropy of the air.

6.2 Water is used as a working fluid in a Carnot cycle whose quality changes from 0.8 to 0.2 during the heat rejection process in a condenser at 25°C. Determine the change in entropy of the working fluid. Also, estimate the change in entropy of the heat sink and total change of entropy for the process.

6.3 The change in entropy of the working fluid of a Carnot cycle during the heat addition process is 0.05 $kW/k$. Determine the amount of heat added if the heat addition process takes place at 400°C. Also, find the entropy change for the heat source and total change in entropy for the process.

6.4 A rigid vessel of 25 $L$ volume contains water at 220 $kPa$ and 180°C. The water exchanges heat with the surroundings till its temperature drops to 45°C. Calculate the change in entropy of the water.

6.5 2 $kg$ water having 80% quality is contained in a piston cylinder system at a constant pressure of 120 $kPa$. Heat is transferred to the system from a constant temperature source at 500°C till its temperature reaches 350°C. Calculate the entropy change of the water. Also, find the entropy change for the heat source and total entropy change during the process.

6.6 A 45 $kg$ copper block is at 90°C and is quenched in 100 $L$ of water at 22°C that is contained in an insulated tank. Determine the change in entropy of the process considering specific heat of copper is 0.38 $kJ/kg-K$.

6.7 Two blocks, one of iron and the other of aluminium, are dropped into a water pond at 20°C. The initial temperature of both the blocks is 75°C. The mass of iron block is 45 $kg$ and mass of aluminium block is 10 $kg$. The specific heats of iron and aluminium are 0.44 $kJ/kg-K$ and 0.90 $kJ/kg-K$, respectively. Determine the change in entropy of the process once the thermal equilibrium is established between blocks and water as the result of heat transfer.

6.8 2 $kg$ air is contained in a piston-cylinder system at 100 $kPa$ and 22°C. The air is then compressed in a polytropic process having $n = 1.25$ till its volume reduces to one fourth of the initial volume. Calculate the change in entropy of air during this compression process.

6.9 Air is compressed isothermally in a piston-cylinder device from 100 $kPa$ and 25°C to 620 $kPa$. Estimate the change in entropy of air and the amount of work done during this process.

6.10 Air is compressed from 120 $kPa$ and 22°C to 620 $kPa$ and 170°C in a steady flow compressor at the rate of 72 $kg/hr$. The power consumption of the compressor is 4.5 $kW$ and it transfers some heat to the surroundings at 20°C during the compression process. Determine the change in entropy of air and total change of entropy during this process. Changes in kinetic and potential energies of air across the compressor can be neglected.

6.11 A rigid vessel contains 5 $kg$ of air at 500 $kPa$ and 35°C. The vessel is insulated and a valve is opened till air pressure drops to 120 $kPa$ within the tank. Determine the final air mass in the tank if air is assumed to undergo a reversible adiabatic expansion.

6.12 A steam power plant operates between the pressure limits of 20 $MPa$ and 30 $kPa$. Water enters the pump as saturated liquid and leaves the turbine as saturated vapour. Determine the work done by the turbine and the work consumed by the pump if the expansion and pumping processes are reversible adiabatic.

6.13 An adiabatic turbine receives steam at 10 $MPa$ and 550°C, discharges at 20 $kPa$. Determine the temperature of steam at turbine exit if its isentropic efficiency is 85%. Also, calculate the power produced by the turbine if the mass flow rate of steam is 5 $kg/s$. Estimate the change in entropy of steam during this process.

6.14 Combustion gases enters the nozzle of a jet engine at 300 $kPa$ and 800°C and leaves at 80 $kPa$. Consider the combustion gases as air and isentropic efficiency of nozzle as 90%. Determine the exit temperature and exit velocity of gases. Also, estimate the change in entropy of gases during this process.

6.15 A geyser installed at IIT Mandi is configured to enter the cold water at 10°C into a well-insulated, double-pipe heat exchanger at the rate of 0.22 $kg/s$ and is heated to 50°C by the hot saturated water (liquid) that enters at 100°C. Find the exit temperature of the hot medium if its flow rate is 2.8 $kg/s$ and the rate of heat transfer between the hot and cold mediums. Also, calculate the rate of entropy generation in the heat exchanger.

6.16 Superheated steam at 250 $kPa$ and 180°C is mixed with with compressed liquid having 250 $kPa$ pressure and 35°C temperature. The mixture leaves the mixing chamber at 250 $kPa$ and 80°C when the compressed liquid enters the mixing chamber at mass flow rate of 2.2 $kg/s$. Calculate the mass flow rate of the superheated steam if the mixing chamber transfers heat to surroundings at 20°C at the rate of 15 $kW$. Also, find the rate of entropy generation during this process.

6.17 250 $kJ$ of heat transfers from a thermal reservoir at 1200 $K$ to an iron vessel at 850 $K$. This vessel transfers the same amount of heat to another thermal reservoir at 490 $K$. Determine the total entropy generation during this process.

6.18 1.2 $kg$ air is contained in a rigid vessel at 120 $kPa$ and 300 $K$. The vessel is provided with an electric resistance heater which supplies energy at the rate of 500 W. Determine the temperature of the air if its pressure measured after one hour is 215 $kPa$. The vessel transfers heat to the surroundings at 300 $K$ during the process. Calculate the amount of heat transferred from the vessel to the surroundings and entropy generation during the process.

6.19 Air is compressed in first stage of a compressor from 100 $kPa$ and 25°C to 2.5 $MPa$. Then an intercooler cools it at constant pressure to 70°C. Air coming from intercooler is again compressed to a pressure of 12 $MPa$. Determine the total specific work and the amount of heat transfer in the intercooler if both the compression stages are reversible adiabatic. Also, find the percentage increase in the specific work if the compression process is achieved in a single state without any intercooling.

6.20 A pump delivers water at the rate of 100 $L/s$ through a vertical distance of 15 $m$ and velocity of 15 $m/s$ at ambient conditions. Determine the work input to the pump if the isentropic efficiency of the combined system is 65%. Also, find the entropy generation during the process.

## PRACTICAL

**Activity 1:** List the causes of irreversibilities associated with the operation of different devices you encounter at your home and workplace. Also, make efforts to quantify the entropy generation in some of these devices if possible.

## REFERENCES AND SUGGESTED READINGS

1. Bejan, A., 2016. Advanced engineering thermodynamics. John Wiley & Sons.
2. Greven, A., Keller, G. and Warnecke, G. eds., 2014. *Entropy* (Vol. 47). Princeton University Press.
3. Cengel, Y.A., Boles, M.A. and Kanoğlu, M., 2011. *Thermodynamics: an engineering approach.* New York: McGraw-hill.
4. Borgnakke, C. and Sonntag, R.E., 2022. *Fundamentals of thermodynamics*. John Wiley & Sons.

# 7 Exergy

## UNIT SPECIFICS

*Through this unit we have discussed the following aspects:*

- *Introduction to the concept of exergy;*
- *Exergy transfer associated with various modes of energy transfer;*
- *Exergy potential of closed systems;*
- *Exergy potential of flowing streams;*
- *Exergy balance equation;*
- *Second law efficiency.*

*Till now you have understood the basic concepts of energy and entropy and associated laws defining the energy conservation and entropy accounting. The concept of entropy was used for predicting the spontaneity of any thermodynamic process. Through the second law of thermodynamics we also know the limit of the conversion efficiency of one type of energy in transit (heat) into another named work. The concept of exergy defines the limiting amount of energy content that can be converted into work in specified ambient conditions where the thermodynamic system is placed.*

## RATIONALE

*The understanding of exergy makes the concepts of thermodynamics closer to reality by simultaneous consideration of system and environment in which the system is expected to produce work. The exergy balance of open and closed systems enables us to estimate the flow of energy into the system which has feasibility for being transferred into work. This concept is practically expressed in terms of second law efficiency of thermodynamic systems.*

## PRE-REQUISITES

*First law of thermodynamics; second law of thermodynamics; entropy*

## UNIT OUTCOMES

*List of outcomes of this unit is as follows:*

*U7-O1: Appreciation of concept of exergy*

*U7-O2: Understating the relationship between exergy transfer and energy transfer*

*U7-O3: Understand the concept of exergy potential*

*U7-O4: Estimation of exergy potentials of closed and open systems*

*U7-O5: Understanding the applications of exergy balance equation*

| Unit-7 Outcomes | EXPECTED MAPPING WITH COURSE OUTCOMES (1- Weak Correlation; 2- Medium correlation; 3- Strong Correlation) | | | | | |
|---|---|---|---|---|---|---|
| | CO-1 | CO-2 | CO-3 | CO-4 | CO-5 | CO-6 |
| U7-O1 | 3 | 2 | 2 | 3 | 3 | 2 |
| U7-O2 | 3 | 2 | 2 | 3 | 3 | 2 |
| U7-O3 | 3 | 2 | 2 | 3 | 3 | 2 |
| U7-O4 | 2 | 2 | 2 | 3 | 3 | 2 |
| U7-O5 | 2 | 2 | 2 | 3 | 3 | 2 |

## 7.1 Introduction to Exergy

We established in the previous chapters that only the amount of energy is not sufficient to know its work potential. Such an argument is specifically discussed for a few devices involving heat and work transformations. Here, we will attempt to generalize this concept for different situations involving energy interactions. Let us first understand the work potential associated with the same amount of heat at different temperatures. Consider $Q$ amount of heat at the surface of the sun that typically has temperature $\simeq 5800K$. The work potential or maximum useful work from this heat can be extracted by operating a Carnot heat engine between this heat source and atmosphere as a heat sink (Fig. 7.1-a). An estimate involving Carnot efficiency suggests that about 95% of this heat can be converted to work. When the same amount of heat reaches the earth surface, it is available at the ambient temperature (Fig. 7.1-b). Now the work potential of this heat is zero. The amount of energy that earth receives from the sun is significantly larger than the global energy demand and it can really solve the global energy crisis if available at higher temperatures. However, when its temperature reaches to the ambient conditions it is simply regarded as the waste heat because it cannot be used to extract energy. This is the reason we attempt to

Scan Me

For more details on exergy.

concentrate the solar radiations on solar collectors and other solar energy based devices to obtain the heat at temperatures above than atmospheric temperature so that it can be used to extract some useful work.

In the previous illustration, we specifically talked about only one form of energy, i.e. heat, however, for a generic source of energy it can be available in different forms such as high pressure gas which can be used to obtain boundary work, chemical energy stored in a fuel that can give heat at very high temperatures and even the combination of many forms. Therefore, a generic property that represents the maximum work potential or maximum useful work that the energy contained in a system can have at a specified state is called **exergy** or **availability** or **available energy**. It is well established in previous chapters that work done by a system during a process depends upon the initial and final states of the system and the process path as indicated in Fig. 7.2 for the boundary work obtained from a closed system between states 1 and 2. Firstly, to obtain the maximum work potential, the process should be reversible, therefore, the irreversibilities are not accounted for in the estimate of exergy. Secondly, the system should not have any potential left to do more work at the end state. For instance, in Fig. 7.2 if end pressure of the system $P_2 > P_0$ where $P_0$ is the pressure of its surroundings. Then, we still have the opportunity to expand the system boundary till it reaches the surrounding conditions. The additional work performed by system from $P_2$ to $P_0$ is shown in Fig. 7.2. Thus, it is quite evident that the end state of the system should be in thermodynamic equilibrium with its surroundings to extract the maximum work from it during a process and such a state of the system is known as **dead state**. A system in dead state is represented in Fig. 7.3. It has chemically inert fluid which is at atmospheric pressure and temperature, thus in chemical, mechanical and thermal equilibrium. Also, the velocity and elevation of the system are zero relative to its surroundings. Any external effects such as surface tension, electrical and magnetic, etc. should also be balanced between the system and surroundings at the dead state of the system if these are of importance. The exergy of a system at its dead state is zero as it cannot undergo a process. The properties at the dead state of a system are usually represented by subscript "0".

When a system undergoes a process, it influences its immediate vicinity of surroundings and the region of surroundings far away from the system remains uninfluenced. Thus, irreversibilities if any are associated with the surroundings in the immediate vicinity of the system. Therefore, the portion of the surroundings in the vicinity of the system that is influenced by a process occurring in the system is known as **immediate surroundings**. The region of surroundings other than immediate surroundings of a system is known as the **environment**. Typically, the environment does not experience any effect of a process occurring in a system. For a typical situation of cooling of a hot water bucket in open

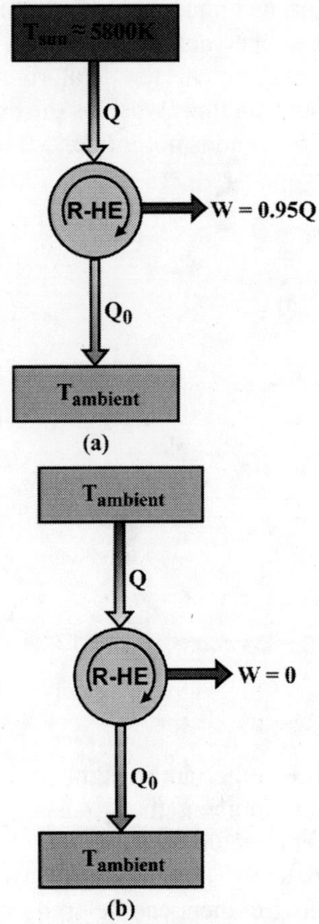


**Fig. 7.1:** A heat engine operating between: (a) sun and atmosphere and (b) two reservoirs at ambient temperature.

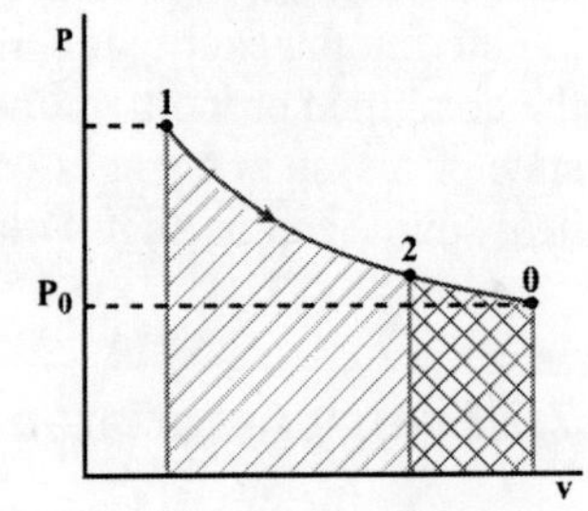


**Fig. 7.2:** Representation of boundary work on $P - v$ diagram during an expansion process.

air, its immediate surroundings and environment are shown in Fig. 7.4. It can be observed from Fig. 7.4 that temperature varies between environment temperature and system temperature in the immediate surroundings whereas the environment remains at a fixed temperature of 25°C and is not influenced by the system.

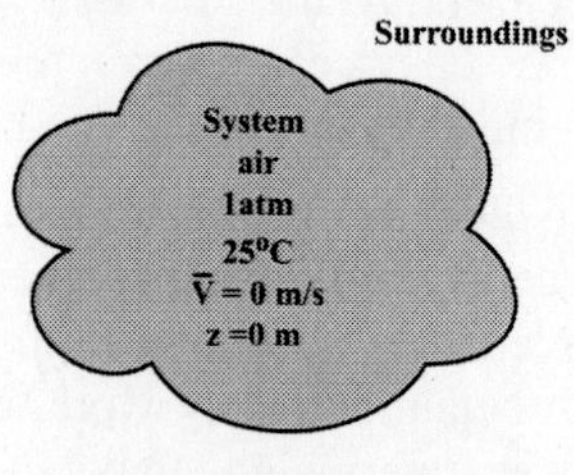


**Fig. 7.3:** Depiction of the dead state of a system.

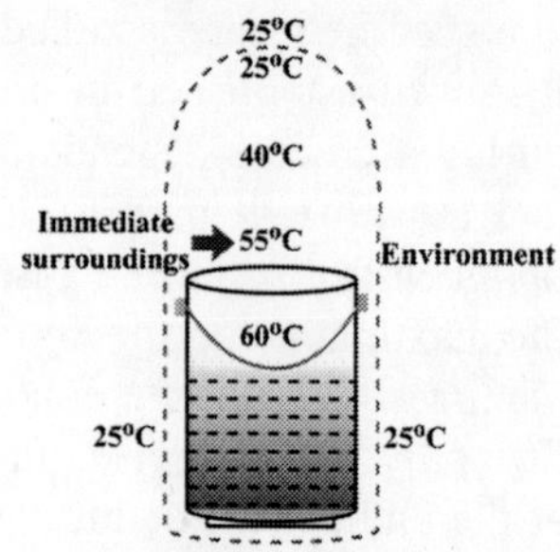


**Fig. 7.4:** Depiction of the immediate surroundings and environment of a system.

It is important to highlight if system temperature is either greater or smaller than the temperature of the environment then it has potential to produce work by operating a heat engine between the two temperature limits. Similarly the deviation in system pressure from the pressure of the environment has potential to do boundary work till pressure is not equalized. The velocity of the system relative to the environment can be used to displace an object or a turbine to extract mechanical work. Only a system in dead state has no potential to do work. The atmosphere around us is in a dead state, that's why it has no potential to do work despite the huge amount of energy it has. A system produces maximum possible work when it undergoes a reversible process from its given initial state to dead state. This maximum possible limit of useful work that a system can deliver without violating the laws of thermodynamics is called exergy and the actual work delivered by a system is always less than its exergy. As the final state in determining exergy is controlled by the environment conditions, therefore, exergy can be regarded as the combined property of the system and environment. The amount of exergy associated with the given state of a system changes with change in conditions of the environment. For example, the exergy of a heat source can be slightly larger in winters than in summers.

## 7.2 Exergy Associated with Different Modes of Energy Transfer

We established in Chapter 2 that energy is stored in a system in different forms such as internal energy, kinetic energy and potential energy, etc. The kinetic and potential energies are the mechanical form of energy whereas the internal energy is associated with the thermodynamic state of a system. The kinetic energy of a system, being the mechanical form of energy, can be fully utilized as useful work irrespective of the thermodynamic state of the system. For example, the velocity of fluid at the exit of a nozzle can be utilized to rotate a turbine and obtain the shaft work. Under the reversible conditions (no losses due to friction and fluid viscosity, etc.), the work produced by the turbine will be equal to the kinetic energy of the fluid. Similarly, the potential energy possessed by a system can be fully utilized to obtain work. For instance, a mass, $m$ lowered through height, $z$ is used to stir a fluid does $mgz$ amount of work on the fluid under reversible conditions. Thus, it can be concluded that the exergy of kinetic energy and potential energy of a system are equal to themselves irrespective of the pressure and temperature of the environment.

The internal energy, which is specified by the thermodynamic state of the system, can change via heat transfer, work transfer and mass transfer where mass transfer is applicable only for control volumes. It is to be noted that the nuclear and chemical reactions are also the other possible ways to change the internal energy of a system, however, we restrict ourselves to thermo-mechanical exergy in the present discussion, therefore, do not consider these effects. We will now establish the amount of exergy associated with heat and work transfer between system and surroundings, whereas the exergy associated with flowing stream is discussed in Section 7.4.

### 7.2.1 Exergy Transfer by Heat

Heat transfer ($Q$) from a constant temperature source at temperature $T$ to its environment at $T_0$ can be achieved reversibly only by employing a Carnot heat engine between the two thermal reservoirs. The maximum work that can be extracted during this heat transfer process is the exergy associated with heat transfer and is expressed as

$$\chi_{heat} = W_{rev} = \left(1 - \frac{T_0}{T}\right) Q \tag{7.1}$$

When the temperature of the heat source changes during the heat transfer process then Eqn. (7.1) can be expressed in integral form as

$$\chi_{heat} = W_{rev} = \int \left(1 - \frac{T_0}{T}\right) \delta Q \tag{7.2}$$

It is to be noted from Eqns. (7.1) and (7.2) that for $T > T_0$ the exergy of the system increases when the heat is added to the system and it decreases when heat is leaving from the system. This can be understood with ease for a variable temperature heat source. If you add heat to such a source its temperature increases and deviates more from the environment temperature, therefore, increases its exergy. Contrary, the removal of heat from such a source will decrease its temperature and bring it more closer to environment temperature, therefore, its exergy decreases. Hence, for a heat source having $T > T_0$, direction of heat transfer and exergy transfer are the same. This is schematically shown in Fig. 7.5-a,b.

It will be interesting to know that how a system having temperature $T < T_0$ possesses exergy. For such a system, the environment acts as the heat source and the system acts as a heat sink for a Carnot heat engine. Now, adding heat to such a system will bring its temperature closer to the environment, therefore, decreasing its exergy. While removing heat from such a cold medium will increase its exergy due to more deviation from the environment. Therefore, heat transfer and exergy transfer take place in opposite direction for a system having $T < T_0$. It is schematically shown in Fig. 7.5-c,d. The exergy associated with heat transfer becomes zero when the temperature of the system reaches environment temperature.

It should also be emphasized that the heat transfer through a finite temperature difference results in entropy generation, hence, the exergy is destructed with such a process. To understand this, consider $Q$ amount of heat is transferred from temperature $T_1$ to $T_2 < T_1$. The associated amount of exergy at temperature $T_1$ is $\chi_{heat,T_1} = \left(1 - \frac{T_0}{T_1}\right) Q$ and exergy at temperature $T_2$ with same amount of heat transfer is $\chi_{heat,T_2} = \left(1 - \frac{T_0}{T_2}\right) Q$. Since $T_1 > T_2$, thus $\chi_{heat,T_1} > \chi_{heat,T_2}$. Therefore, the difference between $\chi_{heat,T_1}$ and $\chi_{heat,T_2}$ is the amount of exergy destroyed during this heat transfer process due to irreversibilities.

## 7.2.2 Exergy Transfer by Work

Work being a high grade or organized form of energy itself represents the maximum potential of a system, however, it becomes essential at this stage to know the exact quantity of useful work that can be extracted from a system. To understand this, lets us consider a piston-cylinder system that produces $W_{rev}$ amount of boundary work while expanding from volume $V_1$ to $V_2$ as shown in Fig. 7.6. During this expansion, the piston has pushed surroundings air that is at atmospheric pressure and the volume of its immediate surroundings has decreased by the amount $V_2 - V_1$. The work associated with changes in immediate surroundings of the system is known as **surroundings work** and is expressed as

$$W_{su} = P_0(V_2 - V_1) \tag{7.3}$$

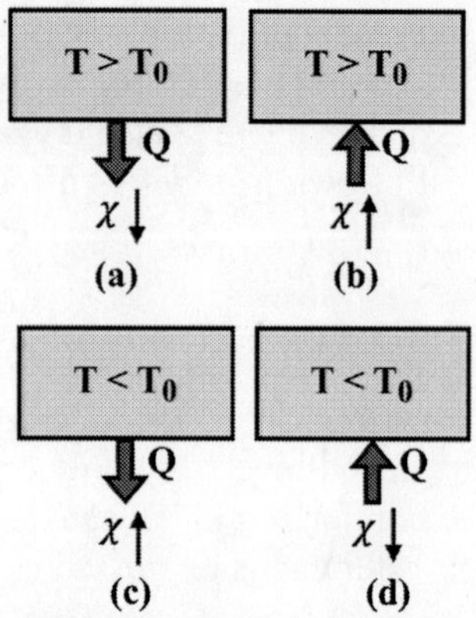


**Fig. 7.5:** Direction of heat and exergy transfer for (a-b) a system at $T > T_0$ and (c-d) a system at $T < T_0$.

This is waste work which cannot be utilized for useful purposes. Therefore, maximum **useful work** that is associated with this expansion process is

$$W_u = W_{rev} - W_{su} = W_{rev} - P_0(V_2 - V_1) \tag{7.4}$$

It must be noted that the surroundings supply an equivalent amount of work to a system during its compression, thus, reducing the input work required for the compression process. Further, the lost potential of work in terms of surroundings work is applicable only for the boundary work whereas the other forms of work such as shaft work in control volumes and electric work, etc. are completely available in useful form as the boundary of the system remains unchanged for such interactions.

The above discussion explains that exergy associated with work transfer can be given as

$$\chi_{work} = \begin{cases} W_{rev} - W_{su} & (for\ boundary\ work) \\ W_{rev} & (for\ other\ forms\ of\ work) \end{cases} \tag{7.5}$$

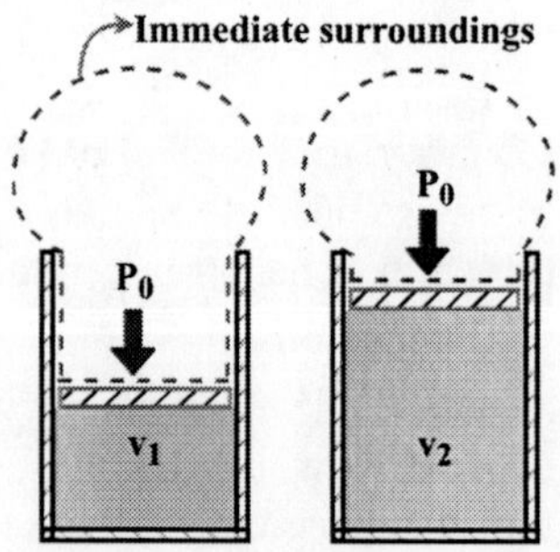


**Fig. 7.6:** Decrease in volume of the immediate surroundings of a system during the expansion of its boundary.

## 7.3 Exergy Potential of a Closed System

Energy is stored in a closed system in the form of internal energy. The only possible modes for changes in internal energy of a closed system are heat transfer and work transfer in the absence of nuclear and chemical reactions. It is established in the previous section that exergy associated with heat transfer is less than the amount of heat exchanged and hence, only a fraction of internal energy of a system can be extracted in the form of useful work when it changes via heat transfer. Contrary, the internal energy change by work transfer can be fully extracted, except for a small waste energy in the form of surroundings work particularly for boundary work done by the system.

To estimate the exergy of a closed system at its given thermodynamic state, consider a piston-cylinder system that contains a fluid at pressure $P$ and temperature $T$. Let $m$ is the mass of the fluid and the pressure and temperature of the environment are $P_0$ and $T_0$, respectively. To transfer the heat from temperature $T$ of the fluid to the environment in reversible manner, a Carnot heat engine is to be connected as shown in Fig. 7.7. Also, due to the pressure difference, the fluid will expand and will do boundary work. Assume that during a differential change in volume, $dV$ of the system $\delta Q$ amount of heat is transferred from the fluid and it does

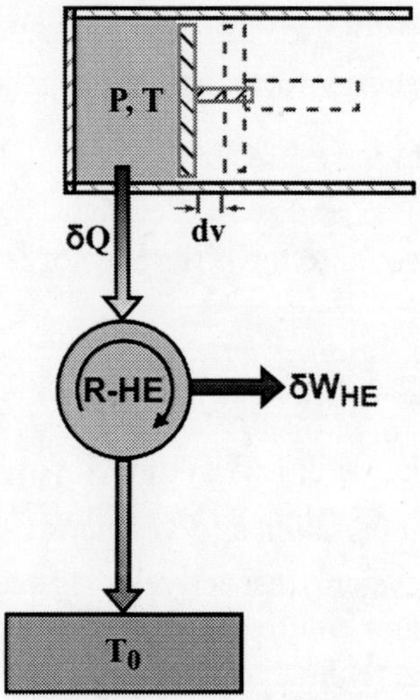


**Fig. 7.7:** A closed system undergoing reversible heat and work interactions.

$\delta W$ amount of boundary work. As per the first law of thermodynamics, we get$-\delta Q - \delta W = dU$ (7.6)

Here, $\delta W = PdV = (P - P_0)dV + P_0 dV = \delta W_u + P_0 dV$ where $\delta W_u$ is useful boundary work. Using Eqn. (7.6), we obtain

$$\delta W_u = -\delta Q - dU - P_0 dV \tag{7.7}$$

The work done by the Carnot heat engine associated with heat transfer $\delta Q$ from the fluid is given as

$$\delta W_{HE} = \left(1 - \frac{T_0}{T}\right)\delta Q \tag{7.8}$$

The total useful work obtained from the system is expressed as

$\delta W_T = \delta W_u + \delta W_{HE} = -\delta Q - dU - P_0 dV + \left(1 - \frac{T_0}{T}\right)\delta Q$. Using the entropy relation, we obtain $dS = -\frac{\delta Q}{T}. \Rightarrow \delta Q = -TdS$. Thus $\delta W_T = -dU - P_0 dV + T_0 dS$. As to extract maximum work from the system, it should reach the dead state. Therefore, integration of $\delta W_T$ from initial state to dead state results in

$$W_T = (U - U_0) + P_0(V - V_0) - T_0(S - S_0) \tag{7.9}$$

The maximum possible total useful work given by Eqn. (7.9) is the exergy possessed by a closed system at the given thermodynamic state. If the system possesses kinetic and potential energies also, then these are directly added to total useful work obtained in Eqn. (7.9) because the kinetic and potential energies are forms of exergy itself. Thus, the exergy of a closed system having mass $m$ becomes

$$\chi = (U - U_0) + P_0(V - V_0) - T_0(S - S_0) + m\frac{V^2}{2} + mgz \tag{7.10}$$

The exergy of a closed system on per unit mass basis ($\phi$) is given as

$$\phi = (u - u_0) + P_0(v - v_0) - T_0(s - s_0) + \frac{V^2}{2} + gz \tag{7.11}$$

All the properties designated by subscript "0" are corresponding to the dead state of the system. Using energy formulation $e = u + \frac{V^2}{2} + gz$, Eqn. (7.11) becomes

$$\phi = (e - e_0) + P_0(v - v_0) - T_0(s - s_0) \tag{7.12}$$

The exergy of a system is zero at its dead state. The change in exergy of a system during a process is equal to the difference of final and initial exergies of the system and is expressed as

$$\Delta\chi = \chi_2 - \chi_1 = m(\phi_2 - \phi_1)$$

$$= m(u_2 - u_1) + m\,P_0(v_2 - v_1) - m\,T_0(s_2 - s_1) + m\left(\frac{V_2^2 - V_1^2}{2}\right) + mg(z_2 - z_1) \tag{7.13}$$

For a closed system involving non-uniform properties, the exergy of the system can be obtained by integration as given here $\chi = \int_m \phi\, dm = \int_V \rho\phi\, dV$. Here, $\rho$ and $V$ are density and volume of the system, respectively. As exergy is a property, therefore, it does not change unless the state of the system and environment is changed during a process. The exergy of a closed system can be either positive or zero and it cannot be negative. A closed system having $P < P_0$ can produce boundary work by allowing atmospheric air to move a piston to do useful boundary work. Similarly, a system at $T < T_0$ can allow operating a heat engine with atmosphere as heat source and produce useful work. Thus, a system having property values less than its dead state also possesses exergy.

## 7.4 Exergy Potential of a Flowing Stream

It is established in Chapter 4 that a flowing stream carries an additional energy, called the flow energy, to maintain its flow in a pipe or duct. The flow energy is given as $w_{flow} = Pv$ where $P$ is pressure and $v$ is the specific volume of the fluid stream. Flow work fundamentally involves the boundary work done by the fluid on its downstream fluid parcels, therefore, the exergy associated with flowing stream is equivalent to exergy associated with boundary work. The exergy with boundary work includes the amount of work in excess of the surroundings work that is the work needed to push atmospheric air at pressure $P_0$. Thus, the exergy associated with a fluid stream due to flow energy can be expressed as $(P - P_0)v$. Therefore, the total exergy of a flow stream is the sum of exergy due to flow and the exergy that it possesses due to its thermodynamic state, kinetic and potential energies (Eqn. 7.11). This is mathematically given as

$$\psi = (u - u_0) + P_0(v - v_0) - T_0(s - s_0) + \frac{V^2}{2} + gz + (P - P_0)v.$$

$$\Rightarrow \psi = (u + Pv) - (u_0 + P_0v_0) - T_0(s - s_0) + \frac{V^2}{2} + gz.$$

$$\Rightarrow \psi = (h - h_0) - T_0(s - s_0) + \frac{V^2}{2} + gz \tag{7.14}$$

Equation (7.14) represents the exergy possessed by a flowing fluid stream and is typically called **flow exergy**. Unlike exergy of a closed system, flow exergy can be negative for fluid stream pressure less than the environment pressure ($P_0$). The change in exergy of a fluid stream when it goes through a process from state 1 to state 2 is given as

$$\Delta\psi = \psi_2 - \psi_1 = (h_2 - h_1) - T_0(s_2 - s_1) + \left(\frac{V_2^2 - V_1^2}{2}\right) + g(z_2 - z_1) \tag{7.15}$$

## 7.5 Decrease of Exergy Principle

In Chapter 6, we established the principle of increase of entropy which states that entropy can be created but it cannot be destroyed. Another variant of the second law of thermodynamics is discussed here which is called the decrease of exergy principle. To demonstrate this, consider an isolated system that has no mass, heat and work interactions (Fig. 7.8). The energy and entropy balance for a process occurring between states 1 and 2 of the isolated system can be expressed as

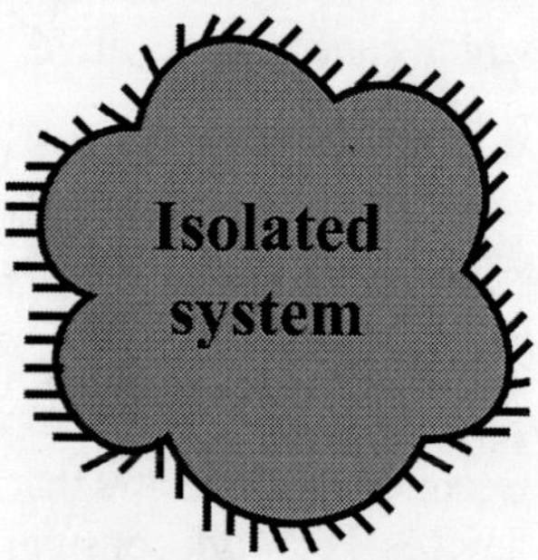

**Fig. 7.8:** Representation of an isolated system.

$$0 = \Delta E_{system} = E_2 - E_1 \quad (7.16)$$

$$0 + S_{gen} = \Delta S_{system} = S_2 - S_1 \quad (7.17)$$

Upon multiplying Eqn. (7.17) by $T_0$ and subtracting from Eqn. (7.16), we get

$$-T_0 S_{gen} = E_2 - E_1 - T_0(S_2 - S_1) \quad (7.18)$$

Using Eqn. (7.13), we know

$$\chi_2 - \chi_1 = (U_2 - U_1) + P_0(V_2 - V_1) - T_0(S_2 - S_1) + m\left(\frac{V_2^2 - V_1^2}{2}\right) + mg(z_2 - z_1).$$

For an isolated system, boundary work will be zero. Thus, $P_0(V_2 - V_1) = 0$. Therefore,

$$\chi_2 - \chi_1 = E_2 - E_1 - T_0(S_2 - S_1) \quad (7.19)$$

where $E = U + m\frac{V^2}{2} + mgz$.

Using Eqns. (7.18) and (7.19), we get

$$\chi_2 - \chi_1 = -T_0 S_{gen} \quad (7.20)$$

As $T_0$ is the absolute temperature of the environment which will always be greater than or equal to zero. Similarly, $S_{gen} \geq 0$. Therefore, change in exergy of an isolated system during a process will always be negative except for a special case of reversible process for which it becomes zero. This states that exergy always decreases during an actual process, this statement is known as **decrease of exergy principle**. For an isolated system, the exergy change during a process is equal to exergy destroyed due to irreversibilities.

Irreversibilities always cause entropy generation, hence, the exergy destruction. The amount of exergy destroyed during a process is directly proportional to the entropy generated during the process and can be expressed as

$$\chi_{destroyed} = T_0 S_{gen} \geq 0 \quad (7.21)$$

Exergy destroyed is always positive for actual processes and it is zero for reversible processes. It represents the lost potential of a system to produce useful work and is also called **lost work** or **irreversibility**. All the systems and surroundings can be put together inside an universe that is an isolated system. Therefore, the decrease of exergy principle and exergy destruction equations are applicable for all the systems of the universe. As no actual process can occur without entropy generation, therefore, the exergy of the universe is decreasing continuously. The decrease of exergy principle should not be confused with the change in exergy of a non-isolated system which can increase when a system moves away from its environment whereas it specifically deals with exergy destruction during a process which cannot be negative.

It is also to be noted that the maximum useful work potential $(W_{u,ideal})$ of a process represents its exergy and in an actual process the useful work $(W_{u,actual})$ produced is less than the maximum useful work potential. Therefore, the difference of the two also specifies the irreversibility or lost work potential. Thus, the actual useful work can also be written as

$$W_{u,actual} = W_{u,ideal} - T_0 S_{gen} \tag{7.22}$$

Equation (7.22) shows that magnitude of actual work will be smaller than ideal work for work producing devices. For work consuming devices such as a compressor, $W_{u,ideal}$ is negative and $W_{u,actual}$ becomes more negative, therefore, having more magnitude.

## 7.6 Exergy Balance Equation

It is well established in the previous section that unlike entropy, the exergy gets destroyed during a process due to irreversibilities within the system. Therefore, the change in exergy of a system during a process is equal to the difference between net exergy transfer across the boundary of the system and exergy destruction within the system boundary due to irreversibilities. Mathematically, it can be expressed as

$$\chi_{in} - \chi_{out} - \chi_{destruction} = \Delta\chi_{system} = \chi_2 - \chi_1 \tag{7.23}$$

The exergy can enter and leave the system by heat, work and mass transfer across its boundaries. For a closed system, there is no mass transfer across the system boundary, thus, exergy is associated with only heat and work transfer. Therefore, exergy balance is given as

$\chi_{heat} - \chi_{work} - \chi_{destruction} = \Delta\chi_{system} = \chi_2 - \chi_1$, which can be further expressed as

$$\left(1 - \frac{T_0}{T}\right) Q - [W - P_0(V_2 - V_1)] - T_0 S_{gen} = \chi_2 - \chi_1 \tag{7.24}$$

In case, there are multiple heat interactions at different temperatures $T_i$ across the boundaries of the system, then

$$\sum_i \left(1 - \frac{T_0}{T_i}\right) Q_i - [W - P_0(V_2 - V_1)] - T_0 S_{gen} = \chi_2 - \chi_1 \tag{7.25}$$

Dividing Eqn. (7.25) by $\Delta t$ and considering a limit of $\Delta t \rightarrow 0$, Eqn. (7.25) can be expressed in rate form as

$$\sum_i \left(1 - \frac{T_0}{T_i}\right) \dot{Q}_i - \left[\dot{W} - P_0 \frac{dV}{dt}\right] - T_0 \frac{dS_{gen}}{dt} = \frac{d\chi}{dt} \tag{7.26}$$

Equations (7.24) to (7.26) are obtained by considering heat transfer to the system and work done by the system as positive quantities. In case of heat transfer from the system and work done on the system, the negative sign should be considered with commensurate terms. Further, the maximum possible reversible work can be obtained from Eqns. (7.24) to (7.26) by considering exergy destruction to be zero. The exergy destruction here is caused due to irreversibilities present within the boundary of the system and does not include the contribution of external irreversibilities present outside the system. Therefore, after subjecting exergy destruction equal to zero, the exergy balance equation holds for an internally reversible process and not always for totally reversible process as there can be unaccounted irreversibilities outside the system boundary. Therefore, the total exergy destroyed during a process can be obtained by together applying the exergy balance for the system and its immediate surroundings where external irreversibilities can be present. It should also be made clear that unlike energy which always remains conserved, the exergy decreases and remains conserved only for a special case of reversible process. This simply means that quality of energy decreases during an actual process due to increase in entropy and decrease of exergy.

The exergy balance for a control volume includes one additional contribution of exergy associated with mass transfer across the boundary of the control volume. Therefore, for a control volume exergy balance can be written as

$$\chi_{heat} - \chi_{work} + \chi_{mass,in} - \chi_{mass,out} - \chi_{destruction} = \Delta\chi_{cv} = \chi_2 - \chi_1 \quad (7.27)$$

Considering heat added to the control volume and work done by the control volume as positive and fixed boundary of the control volume, we can write exergy balance as

$$\left(1 - \frac{T_0}{T}\right) Q - W + m_{in}\psi_{in} - m_{out}\psi_{out} - T_0 S_{gen} = \chi_2 - \chi_1 \quad (7.28)$$

Here, the control volume has only one heat interaction, one inlet and one outlet. For a control volume with multiple heat interactions and multiple inlets and outlets, the exergy balance is given as

$$\sum_i \left(1 - \frac{T_0}{T_i}\right) Q_i - W + \sum m_{in}\psi_{in} - \sum m_{out}\psi_{out} - T_0 S_{gen} = \chi_2 - \chi_1 \quad (7.29)$$

Equation (7.29) in rate form for a control volume becomes

$$\sum_i \left(1 - \frac{T_0}{T_i}\right) \dot{Q}_i - \dot{W} + \sum \dot{m}_{in}\psi_{in} - \sum \dot{m}_{out}\psi_{out} - T_0 \dot{S}_{gen} = \frac{d\chi_{cv}}{dt} \quad (7.30)$$

For a steady flow control volume, the change in property of the control volume with time is zero, i.e. $\frac{d\chi_{cv}}{dt} = 0$, therefore, exergy balance becomes

$$\sum_i \left(1 - \frac{T_0}{T_i}\right) \dot{Q}_i - \dot{W} + \sum \dot{m}_{in}\psi_{in} - \sum \dot{m}_{out}\psi_{out} - T_0 \dot{S}_{gen} = 0 \quad (7.31)$$

A steady flow control volume with single inlet and single outlet has $\dot{m}_{in} = \dot{m}_{out} = \dot{m}$, thus, exergy balance is

$$\sum_i \left(1 - \frac{T_0}{T_i}\right) \dot{Q}_i - \dot{W} + \dot{m}(\psi_{in} - \psi_{out}) - T_0 \dot{S}_{gen} = 0 \quad (7.32)$$

Here, $\psi_{in} - \psi_{out} = (h_{in} - h_{out}) - T_0(s_{in} - s_{out}) + \left(\frac{V_{in}^2 - V_{out}^2}{2}\right) + g(z_{in} - z_{out})$. Equation (7.32) on per unit mass ($\dot{m}$) basis is given as

$$\sum_i \left(1 - \frac{T_0}{T_i}\right) q_i - w + (\psi_{in} - \psi_{out}) - T_0 S_{gen} = 0 \quad (7.33)$$

Here, $q_i = \dot{Q}_i/\dot{m}$ and $w = \dot{W}/\dot{m}$. For an adiabatic steady flow device without any work interaction, we get $T_0 S_{gen} = \psi_{in} - \psi_{out} \Rightarrow \chi_{destruction} = \psi_{in} - \psi_{out}$. This shows that exergy of the flowing stream must decrease as it passes through an adiabatic work free steady flow device and should remain unaltered for a special case of reversible process.

## 7.7 Second Law Efficiency

We defined the performance of different cyclic and non-cyclic devices in previous chapters in terms of their efficiencies and coefficients of performance. These performance parameters were fully derived based on the arguments of energy conservation and hence, the first law of thermodynamics. Thus the thermal efficiency of a heat engine, isentropic efficiency of different steady flow devices is usually termed as the **first law efficiency** also. In many situations, the first law efficiency of a device may not be the true representative of its performance. To understand this concept, consider two heat engines that have 40% first law efficiency and both receive the same amount of heat from the heat source. Consider that first heat engine operates between temperature limits of 1200$K$ and 300$K$ whereas second one operates between 800$K$ and 300$K$ (Fig. 7.9). Despite their different operating temperatures, both will produce the same amount of work output as both have the same first law efficiency and get the same amount of heat input. Now, if we attempt to determine the maximum possible efficiency of a reversible heat engine between the given temperature limits using the concepts of the second law of thermodynamics, then we obtain that

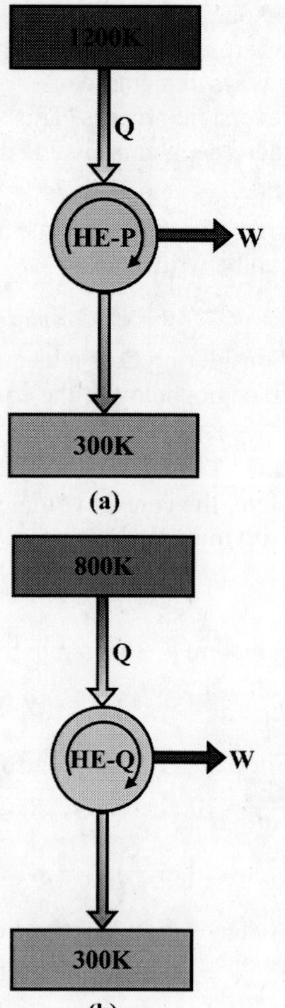

**Fig. 7.9:** Two heat engines P (a) and Q (b) operating between different temperature limits.

$$\eta_{rev,P} = \left(1 - \frac{T_L}{T_H}\right)_P = 1 - \frac{300}{1200} = 75\% \text{ and}$$

$$\eta_{rev,Q} = \left(1 - \frac{T_L}{T_H}\right)_Q = 1 - \frac{300}{800} = 62.5\%.$$

This shows that the first heat engine has the potential to convert 75% of supplied heat into useful work, whereas the second heat engine can convert a maximum of 62.5% of supplied heat into work. It clearly explains that the performance of the first heat engine is poor in comparison to the second one despite their equal first law efficiencies. To overcome this limitation of the first law efficiency, we define the **second law efficiency** as the ratio of actual efficiency of a device to its maximum possible thermal efficiency under the given conditions. It is mathematically expressed as

$$\eta_{II} = \frac{\eta_{actual}}{\eta_{rev}} \tag{7.34}$$

Using this definition, the second law efficiency of above mentioned heat engines is calculated as

$\eta_{II,P} = \frac{0.40}{0.75} = 0.53$ and $\eta_{II,Q} = \frac{0.40}{0.625} = 0.64$. It is clearly evident from the second law efficiency that the second heat engine has converted 64% of available potential into work whereas the first heat engine is only able to convert 53% of its available potential. Thus, the second law efficiency is a real indicator of the performance of a device. This further can be written in the form of work as

$$\eta_{II} = \frac{\eta_{actual}}{\eta_{rev}} = \frac{W_u/Q}{W_{rev}/Q} = \frac{W_u}{W_{rev}} \text{ (work producing devices)} \tag{7.35}$$

Equation (7.35) shows that second law efficiency is also defined as the ratio of actual useful work output to the maximum possible work output that can be obtained in a reversible manner. The value of second law efficiency of a device cannot be more than 100% as the actual device can at most meet the performance of a reversible device under the given conditions if all the irreversibilities are neglected. Equation (7.35) is a more generic definition of the second law of efficiency and it can be applied for non-cyclic devices such as turbines, etc.

The second law efficiency for work consuming cyclic and non-cyclic devices such as refrigerators, pumps and compressors is defined as the ratio of reversible work input to the useful work input of the device. This is given as

$$\eta_{II} = \frac{W_{rev}}{W_u} \quad \text{(work consuming devices)} \quad (7.36)$$

The second law efficiency for work consuming cyclic devices such as refrigerator and heat pump can also be expressed in terms of their COP as

$$\eta_{II} = \frac{COP}{COP_{rev}} \quad \text{(refrigerator and heat pumps)} \quad (7.37)$$

The definitions of second law efficiency described above, involve work interactions. However, the exergy analysis can also be applied to devices (such as heat exchangers, nozzles, mixing chambers, etc.) that involve no work interactions. Therefore, the second law efficiency is required to be presented in more generic form so that it can be applied for all kinds of thermodynamic devices. A general definition of the second law efficiency includes the ratio of exergy recovered to the exergy supplied during a process. It is given as

$$\eta_{II} = \frac{Exergy\ recovered}{Exergy\ supplied} = 1 - \frac{Exergy\ destruction}{Exergy\ supplied} \quad (7.38)$$

Let us illustrate Eqn. (7.38) with the help of the exergy balance equation (Eqn. 7.33) for an adiabatic turbine. Using Eqn. (7.33), we get $w = (\psi_{in} - \psi_{out}) - T_0 S_{gen}$. The maximum possible work from a turbine will be obtained during a reversible process, hence, $w_{rev} = (\psi_{in} - \psi_{out})$. The actual work done by the turbine can be obtained from energy conservation. The energy conservation for an adiabatic turbine with negligible changes in kinetic and potential energies gives $w_u = h_{in} - h_{out}$. Also, $\psi_{in} - \psi_{out} = (h_{in} - h_{out}) - T_0(s_{in} - s_{out})$ when changes in kinetic and potential energies are neglected. Further, for an adiabatic turbine $s_{out} - s_{in} = s_{gen}$, thus,

$w_u = h_{in} - h_{out} = \psi_{in} - \psi_{out} - T_0 s_{gen}$. Using Eqn. (7.35), we get

$$\eta_{II,turbine} = \frac{h_{in}-h_{out}}{\psi_{in}-\psi_{out}} = \frac{\psi_{in}-\psi_{out}-T_0 s_{gen}}{\psi_{in}-\psi_{out}} = 1 - \frac{T_0 s_{gen}}{\psi_{in}-\psi_{out}} \quad (7.39)$$

Equation (7.39) is consistent with the generic definition of the second law efficiency as given in Eqn. (7.38). For a typical adiabatic compressor with negligible kinetic and potential energies, the reversible work becomes

**Fig. 7.10:** Heat exchange between two streams inside a heat exchanger.

$w_{rev} = (\psi_{out} - \psi_{in})$. The actual work supplied is $w_u = h_{out} - h_{in}$. Also,

$\psi_{out} - \psi_{in} = (h_{out} - h_{in}) - T_0(s_{out} - s_{in})$ and $s_{out} - s_{in} = s_{gen}$. Thus, $w_{rev} = \psi_{out} - \psi_{in} = (h_{out} - h_{in}) - T_0 s_{gen}$. Therefore, second law efficiency is given as

$$\eta_{II,compressor} = \frac{\psi_{out}-\psi_{in}}{h_{out}-h_{in}} = \frac{h_{out}-h_{in}-T_0 s_{gen}}{h_{out}-h_{in}} = 1 - \frac{T_0 s_{gen}}{h_{out}-h_{in}} \quad 7.40)$$

Let us now define the second law efficiency for devices that involve no work interaction. Heat exchanger is one such device that involves transfer of heat from a hot stream to a cold stream without their physical mixing as shown in Fig. 7.10. In a heat exchanger hot stream supplies the exergy to a cold stream by the process of heat transfer. Thus, the exergy supplied is the decrease in exergy of the hot stream and exergy recovered is the increase in exergy of the cold medium, considering cold stream is at higher temperature than the temperature of the environment. Thus, using Eqn. (7.38), we obtain

$$\eta_{II,heat\ exch.} = \frac{\dot{m}_c(\psi_{co}-\psi_{ci})}{\dot{m}_h(\psi_{hi}-\psi_{ho})} \quad (7.41)$$

The exergy balance for a heat exchanger results in $\dot{m}_h(\psi_{hi} - \psi_{ho}) + \dot{m}_c(\psi_{ci} - \psi_{co}) - T_0\dot{S}_{gen} = 0$. Therefore,

$\dot{m}_c(\psi_{co} - \psi_{ci}) = \dot{m}_h(\psi_{hi} - \psi_{ho}) - T_0\dot{S}_{gen}$. Substituting this in Eqn. (7.41), we get

$$\eta_{II,heat\ exch.} = \frac{\dot{m}_h(\psi_{hi}-\psi_{ho})-T_0\dot{S}_{gen}}{\dot{m}_h(\psi_{hi}-\psi_{ho})} = 1 - \frac{T_0\dot{S}_{gen}}{\dot{m}_h(\psi_{hi}-\psi_{ho})} \quad (7.42)$$

In Eqn. (7.42), $\dot{S}_{gen}$ can be obtained by applying entropy balance for the heat exchanger as illustrated in Chapter 6. If the temperature of cold stream in a heat exchanger is less than the environment temperature $T_0$, then exergy of the cold stream decreases due to the heat transfer from hot to cold medium. For such a situation, exergy recovered cannot be equated to change in exergy of the cold stream. Therefore, second law efficiency for such a situation can be defined as the ratio of sum of exergies of the outflowing streams to the sum of exergies of the inflowing streams. For a two stream heat exchanger (Fig. 7.10), this becomes

$$\eta_{II,heat\ exch.} = \frac{\dot{m}_h\psi_{ho}+\dot{m}_c\psi_{co}}{\dot{m}_h\psi_{hi}+\dot{m}_c\psi_{ci}} \ (for\ T_{cold\ stream} < T_0) \quad (7.43)$$

In case the heat exchanger is not adiabatic and involves exchange of heat with surroundings, a control volume including the heat exchanger and immediate surroundings should be considered to estimate the second law efficiency of the combined system.

Mixing chamber is another device that involves no work interaction. Consider an adiabatic mixing chamber that involves mixing of hot and cold streams and produces a mixture as the outflowing stream as shown in Fig. 7.11. For such a device, exergy supplied is the addition of exergies of hot and cold streams at their inlet and the exergy recovered is the exergy of the outflowing mixture stream. Thus, the second law efficiency can be expressed as

$$\eta_{II,mixing\ chamber} = \frac{\dot{m}_3\psi_3}{\dot{m}_1\psi_1+\dot{m}_2\psi_2} \quad (7.44)$$

The exergy balance for mixing chamber results in

$\dot{m}_1\psi_1 + \dot{m}_2\psi_2 - \dot{m}_3\psi_3 - T_0\dot{S}_{gen} = 0.$

$\Rightarrow \dot{m}_3\psi_3 = \dot{m}_1\psi_1 + \dot{m}_2\psi_2 - T_0\dot{S}_{gen}$. Thus, Eqn. (7.44) becomes

$$\eta_{II,mixing\ chamber} = \frac{\dot{m}_1\psi_1+\dot{m}_2\psi_2-T_0\dot{S}_{gen}}{\dot{m}_1\psi_1+\dot{m}_2\psi_2} = 1 - \frac{T_0\dot{S}_{gen}}{\dot{m}_1\psi_1+\dot{m}_2\psi_2} \quad (7.45)$$

Here, $\dot{S}_{gen}$ can be obtained by applying entropy balance for the mixing chamber.

**Fig. 7.11:** Mixing of two streams inside a mixing chamber.

## UNIT SUMMARY

This chapter introduces a new property called exergy that represents the maximum useful work potential of a system at the given state of the system. The understanding of exergy requires splitting of surroundings of a system into two regions, known as immediate surroundings and the environment. Environment is the portion of surroundings that remains uninfluenced by the process occurring in a system. A system that is in thermodynamic equilibrium with its environment is known to be in dead state. The system produces maximum useful work when it changes from a given state to its dead state in a reversible process. This maximum useful work produced by the system is known as its exergy. As exergy is dependent on the initial state of the system and final dead state as the conditions of the environment, therefore, it is called a combined property of system and environment. A process can involve heat, work and mass interactions between system and surroundings, therefore, exergy exchange with each of these energy transfer modes is determined.

Further, the principle of decrease of exergy is demonstrated for an isolated system that states that exergy of an isolated system during a process always decreases or remains constant for the special case of a reversible process. This illustrates that exergy gets destroyed during a process due to irreversibilities. Thus, all the spontaneous processes occurring in nature, result in destruction of exergy. Exergy balance equation is presented for closed system and control volume. Moreover, the implication of the exergy balance equation for different steady flow devices is demonstrated. The second law efficiency is obtained as the true representative parameter to assess the performance of different cyclic and non-cyclic devices. Also, the usage of second law efficiency is demonstrated for heat engines and different steady flow devices.

## Solved Examples

**Example 7.1:** Consider a heat engine receives heat at the rate of 800 $kW$ from a heat source at 1200$K$. It rejects a portion of this heat to a heat sink at 300$K$ and produces a work output of 280 $kW$. Determine the maximum power that a heat engine can produce and the rate of irreversibility associated with the process. Also, estimate the waste heat associated with the process that does not have any work potential if surroundings are at $300K$.

**Solution:** Reversible power, rate of irreversibility and waste heat are to be determined.

**Assumptions:** The operation of the heat engine is to be considered reversible between the given temperature limits to find its maximum power.

**Analysis:** For a reversible heat engine $\dot{W}_{max,rev} = \left(1 - \frac{T_{sink}}{T_{source}}\right)\dot{Q_{in}} = 600\ kW$. The useful work delivered by the heat engine is $\dot{W}_u = 280\ kW$.

The irreversibility associated with the heat engine is equal to the difference between reversible work and the actual useful work delivered by the heat engine. Thus, rate of irreversibility is $\dot{I} = \dot{W}_{max,rev} - \dot{W}_u = 600 - 280 = 320\ kW$.

The heat rejected by the reversible heat engine to heat sink is $\dot{Q}_{out} = \dot{Q_{in}} - \dot{W}_{max,rev}$.

$\Rightarrow \dot{Q}_{out} = 800 - 600 = 200\ kW$.

Since, the heat sink and the surroundings both are at the same temperature, therefore, heat available at the heat sink does not have any potential to produce work. Thus, it can be regarded as the waste heat. Therefore, $\dot{Q}_{waste} = \dot{Q}_{out} = 200\ kW$.

**Discussion:** It must be emphasized that the waste heat (200 $kW$) does not have any potential to produce work. Therefore, it is not the part of irreversibility. However, 320 $kW$ energy is wasted due to irreversibilities associated with the operation of the heat engine.

**Example 7.2:** A 100 $kg$ copper block transfers heat to surroundings at 25°C. During this heat transfer process its temperature changes from 150°C to 25°C. Determine the reversible work and the irreversibility associated with the heat transfer process. Assume the average specific heat of copper is $0.38\ kJ/kg - K$.

**Solution:** The reversible work and irreversibility associated with the heat transfer process through a finite temperature difference are to be estimated.

**Assumptions:** In order to estimate the reversible work, a reversible heat engine should be used to transfer heat from the copper block to the surroundings.

**Analysis:** The reversible produced by the assumed heat engine can be given as

$W_{rev} = \int_{T_i}^{T_f} \left(1 - \frac{T_{surr}}{T_{copper}}\right)\delta Q$. Since the temperature of heat source is variable during the heat transfer process, therefore, integral form is used. To perform this integration, a relationship between heat transfer and temperature is required. This relationship can be obtained using energy balance equation applied for the copper block as

$-\delta Q - \delta W = dE$. The heat transfer is given a negative sign because the copper block is losing the heat. $\delta W = 0$ because it does not involve any boundary work interactions. Further, the kinetic and potential energies of the copper block are zero. Thus, $dE = dU$.

$\Rightarrow -\delta Q = dU = mC_{avg}dT \Rightarrow \delta Q = -mC_{avg}dT$. Substituting this relation in integral equation of reversible work gives

$$W_{rev} = \int_{T_i}^{T_f} \left(1 - \frac{T_{surr}}{T_{copper}}\right)\delta Q = \int_{T_i}^{T_f} \left(1 - \frac{T_{surr}}{T_{copper}}\right)(-mC_{avg}dT)$$
$$= -mC_{avg}(T_f - T_i) + mC_{avg}ln\frac{T_f}{T_i}$$

GIven, $m = 100\ kg, C_{avg} = 0.38\ kJ/kg - K, T_f = 25 + 273 = 298K$ and $T_i = 150 + 273 = 423K$.

Thus $W_{rev} = 100 \times 0.38\left[(423 - 298) - ln\frac{298}{423}\right] = 4755.78\ kJ$.

If a reversible heat engine is connected between the copper block and the surroundings during the heat transfer process, it can produce 4755.78 $kJ$ of maximum useful work.

As in reality the heat transfer process is spontaneous and no arrangement is employed to obtain work. Thus, the useful work obtained by this heat transfer process is zero, i.e. $W_u = 0\ kJ$.

Hence, the irreversibility associated with heat transfer process is

$I = W_{rev} - W_u = 4755.75\ kJ$.

**Discussion:** Hence, the entire work potential is lost in this heat transfer process and the cause of this irreversibility is heat transfer through finite temperature difference.

**Example 7.3:** During the expansion process in a piston-cylinder device, 0.02 $kg$ of steam changes its state from 1.2 $MPa$ and 350°C to 300 $kPa$ and 150°C. The steam supplies 1.5 $kJ$ of heat to surroundings during the expansion process which are at at 100 $kPa$ and 25°C. Determine the change in exergy and exergy destruction during the expansion process. Also, calculate the second law efficiency of the process.

**Solution:** The exergy change, exergy destruction and second law efficiency are to be determined for the expansion of steam inside a closed system.

**Assumptions:** There is no mass transfer across the system boundary because the given system is a closed system.

**Analysis:** For a closed system the change in exergy during a process can be estimated as

$\chi_2 - \chi_1 = m(u_2 - u_1) + m\, P_0(v_2 - v_1) - m\, T_0(s_2 - s_1) + m\left(\frac{V_2^2 - V_1^2}{2}\right) + mg(z_2 - z_1)$. Using the superheated steam table B.3, we get,

At $P_1 = 1.2\ MPa$ and $T_1 = 350$°C:

$v_1 = 0.23455\ m^3/kg, u_1 = 2872.7\ kJ/kg, h_1 = 3154.2\ kJ/kg, s_1 = 7.2139\ kJ/kg - K$.

At $P_2 = 0.3\ MPa$ and $T_1 = 150$°C:

$v_2 = 0.63402\ m^3/kg, u_2 = 2571.0\ kJ/kg, h_2 = 2761.2\ kJ/kg, s_2 = 7.0792\ kJ/kg - K$.

The kinetic energy and potential energies are negligible. Thus,

$$\chi_2 - \chi_1 = 0.02[(2571 - 2872.7) + 100(0.63402 - 0.23455) - 298(7.0792 - 7.2139)]$$

$\Rightarrow \chi_2 - \chi_1 = -4.43\ kJ$.

This means that a reversible process between the given initial and final states can produce 4.43 $kJ$ of useful work.

The exergy destruction during the process is $X_{destruction} = T_0 S_{gen}$, therefore, using entropy balance for system, $S_2 - S_1 = m(s_2 - s_1) = \frac{-Q}{T} + S_{gen}$. Consider that the boundary of the system through which heat is transferred to surroundings is at the temperature of surroundings. Hence, $T = T_0 = 298K$.

$$\Rightarrow X_{destruction} = T_0\left[m(s_2 - s_1) + \frac{Q}{T_0}\right] = 298\left[0.02(7.0792 - 7.2139) + \frac{1.5}{298}\right] = 0.6972\ kJ$$

Therefore, 0.6972 $kJ$ useful work could have been extracted if the process would have occurred in a reversible manner.

The second law efficiency of the process can be defined as

$\eta_{II} = 1 - \frac{T_0 S_{gen}}{X_1 - X_2} = 1 - \frac{0.6972}{4.43} = 84.26\%$.

This means that 84.26% of exergy available is obtained in the form of useful work during the process.

**Alternate approach:** The exergy destruction during the process can also be obtained by applying the exergy balance equation for the closed system. This gives,

$\left(1 - \frac{T_0}{T}\right) Q - [W - P_0(V_2 - V_1)] - T_0 S_{gen} = X_2 - X_1$. Since, the boundary of the system which transfers heat to surroundings is assumed at the temperature of surroundings. Thus, $X_{destruction} = T_0 S_{gen} = X_1 - X_2 - [W - P_0(V_2 - V_1)]$. To find the work done by the system, the first law of thermodynamics gives

$$-Q - W = \Delta E = \Delta U = m(u_2 - u_1) \Rightarrow W = -1.5 + 0.02(2872.7 - 2571) = 4.534\ kJ$$

$\Rightarrow X_{destruction} = 4.43 - [4.534 - 100 \times 0.02(0.63402 - 0.23455)] = 0.6949\ kJ$.

Similarly, second law efficiency can be given as

$\eta_{II} = \frac{W_u}{X_1 - X_2} = \frac{W - P_0(V_2 - V_1)}{X_1 - X_2} = \frac{4.534 - 100 \times 0.02(0.63402 - 0.23455)}{4.43} = 84.31\%$.

Here, two different approaches are explained to determine the exergy destruction and the second law efficiency. The small differences between the values obtained using different approaches are due to the truncation and round-off errors involved with calculations.

**Example 7.4:** An adiabatic steam turbine receives 25 $kg/s$ of steam at 1.2 $MPa$ and 350°C. A portion of steam at the mass flow rate of 8 $kg/s$ is taken out of the turbine at 0.4 $MPa$ and 150°C and the remaining steam is allowed to expand till its exit from the turbine at 30 $kPa$ and 95% quality. Determine the actual work output of the turbine and its second law efficiency when the surroundings are at 100 $kPa$ and 25°C.

**Solution:** The actual work output and the second law efficiency of the turbine are to be calculated.

**Assumptions:** The kinetic and potential energies of all streams are negligible.

**Analysis:** Using the superheated steam table B.3, we get,

At $P_1 = 1.2\ MPa$ and $T_1 = 350$°C:

$h_1 = 3154.2\ kJ/kg,\ s_1 = 7.2139\ kJ/kg - K$.

At $P_2 = 0.4\ MPa$ and $T_1 = 150$°C:

$h_2 = 2752.8\ kJ/kg,\ s_2 = 6.9306\ kJ/kg - K$.

Using saturation pressure steam table B.2, we get,

At $P_3 = 30\ kPa$ and $x_3 = 95\%$:

$h_f = 289.27\ kJ/kg, h_{fg} = 2335.3\ kJ/kg, h_3 = h_f + x_3 h_{fg} = 2507.805\ kJ/kg$

$s_f = 0.9441\ kJ/kg - K, s_{fg} = 6.8234\ kJ/kg - K, s_3 = s_f + x_3 s_{fg} = 7.426\ kJ/kg - K$

Using saturation temperature steam table B.1, we get,

At $P_0 = 100\ kPa$ and $T_0 = 25°C$:

$h_0 \simeq h_{f@25°C} = 104.83\ kJ/kg$ and $s_0 \simeq s_{f@25°C} = 0.3672\ kJ/kg - K$.

The exergy associated with a flowing steam is given as

$\psi = (h - h_0) - T_0(s - s_0) + \frac{V^2}{2} + gz$. Since, $\frac{V^2}{2} = gz \simeq 0$. Thus,

$$\psi_1 = (h_1 - h_0) - T_0(s_1 - s_0) = 1009.05\ kJ/kg$$

$$\psi_2 = (h_2 - h_0) - T_0(s_2 - s_0) = 692.08\ kJ/kg$$

$$\psi_3 = (h_3 - h_0) - T_0(s_3 - s_0) = 299.45\ kJ/kg$$

Consider that actual useful work done by the turbine is $W_u$. Using the steady flow energy equation $-\dot{W}_u + \dot{m}_1 h_1 - \dot{m}_2 h_2 - \dot{m}_3 h_3 = 0$. Using mass conservation $\dot{m}_3 = \dot{m}_1 - \dot{m}_2$.

$\Rightarrow \dot{m}_3 = 17\ kg/s$. Now energy balance gives

$\dot{W}_u = \dot{m}_1 h_1 - \dot{m}_2 h_2 - \dot{m}_3 h_3 = 14199.92\ kW$.

The maximum reversible work associated with the turbine can be obtained using exergy balance which is given as

$\sum_i \left(1 - \frac{T_0}{T_i}\right)\dot{Q}_i - \dot{W} + \sum \dot{m}_{in}\psi_{in} - \sum \dot{m}_{out}\psi_{out} - T_0\dot{S}_{gen} = 0$. For the present situation $\dot{Q}_i = 0$ and to extract maximum reversible work $\dot{S}_{gen} = 0$. Thus,

$\dot{W}_{max,rev} = \dot{m}_1\psi_1 - \dot{m}_2\psi_2 - \dot{m}_3\psi_3 = 14598.96\ kW$.

Therefore, the second law efficiency of the turbine is $\eta_{II} = \frac{\dot{W}_u}{\dot{W}_{max,rev}} = \frac{14199.92}{14598.96} = 97.27\%$.

**Example 7.5:** Steam enters a throttling device at 4 $MPa$ and 450°C and leaves at 2 $MPa$. Determine the exergy destruction during the throttling process when the surroundings are at 25°C.

**Solution:** Exergy destruction is to be calculated during a throttling process.

**Assumptions:** Consider the throttling process is adiabatic and changes in kinetic and potential energies of the stream is negligible.

**Analysis:** Using the superheated steam table B.3, we get,

At $P_1 = 4\ MPa$ and $T_1 = 450°C$:

$h_1 = 3331.2\ kJ/kg, s_1 = 6.9386\ kJ/kg - K$.

Since the process in a throttling device is isenthalpic. Thus, $h_2 \simeq h_1 = 3331.2\ kJ/kg$.

At $P_2 = 2\ MPa$ and $h_2 = 3331.2\ kJ/kg$:

$s_2 = 7.2438\ kJ/kg - K$ (using interpolation).

Using saturation temperature steam table B.1, we get,

At $T_0 = 25°C$:

$h_0 \simeq h_{f@25°C} = 104.83\ kJ/kg$ and $s_0 \simeq s_{f@25°C} = 0.3672\ kJ/kg - K$.

The exergy associated with a flowing steam is given as

$\psi = (h - h_0) - T_0(s - s_0) + \frac{V^2}{2} + gz$. Since, $\frac{V^2}{2} = gz \simeq 0$. Thus,

$$\psi_1 = (h_1 - h_0) - T_0(s_1 - s_0) = 1268.09\ kJ/kg$$

$$\psi_2 = (h_2 - h_0) - T_0(s_2 - s_0) = 1177.14\ kJ/kg$$

The exergy balance for a steady flow device is given as

$$\sum_i \left(1 - \frac{T_0}{T_i}\right) q_i - w + (\psi_{in} - \psi_{out}) - T_0 S_{gen} = 0.$$

Thus, exergy destruction is $\chi_{destruction} = T_0 S_{gen} = \psi_1 - \psi_2 = 90.95\ kJ/kg$.

**Example 7.6:** An engineer is given the task to design a heating system for a room that is to be maintained at 23°C in a surroundings of 5°C. At a first glance, the engineer proposes to use an electric resistance heater which will supply one unit of heat for a unit of electric work it will consume. Assess the effectiveness of the proposed heating arrangement.

**Solution:** The effectiveness of an electric heater to heat a room is to be assessed.

**Assumptions:** As the heater provides one unit of heat (output) to each unit of work it consumes (input), therefore, its first law efficiency is 100%.

**Analysis:** An alternate approach can be considered to supply this electric work input to a heat pump and use it to heat the room. Let us now assess the performance of this reversible heat pump. The COP of the reversible heat pump operating between given temperature limits is

$$\beta'_{rev} = \frac{T_H}{T_H - T_L} = \frac{296}{296 - 278} = 16.44 = \frac{Q_H}{W}.$$

Thus, a reversible heat pump can supply 16.44 units of heat for each unit of electric work it consumes.

Further, the second law efficiency of the electric resistance heater can be estimated as

$$\eta_{II} = \frac{\beta'_{electric-heater}}{\beta'_{rev}} = \frac{1}{16.44} = 6.083\%.$$

As the second law efficiency of the proposed heater is very less, therefore, it is recommended to use a heat pump instead of using an electric resistance heater for heating the room.

# EXERCISES

## Multiple Choice Questions

| | | | | |
|---|---|---|---|---|
| 7.1 | Exergy is a property of | | | |
| | A. system | B. surroundings | C. system and its immediate surroundings | D. system and environment |
| 7.2 | A region that is influenced by a process occurring within a system is called | | | |
| | A. surroundings | B. immediate surroundings | C. environment | D. subsystem |
| 7.3 | In order to extract the maximum useful potential of a system during a process, the system must finally reach to the state of its | | | |
| | A. surroundings | B. immediate surroundings | C. environment | D. subsystem |
| 7.4 | The maximum work potential is obtained from a system when it changes from its given initial state to dead state through an | | | |
| | A. internally reversible process | B. externally reversible process | C. irreversible process | D. internally irreversible process |
| 7.5 | The exergy associated with heat transfer $\delta Q$ from a system at temperature $T$ in the environment having temperature $T_0$ is | | | |
| | A. $\int \left(\frac{T_0}{T}\right)\delta Q$ | B. $\int T_0\delta Q$ | C. $\int \left(1-\frac{T_0}{T}\right)\delta Q$ | D. $\int \left(1-\frac{T}{T_0}\right)\delta Q$ |
| 7.6 | The exergy associated with boundary work in an environment having pressure $\square_0$ is | | | |
| | A. $\int PdV$ | B. $\int P_0dV$ | C. $\int VdP$ | D. $\int (P-P_0)dV$ |
| 7.7 | Spontaneous process always occurs in the direction of<br>i. decreasing exergy | | | |

| | | | | |
|---|---|---|---|---|
| | ii. decreasing energy<br>iii. decreasing entropy<br>iv. increasing entropy | | | |
| | A. i and ii | B. i and iii | C. i and iv | D. ii and iv |
| 7.8 | The irreversibility or exergy destruction $(T_0 S_{gen})$ during a process is always | | | |
| | A. $< 0$ | B. $> 0$ | C. $\geq 0$ | D. $\leq 0$ |
| 7.9 | The second law efficiency of a work consuming device is | | | |
| | A. $\eta_{II} = \frac{w_u}{w_{rev}}$ | B. $\eta_{II} = \frac{w_{rev}}{w_u}$ | C. $\eta_{II} = \frac{\eta_{actual}}{\eta_{rev}}$ | D. $\eta_{II} = \frac{w_{rev}}{w_{isentropic}}$ |
| 7.10 | The exergy of a fluid stream passing through an adiabatic, work free, single inlet, single outlet, steady flow device during an irreversible process | | | |
| | A. increases | B. remains constant | C. decreases | D. becomes infinite |
| 7.11 | The exergy of a system having temperature less than environment temperature ____________upon addition of heat to the system. | | | |
| | A. increases | B. decreases | C. remains unchanged | D. becomes infinite |
| 7.12 | The transfer of heat through a finite temperature difference results in | | | |
| | A. energy destruction | B. heat destruction | C. entropy destruction | D. exergy destruction |

**Answers of Multiple Choice Questions**

**7.1 D, 7.2 B, 7.3 C, 7.4 A, 7.5 C, 7.6 D, 7.7 C, 7.8 C, 7.9 B, 7.10 C, 7.11 B, 7.12 D**

**Short and Long Answer Type Questions**

7.1 What should be the final state of a system during a process to produce maximum useful work?

7.2 What is useful work? Explain. How is it different from reversible work?

7.3 Differentiate between actual work and useful work.

7.4 Can the exergy of a system change during a process if no work is involved?

7.5 Two geothermal sites are located in different regions having the same energy content. Comment on their exergy content.

7.6 Does the amount of internal energy stored in a system equal to the amount of exergy associated with it? Explain.

7.7 A flowing stream includes exergy due to its thermodynamic state, kinetic energy and potential energy. Can all these components of exergy be negative?

7.8 Does the exergy equation for a system show any dependence on its energy and entropy equations? Elucidate.

7.9 What happens to the exergy of a system having temperature less than the temperature of the environment when heat is added to it?

7.10 Explain the differences between first and second law efficiencies.

7.11 Can a refrigerator having higher COP always have higher second law efficiency? Explain.

7.12 Define the second law efficiency for a nozzle.

7.13 Do a process with no entropy generation is always reversible? Elaborate.

7.14 Can the second law efficiency of a system be greater than its first law efficiency? Explain.

7.15 Can a process with no entropy generation have 100% second law efficiency?

**Numerical Problems**

7.1 A heat engine gets 2 $kW$ heat from a source at 1300$K$ and 5 $kW$ heat from another source at 850$K$. It operates in a reversible manner and rejects heat to a heat sink at 400$K$. Determine the power produced by the heat engine. Also, estimate the exergy associated with the given heat sources if the environment temperature is 300$K$.

7.2 The freezer of a household refrigerator is maintained at -10°C and the cold space beneath the freezer is at 10°C. The refrigerator removes heat from the freezer and the cold space at the rate of 2 $kW$ and 1.5 $kW$, respectively. Determine the minimum power input to the refrigerator if the environment is at 25°C.

7.3 Atmospheric air at 100 $kPa$ and 27°C is passed through a compressor which gives pressurized air at 670 $kPa$ and 350°C. Determine the minimum work input for unit mass of air to the compressor.

7.4 0.02 $kg$ air at its dead state of 100 $kPa$ pressure and 27°C temperature is contained in a piston cylinder system. It is compressed to 450 $kPa$ and 170°C and the useful work involved in the process is 15 $kJ$. Determine the exergy of the compressed air at its final state and the minimum work required for the compression process. Also calculate the second law efficiency of the compression process.

7.5 A rigid tank of 25 $L$ volume contains superheated steam at 220 $kPa$ and 250°C. The steam supplies heat to a room maintained at 23°C till its temperature drops to 130°C. Determine the amount of heat transferred to the room. Now, consider a heat engine that operates between the rigid tank and the surroundings at 8°C which supplies its work output to operate a heat pump between the surroundings and the room. Calculate the maximum amount of heat supplied by the heat pump to the room.

7.6 A 50 $L$ vessel is divided into two halves using a partition such that one portion contains superheated vapour at 250 $kPa$ and 400°C and the other portion is evacuated. The partition ruptures and the steam fills the entire volume with a final pressure of 120 $kPa$. Determine the exergy destruction during the process if the environment is at 100 $kPa$ and 27°C.

7.7 A well insulated tank of 120 $L$ volume contains water at 27°C. An iron block of specific heat 0.44 $kJ/kg-K$ is dropped into the water. The water is stirred to have uniform temperature and reaches to thermal equilibrium with iron at 35°C. 500 $kJ$ work is done on the water by the stirrer. Determine the mass of the iron block. Also calculate the exergy destruction during the process if the surroundings temperature is 22°C.

7.8 Air flows through a steady flow nozzle where its velocity is increased from 45 $m/s$ to 280 $m/s$. The pressure and temperature of air at the inlet of nozzle is 260 $kPa$ and 95°C, respectively. The pressure of air at nozzle exit is 100 $kPa$. Determine the temperature of air at nozzle exit if a heat loss of 5 $kJ/kg$ takes place from the nozzle. Also calculate the exergy destruction in the process if the surroundings temperature is 22°C.

7.9 Steam expands in a 2.5 $MW$ adiabatic turbine from 5 $MPa$, 550°C to 80 $kPa$ and 120°C. The changes in kinetic energy of steam across the turbine can be neglected. Determine the maximum power output and the second law efficiency of the turbine.

7.10 Compressed water enters a mixing chamber at the rate of 2 $kg/s$ and at a pressure and temperature of 250 $kPa$ and 25°C. It is heated using superheated steam entering mixing chamber at 250 $kPa$ and 350°C. The mixing chamber loses heat at the rate of 15 $kJ/s$ to surroundings at 25°C during the mixing process. The mixture leaves the mixing chamber at 250 $kPa$ and 80°C. Determine the required mass flow rate of the steam, exergy destruction during the mixing process and the second law efficiency of the mixing chamber.

7.11 Air flowing through a heat exchanger at the rate of 0.1 $kg/s$ changes its temperature from 1300°C to 300°C at a constant pressure of 200 $kPa$. The heat exchanger is connected to a heat engine that receives the heat from the air being cooled in the heat exchanger. Determine the amount of heat delivered to the heat engine and the maximum work that the heat engine can produce in a surroundings of 100 $kPa$ and 25°C.

7.12 Air enters the compressor of a turbocharged automotive engine at 100 $kPa$ and 25°C and leaves at 300 $kPa$. It is then cooled in an intercooler by 65°C before it is allowed to enter the engine. Determine the temperature of air entering the engine if the isentropic efficiency of the compressor is 80%. Also calculate the irreversibility associated with the combined compression and intercooling process.

7.13 Hot water is extracted from a geothermal source at the rate of 25 $kg/s$ at 650 $kPa$ and 160°C. Then it is throttled to a pressure of 220 $kPa$ in a flash chamber. The saturated liquid and saturated vapour are taken out separately from the flash chamber. Determine the rate of irreversibility associated with the process.

7.14 A pressure cooker of 5 $L$ volume operates at a pressure of 200 $kPa$. The cooker initially contains equal volumes of saturated liquid and saturated vapour at its operating pressure. Then it is placed on an electric heater of 1 $kW$ capacity that is kept on for 10 minutes. Determine the amount of water in the cooker at the end of heating. Also calculate the irreversibility associated with the entire process if the surroundings are at 100 $kPa$ and 25°C.

7.15 A room is maintained at 35°C using a heat pump that consumes 4 $kW$ power. The room loses 12 $kW$ of heat to the colder surroundings at 12°C. Determine the COP of the heat pump. Also calculate the rate of exergy destruction within the heat pump and rate of exergy destruction associated with the heat loss process.

## PRACTICAL

**Activity 1:** Find the maximum work potential of different energy conversion devices that you see in your surroundings. Compare it with their actual useful work, estimate the irreversibility and the second law efficiency.

## REFERENCES AND SUGGESTED READINGS

1. Bejan, A., 2016. Advanced engineering thermodynamics. John Wiley & Sons.
2. Moran, M.J., Shapiro, H.N., Boettner, D.D. and Bailey, M.B., 2010. *Fundamentals of engineering thermodynamics*. John Wiley & Sons.
3. Cengel, Y.A., Boles, M.A. and Kanoğlu, M., 2011. *Thermodynamics: an engineering approach*. New York: McGraw-hill.
4. Dewulf, J., Van Langenhove, H., Muys, B., Bruers, S., Bakshi, B.R., Grubb, G.F., Paulus, D.M. and Sciubba, E., 2008. Exergy: its potential and limitations in environmental science and technology. *Environmental Science & Technology*, *42*(7), pp.2221-2232.
5. Szargut, J., 2005. *Exergy method: technical and ecological applications* (Vol. 18). WIT press.
6. Dincer, I. and Rosen, M.A., 2012. *Exergy: energy, environment and sustainable development*. Newnes.

# 8 Basic Power and Refrigeration Cycles

## UNIT SPECIFICS

*Through this unit we have discussed the following aspects:*

- *Introduction of realistic power cycles for enabling students for estimating power production;*
- *Analysis of Rankine cycle based steam power plants;*
- *Analysis of Brayton cycle based gas turbine powerplants;*
- *Analysis of Vapour compression refrigeration cycles;*
- *Introduction to Otto and Diesel cycles for representing the power production in reciprocating engines.*

*After completing the discussion on the fundamentals of thermodynamics in the initial chapters of this book, this chapter applies those concepts to the thermodynamic cycle based representation of various real world systems such as power plants, refrigerators, automobile engines, etc.*

## RATIONALE

*Contents of this chapter are applied in nature and it will help the students to appreciate the application of thermodynamic concepts in estimating the operating efficiencies of various real world systems. For providing continuous power, the systems need to ensure repeatable cyclic operation. For the analysis of various cycles representing power generating systems, individual processes are represented with best matching basic thermodynamic processes and the entire cycle is analysed through the combined analysis of various constituent processes.*

## PRE-REQUISITES

*Thermodynamic processes, First law of thermodynamics*

## UNIT OUTCOMES

*List of outcomes of this unit is as follows:*

*U8-O1: Appreciation of requirement of representing the systems as thermodynamic cycles*

*U8-O2: Developing capability to analyse steam power plants by using Rankine cycle*

*U8-O3: Developing capability to analyse gas turbine power plants by using Brayton cycle*

*U8-O4: Developing capability to analyse vapor compression refrigeration cycle based refrigeration systems*

*U8-O5: Developing capability to analyse reciprocating engines by using Otto and diesel cycles*

| Unit-8 Outcomes | EXPECTED MAPPING WITH COURSE OUTCOMES (1- Weak Correlation; 2- Medium correlation; 3- Strong Correlation) | | | | | |
|---|---|---|---|---|---|---|
| | CO-1 | CO-2 | CO-3 | CO-4 | CO-5 | CO-6 |
| U8-O1 | 2 | 2 | 1 | 2 | 2 | 3 |
| U8-O2 | 2 | 2 | 1 | 2 | 2 | 3 |
| U8-O3 | 2 | 2 | 1 | 1 | 2 | 3 |
| U8-O4 | 2 | 2 | 1 | 1 | 2 | 3 |
| U8-O5 | 2 | 2 | 1 | 1 | 2 | 3 |

## 8.1 Introduction to Power Cycles

The term power cycles refers to a series of energy conversion devices that produce power. For instance, a steam thermal power plant provides mechanical power to rotate an electric generator that produces electricity. A jet propulsion engine produces sufficient thrust to propel the aircraft. A diesel locomotive engine produces power to pull a train. A car or bus engine develops mechanical power to make its wheel rotate. All these examples include a list of engineering devices that produce mechanical power. Considering mechanical power as the common outcome in all devices, the human body can also be considered a power producing device that develops sufficient mechanical power to walk, lift different objects and carry out different routine activities that need mechanical power. It must be emphasized that the human body is a superior power producing device that provides instant power with precise variation as per the requirement of the task. For example, to lift a cup of tea and a bucket of water, hand automatically and instantly meters the efforts required. However, to perform the same task engineering devices require complicated control systems and still cannot be as efficient as the human body. We leave it to the readers to understand the principle of power generation and mechanism of actuation to precisely and instantly provide the required power by the human body. Here, we will mainly attempt to introduce the basic power cycles that convert heat into mechanical power and the methods to analyze their performance.

A steam power plant uses water as the working fluid which passes through a series of steady flow devices in a cyclic loop and completes a **closed cycle** (Fig. 8.1-a). Contrary, a gas turbine engine (Fig. 8.1-b) uses gaseous working fluid and takes atmospheric air at the inlet and discharges products of combustion to the ambient at its outlet. In such a power producing device the inlet and exit states of the working fluid are different and it always takes ambient air as intake. Such devices are often called **open cycles**. However, when we analyze a power producing device it is always convenient to idealize it to a closed cycle that closely approximates it and then present the thermodynamic analysis for the same. The operation of actual power cycles includes a lot of complexities such as friction, chemical reactions, heat transfer through finite temperature difference and the speed of expansion and compression processes such that equilibrium cannot be established during the intermediate states. To make the thermodynamic analysis of actual cycles feasible, the irreversibilities within them are usually avoided which makes all the processes that form the cycle internally reversible. A cycle that is represented by all internally reversible processes is known as an **ideal cycle** and it can be analyzed thermodynamically. An actual cycle with its corresponding ideal cycle is shown on $P - v$ diagram in Fig. 8.2. The analysis of ideal cycles helps the engineers to know the influence of different parameters on the performance of the cycle without much complications. In reality, the efficiency of actual cycles will always be less than the ideal cycles, however, the general characteristics of their performance remains the same. For example, the analysis of the ideal steam power plant cycle shows that its efficiency increases with increase of boiler pressure. This effect is typically observed in actual operation of steam power plants as well.

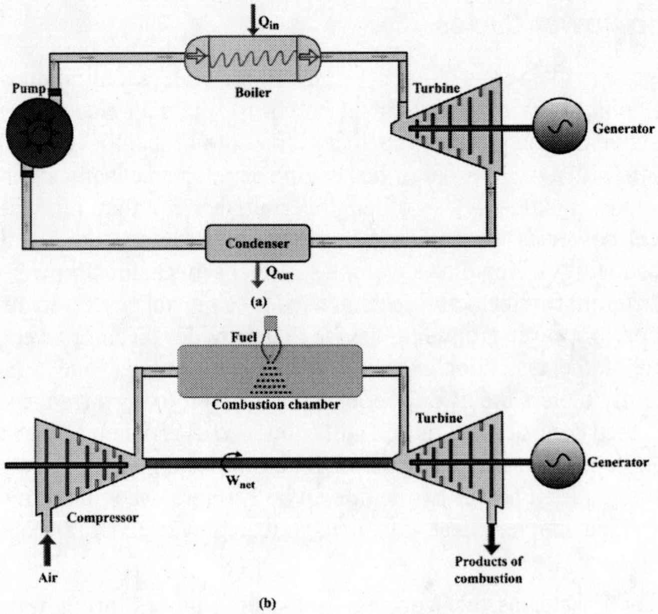


**Fig. 8.1:** Representation of: (a) steam power plant as a closed cycle and (b) a gas turbine engine as an open cycle.

In this chapter, we will mainly discuss the power cycles that fundamentally utilizes conversion of heat into work. We know from Chapter 5 that such a device is typically called a heat engine and the idealized heat engine that includes all totally reversible processes is Carnot cycle. The Carnot cycle has the maximum efficiency among all heat engines operating between the same temperature limits. A question may arise that why we are studying different ideal cycles when the most idealized version, the Carnot cycle, is already available for a heat engine. The answer to this question is that achieving a totally reversible cycle in practice is not feasible, that's why a practical work producing device cannot be made to completely follow the Carnot processes. An actual work producing device can be made to work in different ways and some of which are far from the Carnot processes. Thus, a more closer internally reversible version of these real work producing devices form ideal cycles that can have significant differences from the Carnot cycle. Thus, it is realistic to approximate each work producing device to its commensurate idealized version and perform its thermodynamic analysis. Note that the efficiency of the Carnot cycle is higher than the efficiency of an ideal cycle that is internally reversible and operating under the similar conditions. The ideal cycle takes care of external irreversibilities. Moreover, the efficiency of the actual cycle is less than the efficiency of the ideal cycle. An ideal cycle that is formed by all internally reversible processes is based on the following main assumptions:

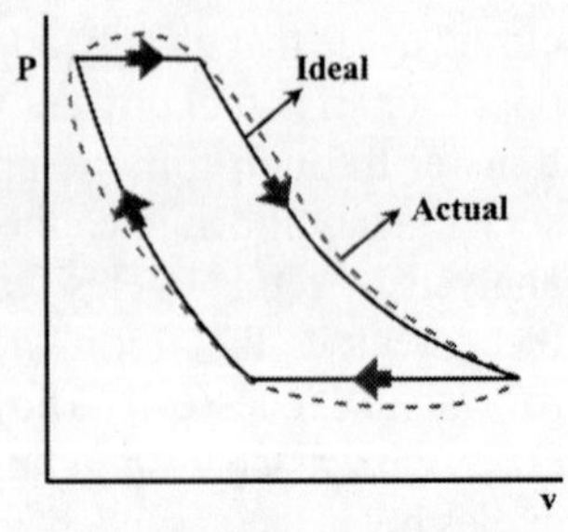


**Fig. 8.2:** Representation of an ideal cycle for an actual cyclic process.

- The friction is negligible for the cycle. Thus, there is no pressure drop due to flow of working fluid through different components and pipes connecting the subsequent components.
- The expansion and compression processes are quasi-equilibrium.
- The heat transfer through the pipes connecting different components is negligible. This means that pipes are well insulated and the thermodynamic state of the working fluid at the exit of one component is equal to the thermodynamic state of the working fluid at the intake of the next adjacent component in the cycle.

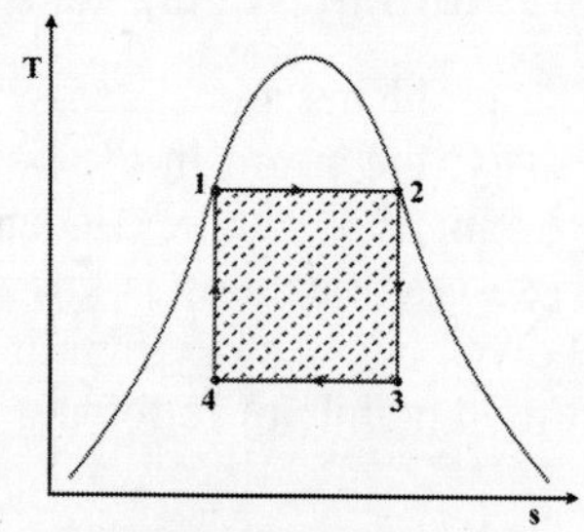

**Fig. 8.3:** Representation of a Carnot vapour cycle on $T-s$ diagram.

## 8.2 Limitations of Carnot Vapour Cycle

The vapour power cycles mainly operate in the two-phase regime and involve conversion of liquid to vapour and vice-versa in different components that form the cycle. Water is the most common working fluid for most vapour power cycles. We will now first consider the Carnot vapour cycle as the ideal cycle for steam power plants and then see why it cannot be used in reality. Consider a Carnot vapour cycle involving an isotherm heat addition process (1-2), isentropic expansion in turbine (2-3), isothermal heat rejection (3-4) and isentropic compression (4-1) as shown in Fig. 8.3.

It is to note that having an isothermal process of heat transfer is not challenging for working fluids that involve phase change during the heat addition or removal. If we maintain a boiler at the constant pressure conditions during the heat addition process 1-2, isothermal conditions will automatically prevail within the liquid-vapour dome which maintains the process isothermal as well. Whereas the addition of heat in the single phase regime above the critical point of the working fluid again brings the difficulty of maintaining isothermal conditions. The expansion of steam in a turbine can also be maintained with sufficient accuracy as an isentropic process. However, the expansion process results in formation of a significant amount of liquid within the turbine in process 2-3. The liquid formation in the form of high speed drops impacting the surface of the turbine blades causes erosion. As this becomes a major source of wear in the turbine, thus a turbine cannot operate with steam quality less than about 88% at its exit in steam power plants. This problem can be avoided for working fluids having steeper saturated vapour curve. The process 4-1 involves the compression of saturated liquid-vapour mixture to saturated liquid state. Practically, separate devices are available for compressing pure liquids and pure gaseous substances. However, it is not possible for a compressor to compress the mixture of liquid and vapour. Another issue with process 4-1 is that precisely controlling state 4 is not possible in practical scenarios within a condenser due to the complex interfacial dynamics of the liquid-vapour mixture involved during the heat transfer process. Thus, the Carnot cycle cannot be considered an ideal cycle for vapour power cycles.

## 8.3 Rankine Cycle as an Ideal Cycle for Steam Power Plants

The problems associated with the Carnot cycle can be mainly avoided by doing two modifications. Firstly, the steam in the boiler can be superheated before it is expanded in a turbine. Secondly, the working fluid can be condensed in the turbine till it becomes saturated liquid. An ideal cycle including these modifications is known as **Rankine cycle**. It consists of all internally reversible processes as shown on $T - s$ diagram in Fig. 8.4. The Rankine cycle is an ideal cycle for vapour power plants and consist mainly of following four processes:

1-2: Isentropic compression in a pump

2-3: Isobaric heat addition in a boiler

3-4: Isentropic expansion in a turbine

4-1: Isobaric heat rejection in a condenser

The working fluid in a steam power plant is typically water. Water enters the pump as saturated liquid and is isentropically compressed from condenser pressure to boiler pressure. The temperature of water increases slightly during this compression process due to a small decrease in its specific volume. This isentropic compression of water is shown on $T - s$ diagram using process 1-2. States 1 and 2 are very close for process 1-2 that is why a zoomed in inset view is shown in Fig. 8.4 to clearly represent process 1-2.

Water at state 2 is passed through a boiler where it is heated under the constant pressure condition. Water is in subcooled regime at state 2, therefore, first it goes through a sensible heating process till it reaches the saturation liquid state. Further heating at this point results in a change of its phase from saturation liquid to saturation vapour state and hence a latent heating process. The saturated vapour is further heated to reach the superheated state at the exit of the boiler. The process 2-3 on $T-s$ diagram in Fig. 8.4 depicts the constant pressure heating inside the boiler. The water receives the necessary heat inside the boiler from the burning of fuels in thermal power plants whereas this heat is supplied by nuclear fission reactions in a nuclear reactor. The boiler is typically known as a steam generator as well.

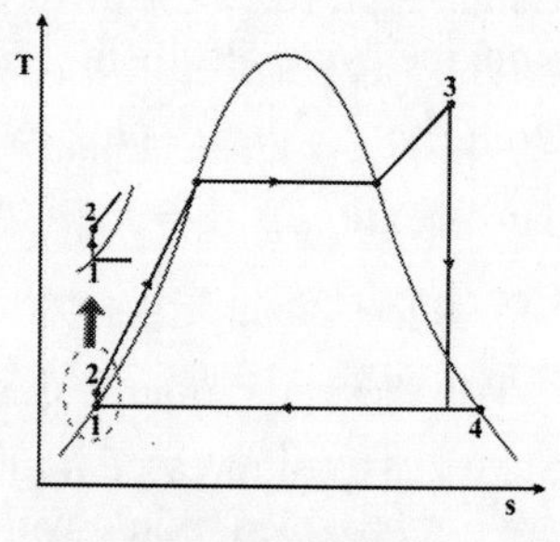

**Fig. 8.4:** Representation of an ideal Rankine cycle on $T-s$ diagram.

The superheated steam exiting boiler is passed through the turbine where it is isentropically expanded and rotates the turbine shaft. This turbine shaft is further connected to an electric generator which supplies electricity as an output from a steam power plant. The expansion of superheated steam in the turbine helps in maintaining the required quality at its exit. Process 3-4 represents the isentropic expansion in a turbine on $T-s$ diagram.

Steam at state 4 is passed through a condenser where it is cooled at constant pressure conditions and is obtained as a saturated liquid at its outlet to complete the cycle. Typically atmospheric water is used as a cooling medium in a condenser. However, in water scarce areas, air cooled condensers can be used. Usually an air cooled condenser has a considerably larger size than a water cooled condenser. The four processes that form the ideal Rankine cycle are performed in different steady flow devices as schematically shown in Fig. 8.5. It is to be noted that the changes in kinetic and potential energies across each steady flow device are very small as compared to the changes in enthalpy of the working fluid. Therefore, these can be neglected during the thermodynamic analysis.

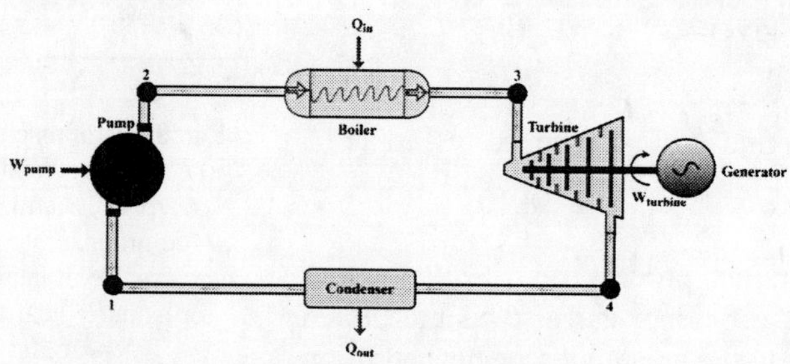

**Fig. 8.5:** Schematic of an ideal Rankine cycle based steam power plant.

For a steady flow device with single inlet and single outlet streams, the energy conservation can be written as

$$q_{in} - q_{out} - (w_{out} - w_{in}) + h_{in} - h_{out} = 0 \quad (8.1)$$

Using Eqn. (8.1) for different steady flow devices involved in an ideal Rankine cycle results in

Boiler: $q_{in} = h_{out} - h_{in} = h_3 - h_2$ (8.2)

Turbine: $w_{turbine} = h_{in} - h_{out} = h_3 - h_4$ (8.3)

Condenser: $q_{out} = h_{in} - h_{out} = h_4 - h_1$ (8.4)

Pump: $w_{pump} = h_{out} - h_{in} = h_2 - h_1$ (8.5)

State 2 is a compressed liquid state. When the compressed property tables are not provided, pump work can also be expressed as

$w_{pump} = -\int_1^2 vdP$. The specific volume of water can be considered almost constant due to the incompressible nature of water. Thus, $v_2 \simeq v_1 = v_{f@P_1}$. This gives

$$w_{pump} = -v_{f@P_1}\int_1^2 dP = -v_{f@P_1}(P_2 - P_1) \quad (8.6)$$

Here, a negative sign represents that work is done on the system. If we compare the magnitude of pump work using Eqns. (8.5) and (8.6), we get

$$h_2 = h_1 + v_{f@P_1}(P_2 - P_1) \quad (8.7)$$

Equation (8.7) is very useful in absence of compressed steam tables. The net work output produced by the ideal Rankine cycle can be expressed as

$$w_{net} = w_{turbine} - w_{pump} \quad (8.8)$$

It must be noted that the work produced by the turbine is considerably larger than the work input to the pump operating between the same pressure limits. This is because the working fluid in the turbine is in vapour state having high specific volume whereas the specific volume of working fluid in the pump is very small being it in liquid state. As the work done in a steady flow device is proportional to specific volume, therefore, pump work can even be neglected in comparison to the turbine work. Since for a cyclic process, $\oint \delta q = \oint \delta w$, thus, $q_{net} = w_{net}$. Therefore, Eqn. (8.8) can also be given as

$$w_{net} = q_{net} = q_{in} - q_{out} \quad (8.9)$$

The thermal efficiency of the ideal Rankine cycle can be expressed as

$$\eta_{Rankine} = \frac{w_{net}}{q_{in}} = 1 - \frac{q_{out}}{q_{im}} \quad (8.10)$$

As the area under an internally reversible process on $T - s$ diagram represents the heat transfer, thus thermal efficiency can also be interpreted as the ratio of area enclosed by the cycle to the area under the heat addition process on $T - s$ diagram.

The thermal efficiency of an ideal Rankine cycle can also be compared with efficiency of a Carnot cycle. To do so, we need to first know the concept of **mean temperature of addition** during a process. Consider the heat addition process of an ideal Rankine cycle as shown in Fig. 8.6-a.

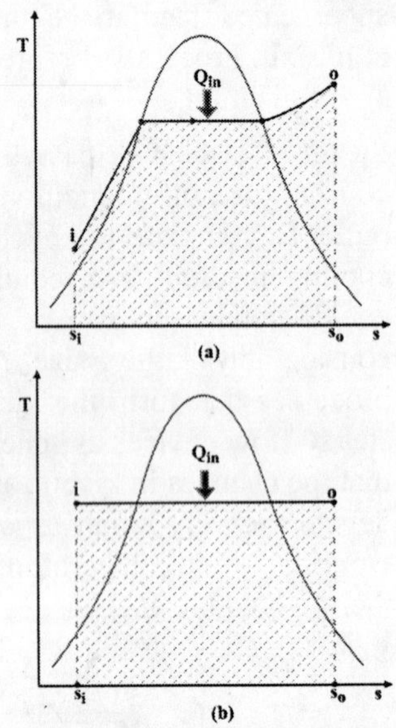


**Fig. 8.6:** Representation of (a) an isobaric heat addition process and (b) commensurate mean temperature of heat addition for isobaric heat addition process.

The area under this process between states $i$ and $o$ represents the amount of heat added $(q_{in})$. Now if we fix the end states and draw a horizontal line on $T-s$ diagram such that the area enclosed is equal to the amount of heat added during the actual process (Fig. 8.6-b), then the temperature indicated by the horizontal process line in Fig. 8.6-b is known as mean temperature of heat addition. Thus, it is the hypothetical average temperature during a heat transfer process. It can be expressed mathematically as

$$q_{in} = T_m(s_o - s_i) \Rightarrow T_m = \frac{q_{in}}{s_o - s_i} \quad (8.11)$$

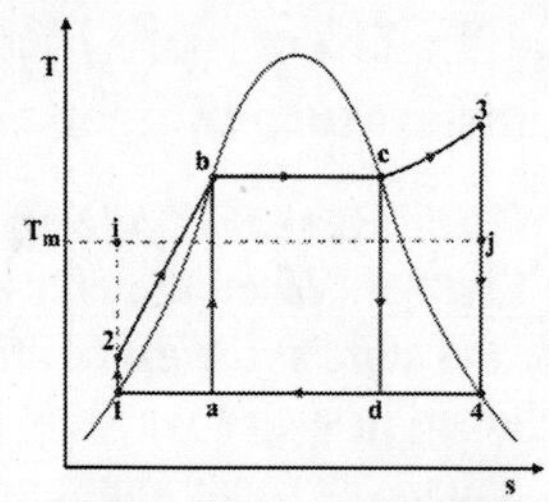


**Fig. 8.7:** Comparison between a Carnot vapour cycle and an ideal Rankine cycle.

Now let us use this concept to compare the thermal efficiency of a Carnot cycle and an Ideal Rankine cycle. An ideal Rankine cycle is shown on $T-s$ diagram in Fig. 8.7 by the process loop 1-2-3-4-1. A commensurate Carnot cycle for on the same $T-s$ diagram is shown by processes $a-b-c-d-a$. The mean temperature of heat addition for the ideal Rankine cycle is shown by dashed line $i-j$ in Fig. 8.7. The mean temperature of heat addition for Rankine cycle is less than the temperature of heat addition for the commensurate Carnot cycle whereas both the cycles have same temperature of heat rejection. The mean temperature of addition for Rankine cycle has decreased due to the low temperatures during the heating of liquid in the subcooled region. As the efficiency of the Carnot cycle is expressed as $\eta_{Carnot} = 1 - \frac{T_{heat\ rejection}}{T_{heat\ addition}}$, it can be written for ideal Rankine cycle using the concept of mean temperature of heat addition as

$$\eta_{Rankine} = 1 - \frac{T_{condenser}}{T_m} \quad (8.12)$$

This clearly explains that efficiency of an ideal Rankine cycle is less than the efficiency of corresponding Carnot cycle operating between similar conditions.

## 8.4 Factors Influencing the Efficiency of Rankine Cycle

Steam power plants are the major source of electricity production across the globe. Due to this reason, a slight improvement in their efficiency can result in huge savings in cost of fuels and controlling the environmental pollution. The fundamental idea to improve the efficiency of an ideal Rankine cycle is to devise ways to increase its mean temperature of heat addition and decrease the temperature of heat rejection. Doing so poses several practical constraints which also should be tackled while modifying the cycle parameters. We demonstrate below the effect of different cycle parameters on its thermal efficiency and associated practical limitations.

### 8.4.1 Effect of increasing boiler pressure (or mean temperature of heat addition)

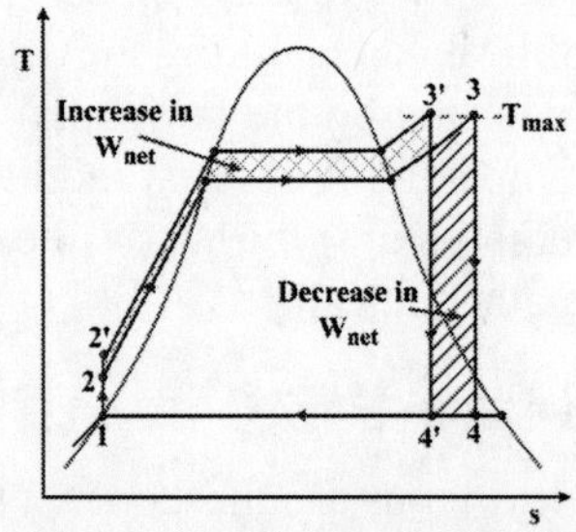


Fig. 8.8: Effect of increasing boiler pressure on an ideal Rankine cycle.

An ideal Rankine cycle operating between pressures $P_1$ and $P_2$ is shown in Fig. 8.8. When the pressure of the boiler is increased from $P_2$ to $P_{2'} > P_2$, the state at the exit of the boiler changes from 3 to 3′. Note that 3′ is selected in such a way that its temperature is equal to temperature of state 3. This is because the highest temperature of the steam in the boiler is limited by the metallurgical limits of its materials and the materials of the turbine. The temperatures close to 600°C are presently tolerated by the turbine and the recent advancements in materials technology can further increase this value. The new cycle $1 - 2' - 3' - 4'$ obtained by increasing boiler pressure results in a decrease of some amount of work due to shift of expansion process in turbine toward left (Fig. 8.8). However, it also results in an increase in some work enclosed between the earlier boiler pressure and modified boiler pressure. The overall effect is an increase in net work output which increases the efficiency of the cycle. The main reason for increase in efficiency of the cycle is increase in mean temperature of heat addition upon increasing the boiler pressure. The shifting of expansion process towards the left also results in increasing moisture content at the exit of the turbine which decreases its efficiency and leads to erosion of its surface. Thus increasing boiler pressure is good for efficiency of the cycle, however, detrimental for the operation of the turbine. Reheat Rankine cycle is a modified version of ideal Rankine cycle that makes it possible to add heat at higher boiler pressure, hence, at high mean temperature of heat addition without causing any harm to the turbine. The readers are directed to refer to books on applied thermodynamics to read details on reheat Rankine cycle as it is not in the scope of the present book.

### 8.4.2 Effect of decreasing condenser pressure (or temperature of heat rejection)

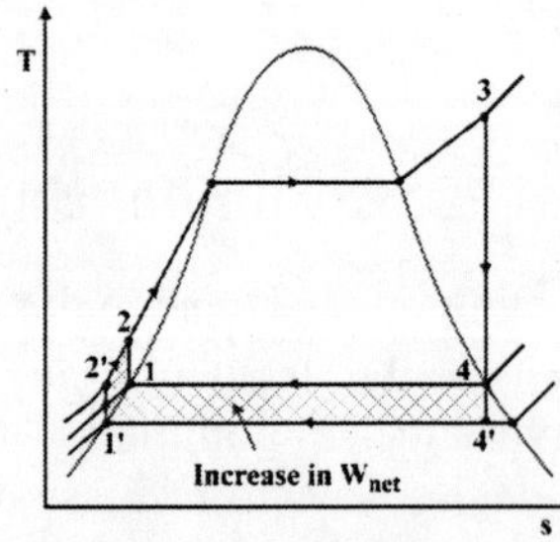


Fig. 8.9: Effect of decreasing condenser pressure on an ideal Rankine cycle.

The working fluid in a condenser mostly stays in the saturated liquid-vapour dome. Thus, decreasing condenser pressure also results in decrease of its temperature. Decreasing condenser pressure from $P_4$ to $P_{4'} < P_4$ results in an increase in net work output as shown in Fig. 8.9. Though it also increases the amount of heat added as indicated by the increase in area under the heat addition process from state 2′ to state 2. However, the increase in net work output is more in comparison to increase of heat addition, thus the efficiency of the cycle increases. This can also be seen in terms of considerable decrease in temperature of heat rejection which in turn increases the cycle efficiency. Moreover, decreasing condenser pressure also decreases the quality of steam at turbine exit, hence, it is not good for its operation. Since the working fluid in the condenser transfers heat to water or air at ambient conditions. Thus its temperature is limited by the temperature of the surrounding atmosphere. Since, the saturation pressure corresponding to atmospheric temperature is less than the atmospheric pressure for water, therefore, the condenser of a typical steam power plant is operated below the atmospheric pressure. It can be easily operated due to the closed loop nature of the cycle, however, it poses a difficulty of air leakages into the condenser.

### 8.4.3 Increasing mean temperature of heat addition by superheating steam

Increasing the temperature of the steam at the boiler exit even at fixed boiler pressure also increases the mean temperature of heat addition. Increasing the temperature of working fluid from state 3 to 3′ shifts the expansion process towards right on $T - s$ diagram as shown in Fig. 8.10 and increases the net work output of the cycle. Though heat addition indicated by area under process $3 - 3'$ also increases, overall effect produces more increase in net work output than increase of heat addition, hence efficiency of the cycle increases. The increase in mean temperature of heat addition upon increasing the temperature of superheat is the reason for increase in efficiency of the cycle. Superheating steam to higher temperature is also beneficial in terms of improvement of steam quality at the exit of the turbine, however, its highest value is limited by the metallurgical limits. Moreover, a very high value of superheat temperature can also lead to turbine exit at the superheated state which in turn will increase the average temperature of heat rejection and hence decrease in cycle efficiency.

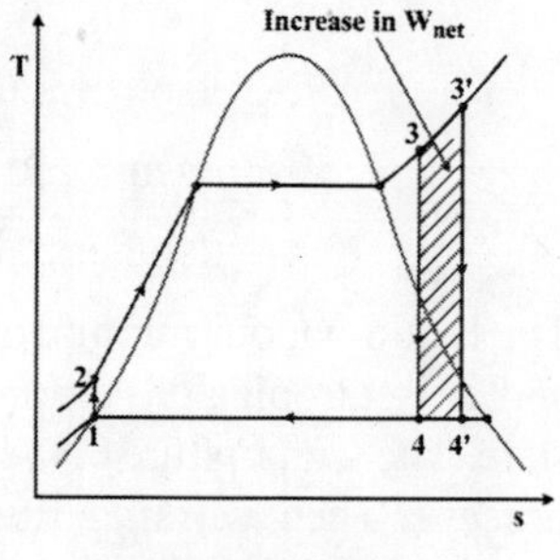


**Fig. 8.10:** Effect of superheating on an ideal Rankine cycle.

It must be noted from the above discussion that variation in each cycle parameter has some positive and some negative effects on the cycle. Therefore, using this fundamental understanding the combination of different parameters is used to improve the cycle efficiency in practical scenarios. The reheat and regenerative Rankine cycles are used in practice for power production using steam power plants and have better efficiency than the ideal Rankine cycle. The idea of the present book is to introduce the readers to practical applications of the fundamentals of thermodynamics. Thus, we discussed only the basic power cycles in this book and readers are suggested to refer to the books of applied thermodynamics to know more about advanced power cycles.

Scan Me

For more details on reheat and regenerative Rankine cycles.

## 8.5 Ideal Vapour Compression Refrigeration Cycle

We established in Chapter 5 that the reverse of a heat engine which facilitates the transfer of heat from low temperature heat source to a high temperature heat sink by consuming work is known as a refrigerator or a heat pump. The idea of using a refrigerator is to maintain the temperature of a space below the atmospheric temperature.

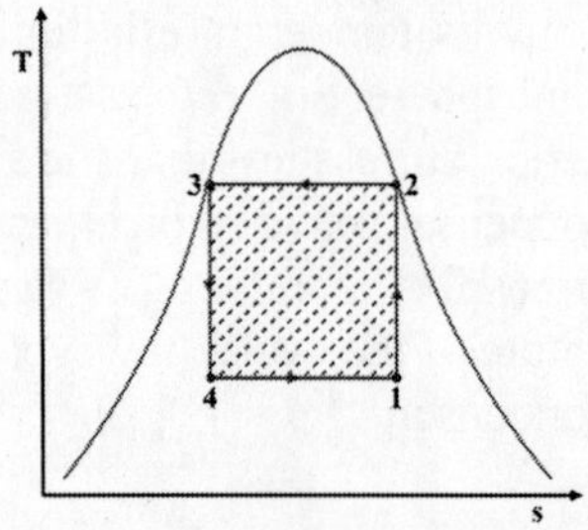


**Fig. 8.11:** Representation of a reverse Carnot refrigeration cycle on $T - s$ diagram.

This is usually achieved by transferring heat from the desired space to the atmosphere. A Carnot heat engine that consists of all reversible processes can be reversed completely to make it work as a refrigerator. The $T - s$ diagram for a reverse Carnot refrigerator using a working fluid in the two-phase regime is shown in Fig. 8.11. It consists of four internally reversible processes:

1-2: Isentropic compression in a compressor

2-3: Isothermal heat rejection in a condenser

3-4: Isentropic expansion in a turbine

4-1: Isothermal heat addition in an evaporator

The liquid-vapour mixture of working fluid at state 1 enters a compressor and is isentropically compressed to state 2 which is a saturated vapour state. The temperature of working fluid increases during the compression process. Then working fluid rejects heat to high temperature sink at temperature $T_H$ and it changes from saturated vapour to saturated liquid in a condenser. This isothermal heat rejection process is represented by process 2-3 in Fig. 8.11. Further saturated liquid at state 3 is passed through a turbine where it is isentropically expanded to state 4 and its temperature decreases during the expansion process. The low temperature refrigerant at state 4 then passes through an evaporator where heat is added under isothermal conditions and its quality increases. The heat addition process in the evaporator is continued till the working fluid reaches state 1 to complete the cycle.

It is to be mentioned that isothermal heat addition and heat rejection processes are possible within the saturated liquid-vapor regime whereas the compression of saturated liquid-vapour mixture at state 1 is not practically possible in a compressor. Moreover, the expansion process in the turbine involves handling of high moisture content working fluid. Due to these reasons, the actual refrigerator cannot be approximated as a reversible Carnot cycle.

The two modifications are done to the reverse Carnot cycle that make it an ideal cycle for the refrigerating devices. Firstly, the working fluid is heated to saturated vapour state in the evaporator and then pure vapour is compressed in a compressor. Secondly, the expansion of saturated liquid in a turbine is replaced with a throttling process that is performed in an expansion valve or a capillary tube. The cycle including these modifications is termed as an **ideal vapour compression refrigeration cycle** and the processes involved in such a cycle are as follows:

1-2: Isentropic compression in a compressor

2-3: Isobaric heat rejection in a condenser

3-4: Isenthalpic expansion in a throttling device

4-1: Isobaric heat addition in an evaporator

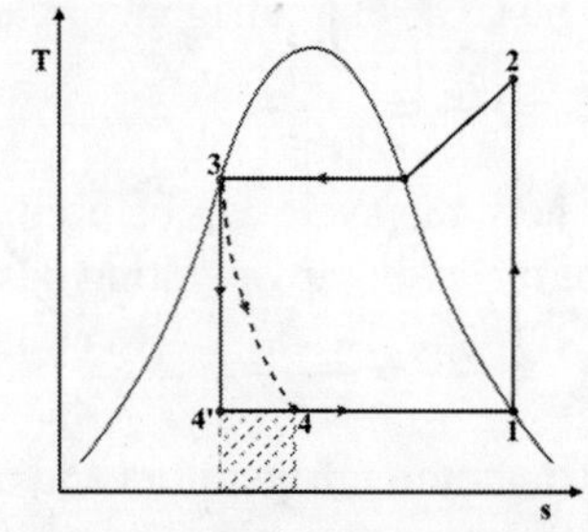


**Fig. 8.12:** Representation of an ideal vapour compression refrigeration cycle on $T-s$ diagram.

The $T-s$ diagram involving four processes of ideal vapour compression refrigeration cycle is shown in Fig. 8.12. The working fluid in saturated vapour state 1 enters a compressor where it is isentropically compressed to a superheated state 2. The temperature of the working fluid at the exit of the compressor is greater than the ambient temperature. The working fluid then passes through a condenser where it rejects heat to the atmosphere under the constant pressure conditions till it becomes saturated liquid. Generally air cooled condensers are used in refrigerators. Then working fluid is passed through a throttling device where it is expanded in an isenthalpic process from saturated liquid state 3 to state 4. The pressure and temperature of the working fluid decreases in a throttle. Typically the temperature of working fluid becomes less than the temperature of the refrigerated space. Further working fluid is passed through an evaporator where it is heated in a constant pressure process and comes out of the evaporator as an saturated vapour to complete the cycle. It must be stated that ideal vapour compression refrigeration cycle is not an internally reversible cycle because isenthalpic expansion process in a throttling device is not internally reversible. Thus, it is represented by dashed line on $T-s$ diagram in Fig. 8.12. A schematic representation of different components of an ideal vapour compression refrigeration cycle is shown in Fig. 8.13.

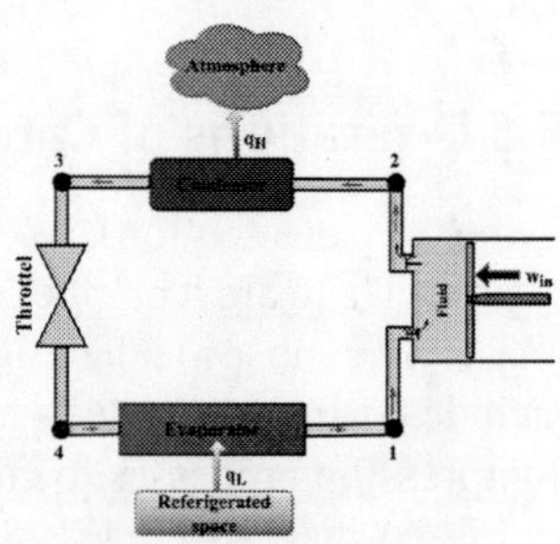


**Fig. 8.13:** Schematic representation of an ideal vapour compression refrigeration cycle based refrigerator.

It is interesting to note that if a turbine is used in place of a throttling device for the expansion process then the cycle becomes internally reversible. The isentropic expansion of working fluid is shown by process

$3-4'$ in Fig. 8.12. It is evident from Fig. 8.12 that expansion in the turbine increases the refrigerating effect by an amount shown with shaded area in Fig. 8.12 for the same amount of compressor work. Moreover, the turbine can also produce some work output during the expansion process. Despite these benefits the turbine is not used in vapour compression refrigeration systems because of complexities in its operation and high cost that does not justify the added benefits. Contrary the throttling device is very simple and cheap.

The thermodynamic analysis of vapour compression refrigeration systems is also based on the steady flow energy equation (Eqn. 8.1) considering changes in kinetic and potential energies negligible compared to corresponding enthalpy changes. Thus, using Eqn. (8.1), we get

Compressor: $w_{in} = h_{out} - h_{in} = h_2 - h_1$ (8.13)

Condenser: $q_H = h_{in} - h_{out} = h_2 - h_3$ (8.14)

Throttling device: $h_{in} \simeq h_{out} \Rightarrow h_3 \simeq h_4$ (8.15)

Evaporator: $q_L = h_{out} - h_{in} = h_1 - h_4$ (8.16)

The COP of a refrigerator can be defined as

$$\beta = \frac{q_L}{w_{in}} = \frac{h_1 - h_4}{h_2 - h_1} \tag{8.17}$$

The same cycle can be used as a heat pump by just changing the heat source and heat sink. Thus, COP of a heat pump operating based on ideal vapour compression refrigeration cycle is given as

$$\beta' = \frac{q_H}{w_{in}} = \frac{h_2 - h_3}{h_2 - h_1} \tag{8.18}$$

The home refrigerators and air conditioning systems work on the basis of an ideal vapour compression refrigeration cycle. The tubes passing through the freezer of a home refrigerator acts as an evaporator and the tubes on its back work as a condenser. This is the reason the back portion of a home refrigerator is typically warmer than the ambient. The working fluid used in a refrigerator is known as refrigerant. A large number of refrigerants can be used in refrigeration systems depending upon the system requirements and environmental constraints.

## 8.6 Limitations of Carnot Gas Cycle

In the previous sections, we discussed power cycles that use working fluid in the saturated liquid-vapour regime. There are a number of power cycles that use working fluid throughout the cycle in complete gas phase. The most idealized Carnot cycle for power producing devices that use gaseous working fluid includes isothermal heat addition, isentropic expansion, isothermal heat rejection and isentropic compression processes to complete the thermodynamic cycle. The Carnot gas cycle is shown on $P - v$ and $T - s$ diagrams in Fig. 8.14. The liquid-vapour dome in Fig. 8.14 is shown to highlight the operating zone of gas power cycles on property diagrams. Using gaseous working substance, the Carnot cycle can be achieved either in a piston-cylinder device as demonstrated in Chapter 5 or using a number of steady flow devices in a closed loop. Here, we present steady flow devices based Carnot gas cycle that uses two turbines and two compressors as shown in Fig. 8.15. The first turbine allows isothermal heat addition and simultaneous expansion of the gas. Further gas is expanded isentropically in the second turbine. After this an isothermal heat rejection and compression of gaseous working fluid is performed in a compressor and finally the working fluid is compressed isentropically to complete the cycle. This cycle is not practically possible because the isothermal heat addition and heat rejection processes are extremely slow and require very large surface area to exchange a finite amount of heat. Contrary, the isentropic expansion and compression processes should be rapid to avoid any heat exchange with the surroundings. That is why operating different steady flow devices at different speeds in a closed loop thermodynamic cycle is not feasible. Thus, Carnot gas cycle cannot be used to closely approximate the gas power cycles and used as a standard to compare the performance of actual devices.

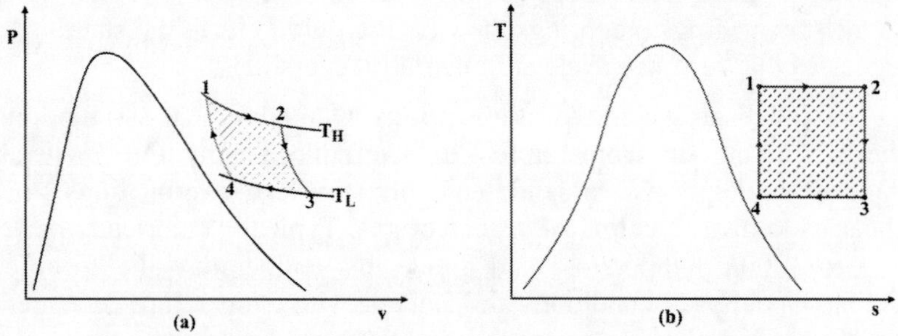


**Fig. 8.14:** Representation of Carnot gas cycle on (a) $P - v$ diagram and (b) $T - s$ diagram.

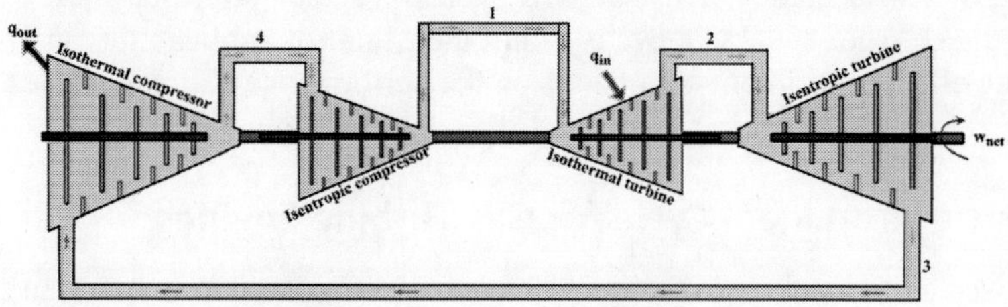


**Fig. 8.15:** Schematic of a Carnot gas cycle.

## 8.7 Air Standard Assumptions

The gas turbine engines and the reciprocating engines are the main power producing devices that use gas as the working fluid. In all these engines, air taken from the atmosphere is mixed with fuel and combustion is performed within the system boundaries to produce energy. This is the reason these are also called **internal combustion engines**. Contrary, a device in which the combustion takes place outside the system boundary and products of combustion are used as input energy is called an **external combustion engine**. A steam power plant uses the heat from hot gases to produce steam that are obtained from combustion of fuel in an external furnace, hence, it can be regarded as an external combustion engine. The composition of gaseous medium changes in a gas turbine and a reciprocating engine throughout the cycle, therefore, these are difficult to analyze. Considering the fact that air contains a significant amount of nitrogen and it is not participating in the chemical reactions, it can be said that the products of combustion will also be nitrogen dominant. Thus, the working fluid remains closer to air like substance throughout the cycle. Another difficulty with these engines is that these operate in an open cycle as air taken as intake leaves in the form of products of combustion. The characteristic of operating in an open cycle is shown by all the internal combustion engines.

The easiest approach to thermodynamically analyze the internal combustion engines is to use an approximate closed cycle that resembles closest to the actual open cycle operation. Such an approximated closed cycle is called **air standard cycle** and is based on following assumptions:

1. A fixed mass of air is considered as the working fluid that operates in a closed loop without any intake and exhaust. The air also behaves as an ideal gas throughout the cycle.
2. The combustion process is replaced with a heat addition process from an external source.

3. The exhaust of products of combustion in an actual engine is replaced by a heat rejection process to the surroundings that brings the working fluid to its initial state.
4. All the processes that form the cycle are internally reversible.

These assumptions are called **air standard assumptions**. One additional assumption of considering constant specific heats of air at room temperature is also introduced many times. When this assumption is used, the air standard assumptions are called **cold air standard assumptions**. An ideal gas with constant specific heats is known as **calorically perfect gas**. Typically calorically perfect behaviour is shown by gases up to $1000K$ temperature and above this temperature the specific heats become temperature dependent. Under such conditions an ideal gas with temperature dependent specific heats is known as **thermally perfect gas**. For a thermally perfect gas, the property tables should be used to find internal energy, enthalpy and entropy values. If the temperature of a gas becomes higher than $\simeq 4000K$, then dissociation of gas molecules may start and air standard assumptions may not be applicable with sufficient accuracy. Air standard cycles provide simplified thermodynamic analysis without deviating much from the actual cycle within their limit of applicability and help in qualitative investigation of the effect of different parameters on the performance of actual cycles.

## 8.8 Brayton Cycle as an Ideal Cycle for Gas Turbine Engines

Brayton cycle is used as an ideal cycle for gas turbine engines. A typical gas turbine engine consists of a compressor, a combustion chamber and a turbine as shown in Fig. 8.16. The compressor takes in the atmospheric air and increases its pressure and temperature. The high pressure and high temperature gas then mixes with fuel in the combustion chamber where combustion of fuel takes place. The heat released by the fuel further increases the pressure and temperature of the products of combustion which are then expanded in the turbine to obtain the shaft work. The products of combustion after their expansion in the turbine are exhausted to the atmosphere. Thus, the actual operation of a gas turbine forms an open cycle. To thermodynamically analyze a gas turbine engine, we impose the air standard assumptions which gives a closed cycle as shown in Fig. 8.17. The expansion and the compression processes in the closed loop cycle remain the same. The changes occur in the combustion process which is replaced by a heat addition process from the external source. Also, the working fluid at the exit of the turbine is passed through another heat exchanger where it gives heat to surroundings in order to reach the initial state to complete a closed loop cycle. Such a closed loop thermodynamic cycle is known as **ideal Brayton cycle** which includes the following ideal processes:

1-2: Isentropic compression in a compressor

2-3: Isobaric heat addition in a heat exchanger

3-4: Isentropic expansion in a turbine

4-1: Isobaric heat rejection in a heat exchanger

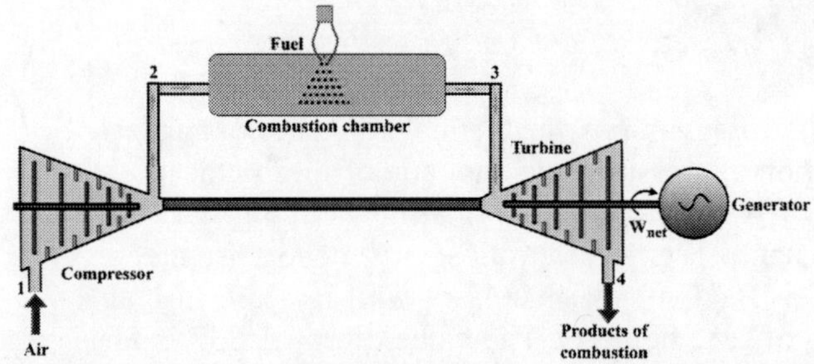


**Fig. 8.16:** Schematic representation of a gas turbine engine as an open cycle.

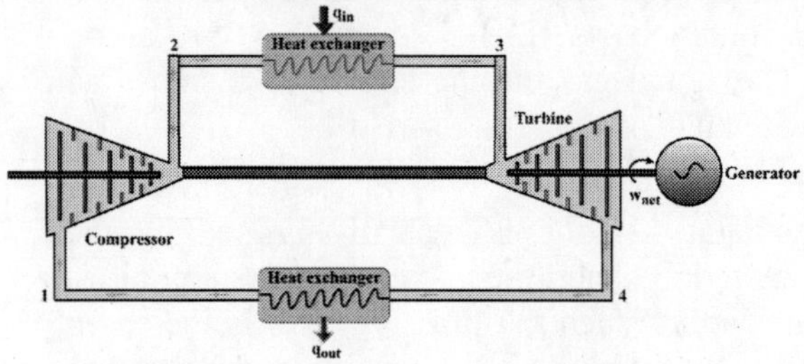


**Fig. 8.17:** Schematic representation of an ideal Brayton cycle for a gas turbine engine.

The $P - v$ and $T - s$ diagrams for an ideal Brayton cycle are shown in Fig. 8.18. The thermodynamic analysis of such a cycle can be performed by analyzing the individual steady flow devices by considering the negligible changes in kinetic and potential energies of the working fluid across each device. Thus, using Eqn. (8.1) and considering cold air standard assumption, we get

Compression: $w_{comp} = h_{out} - h_{in} = h_2 - h_1 = C_p(T_2 - T_1)$ (8.19)

Heat addition: $q_{in} = h_{out} - h_{in} = h_3 - h_2 = C_p(T_3 - T_2)$ (8.20)

Expansion: $w_{turbine} = h_{in} - h_{out} = h_3 - h_4 = C_p(T_3 - T_4)$ (8.21)

Heat rejection: $q_{out} = h_{in} - h_{out} = h_4 - h_1 = C_p(T_4 - T_1)$ (8.22)

The efficiency of an ideal Brayton cycle is given as

$$\eta_{Brayton} = \frac{w_{net}}{q_{in}} = 1 - \frac{q_{out}}{q_{in}} = 1 - \frac{C_p(T_4 - T_1)}{C_p(T_3 - T_2)} = 1 - \frac{T_1\,(T_4/T_1 - 1)}{T_2\,(T_3/T_2 - 1)} \quad (8.23)$$

Processes 2-3 and 4-1 are isobaric, thus, $P_2 = P_3$ and $P_4 = P_1$, hence, $\frac{P_2}{P_1} = \frac{P_3}{P_4} = r_p$ where $r_p$ is known as **pressure ratio**. For isentropic compression and expansion processes, we obtain $\frac{T_2}{T_1} = \left(\frac{P_2}{P_1}\right)^{\frac{k-1}{k}} = r_p^{\frac{k-1}{k}}$ and $\frac{T_3}{T_4} = \left(\frac{P_3}{P_4}\right)^{\frac{k-1}{k}}$, where $k$ is the ratio of specific heats. Thus, we get $\frac{T_2}{T_1} = \frac{T_3}{T_4} \Rightarrow \frac{T_4}{T_1} = \frac{T_3}{T_2}$. Therefore, Eqn. (8.23) can be expressed as

$$\eta_{Brayton} = 1 - \frac{1}{r_p^{(k-1)/k}} \quad (8.24)$$

Using the cold air standard assumptions, the thermal efficiency of an ideal Brayton cycle is a function of pressure ratio and ratio of specific heats. A plot of thermal efficiency based on Eqn. (8.24) is shown in Fig. 8.19 for different pressure ratios and specific heat ratios. It is evident from Fig. 8.19 that efficiency of a Brayton cycle increases with increase in both pressure ratio and ratio of specific heats. The value of $k = 1.667$ is for monoatomic gases such as helium, argon, etc. and shows the highest thermal efficiency at all the pressure ratios. The value of specific heat ratio at cold air standard conditions is 1.2 for ethane, 1.3 for carbon dioxide and 1.4 for air. The actual gas turbine engine also shows an increase in its efficiency with an increase of pressure ratio.

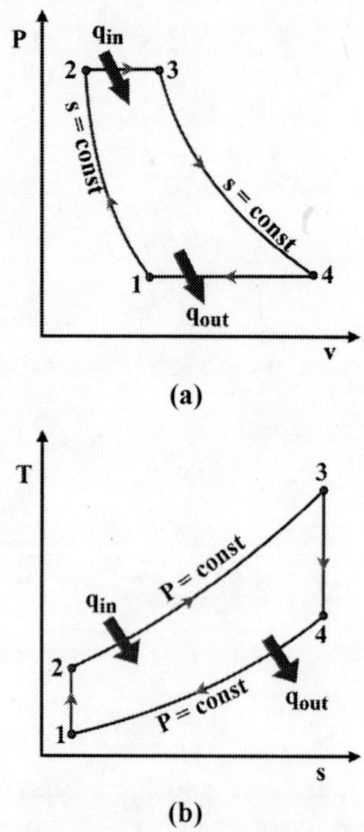


**Fig. 8.18:** Representation of an ideal Brayton cycle on (a) $P-v$ diagram and (b) $T-s$ diagram.

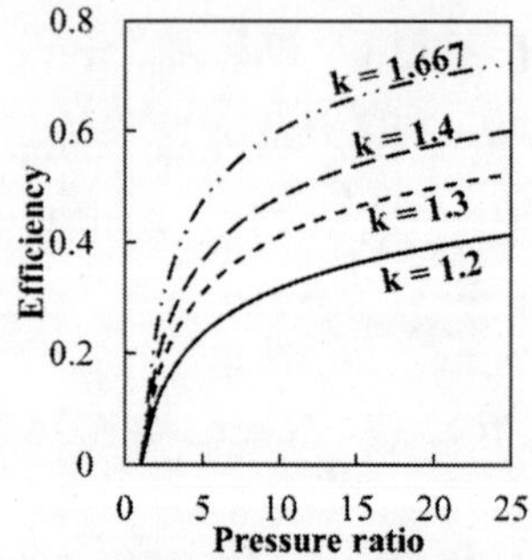


**Fig. 8.19:** Thermal efficiency of an ideal Brayton cycle as function of pressure ratio and ratio of specific heats.

The maximum temperature in the Brayton cycle is obtained at the end of the heat addition process (state 3) and is typically limited by the metallurgical limits of the materials. Thus, the maximum pressure ratio is limited by this temperature limit. When the pressure ratio is increased from very low values, gradually by keeping the maximum temperature ($T_3$) fixed, the net work done by the cycle first increases, reaches a maximum value and then starts decreasing. This is represented in Fig. 8.20 for three different pressure ratios where the area enclosed by the cycle represents its net work output. In order to achieve the high net work output at high pressure ratios (high thermal efficiency) the mass flow rate of air should be large. This increases the size of the equipment which causes difficulties in handling the system and the high capital cost. Thus, this approach is not economical and a compromise between pressure ratio and net work output is established which gives a typical pressure ratio range of 10 to 18 for operating a gas turbine engine. It is also important to mention that the compressor work in a gas turbine is comparable to the work output of the turbine. This is because the working fluid remains in gaseous phase in both the compressor and the turbine, therefore, has comparable specific volume in both devices. Thus, a large portion of turbine work is utilized to drive the compressor and the remaining portion is obtained as net work output. This is the reason for having a larger size turbine in a gas turbine engine in comparison to the turbine of a steam power plant where the power required by the pump is very small. The ratio of compressor work to turbine work is known as **back work ratio** and it can be even more than 0.5 when the isentropic efficiencies of the compressor and the turbine are low.

It is interesting to note that air in a gas turbine is used for two purposes, first to burn the fuel and second to cool the turbine and other components to avoid their overheating. The mass of air supplied for combustion of unit mass of fuel is known as **air-fuel ratio**. Due to the large amount of air used for cooling in a gas turbine engine, its air-fuel ratio typically ranges from 50 to 120. The gas turbine engines are mainly used for electricity production and aircraft propulsion. In case of electricity generation, the turbine shaft is connected to an electric generator to produce electricity. The gas turbine power plants can be used independently for electricity generation, however, many times these are used in conjunction with steam power plants where the exhaust gases from the gas turbine are used to produce steam. This is because the exhaust of a gas turbine takes place at a higher temperature than the maximum temperature of a steam power plant. Such plants are based on the **combined gas-vapour power cycle**. The use of gas turbines for aircraft propulsion involves partial expansion of gases in the turbine that is sufficient to run the compressor and provide power for running auxiliary equipment. The gases then are expanded through a nozzle to produce the required thrust. The thermodynamic cycle used for aircraft propulsion is also known as the **ideal jet propulsion cycle**.

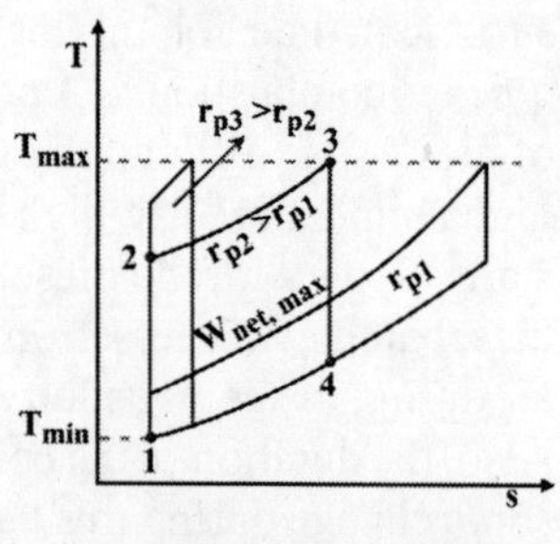


**Fig. 8.20:** Representation of Brayton cycle for fixed minimum and maximum temperatures at three different pressure ratios.

Scan Me

For more details on reciprocating engines.

## 8.9 Fundamentals of Reciprocating Engines

The reciprocating engines powering the majority of cars, two wheelers, railway locomotives, etc. are thermodynamically analysed by considering them closed systems unlike the steady flow systems discussed in this chapter till now. The system volume keeps on changing based on piston motion that enables the intake of fresh air/air-fuel mixture (Fig 8.21-a), compression of charge (air and fuel mixture) (Fig. 8.21-b), combustion of charge followed expansion (Fig. 8.21-c), and exhaust of burned gases from the cylinder (Fig. 8.21-d) with proper synchronization of piston motion and valves opening and closing timings. Process represented in Fig. 8.21(a-d) is the representation of typical four stroke engines. In one cycle of the reciprocating engines, a fixed mass of charge is inducted inside the system and it undergoes compression, combustion, expansion and exhaust of the system. Hence, during analysis of one cycle the mass remains constant and inflow and outflow from the thermodynamic system is not allowed. For proper combustion of the fuel, mixing of the fuel and air is a critical parameter. Depending upon the evaporation and diffusion properties of the fuels suitable air fuel mixing mechanisms are used in different types of the engines. After the large-scale popularisation of petroleum based liquid fuels for the automobile engines, based on the distillation characteristics of petroleum diesel and petrol (gasoline) are two most popular fuels for reciprocating engines. It is easier to evaporate petrol than diesel, hence generally petrol is injected in low pressure air outside the cylinder by deploying low pressure ($\sim 5\ bar$) fuel injectors or other mechanisms named as carburettors. The intake and compression strokes provide sufficient time for mixing of fuel and air. This homogeneously mixed mixture rapidly burns. This quicker combustion is represented as constant volume combustion on a thermodynamic cycle. In case of diesel, it is difficult to evaporate the diesel. Hence, diesel is injected at high pressure (~1000 $bar$) in high temperature and high pressure gases after the completion of the compression process. Combustion starts along with the diesel injection process. Due to difficulty in the evaporation and diffusion of the diesel along with lesser time available for the mixing of diesel and air, the mixture condition of diesel-air during the combustion of the diesel is heterogeneous. Hence, it is called heterogeneous combustion. Also, the duration of the combustion (heat addition) process is longer and the cylinder volume undergoes some change during this heat addition process. Hence, the heat addition process of the diesel engines is represented as a constant pressure process in the thermodynamic cycles.

Due to homogeneous mixing of gasoline and air generally spark plugs are used for igniting the gasoline air mixtures. Hence, gasoline engines are also referred to as spark ignition (SI) engines. The autoignition temperature of diesel-air mixture is lower than gasoline-air mixture. For taking advantage of this property of fuel-air mixture along with certain other parameters affecting the combustion (not in the scope of this book), in diesel fuelled engines the combustion starts by raising the temperature though compression. Hence, diesel engines are also referred to as compression ignition (CI) engines.

Similar to the description of open and closed cycles earlier in section 8.8, the real world cycles occurring in the reciprocating engines are also open cycles but for analytical simplicity they are approximated as closed cycles. In most of the modern automobile engines, turbochargers are deployed for increasing the output power of the engines by increasing the intake air pressure during induction (8.21-a) so that more fuel can be burned inside the cylinder that increases the amount of power produced by the engine. Turbocharges are a combination of turbine and compressor. Turbines are powered by hot exhaust gases coming out from the exhaust valve of the reciprocating engine cylinder. Other details of these turbochargers are similar to descriptions of turbines and compressors used in Brayton cycle based gas turbines as described in section 8.8. Based on the time taken by the heat addition process which in turn depends on the properties of the air and fuel used for producing this heat, the heat addition process can be approximated as constant volume heat addition, constant pressure heat addition or combination of these processes. The next section describes two popular thermodynamic cycles applicable for reciprocating engines named as Otto and Diesel cycles.

## 8.10 Otto and Diesel Cycles as Ideal Cycles for reciprocating engines

Otto and Diesel cycles are used as the ideal cycles for reciprocating spark ignited and compression ignited engines, respectively.

The actual operation of a reciprocating engine forms an open cycle. To thermodynamically analyze a reciprocating engine, we impose the air standard assumptions which gives a closed cycle 1-2-3(3')-4 as shown in Fig. 8.22. Such a closed loop thermodynamic cycle is known as **ideal Otto (Diesel) cycle** which includes the following ideal processes:

1-2: Isentropic compression through the piston

2-3(3'): Isochoric (Isobaric) heat addition

3(3')-4: Isentropic expansion of the piston

4-1: Isochoric heat rejection through the sudden opening of the exhaust valve

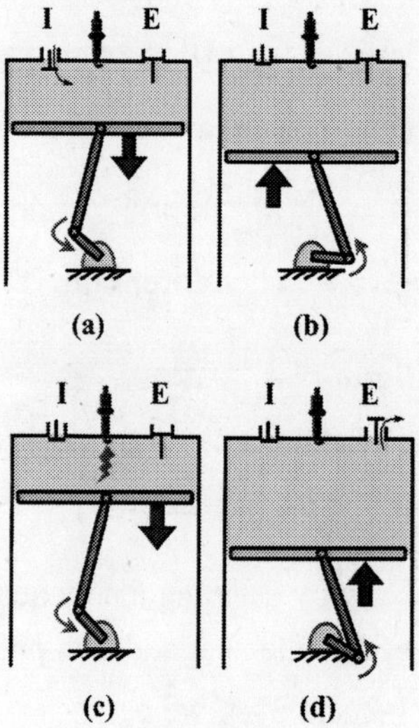


**Fig. 8.21:** Piston motion and valve openings and closing for different steps of reciprocation engine cycles: (a) intake, (b) compression, (c ) combustion and expansion and (d) exhaust.

The $P - v$ diagram for the ideal Otto (Diesel) cycle shows the same compression ratios for both Otto as well as Diesel cycles. But in actual practice the compression ratio of petrol engines (about 12) is much lower than the compression ratio of diesel engines (about 20). The thermodynamic analysis of such a cycle can be performed by analyzing the individual energy changes for individual processes during these cycles. Thus, using first law of thermodynamics for a closed system and considering cold air standard assumption, we get

Compression process:

$$w_{comp} = u_2 - u_1 = C_v(T_2 - T_1) \quad (8.25)$$

Heat addition for Otto Cycle:

$$q_{in} = u_3 - u_2 = C_v(T_3 - T_2) \quad (8.26)$$

Heat addition for Diesel Cycle:

$$q_{in} = h_{3'} - h_2 = C_p(T_{3'} - T_2) \quad (8.27)$$

Expansion for Otto cycle:

$$w_{expansion} = u_3 - u_4 = C_v(T_3 - T_4) \quad (8.28)$$

Expansion for Diesel cycle:

$$w_{expansion} = u_{3'} - u_4 = C_v(T_{3'} - T_4) \quad (8.29)$$

Heat rejection: $q_{out} = u_4 - u_1 = C_v(T_4 - T_1)$ (8.30)

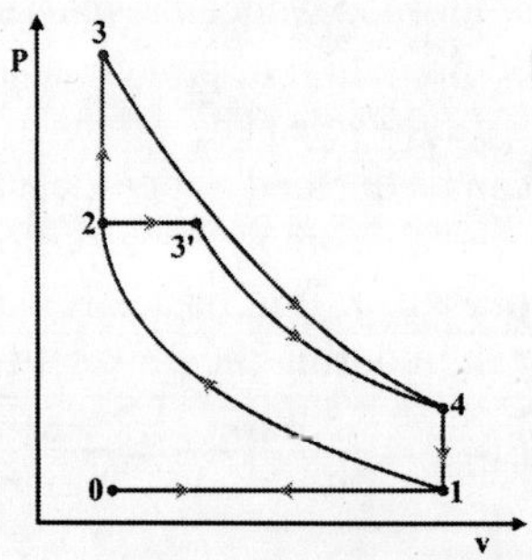


**Fig. 8.22:** Representation of an ideal Otto cycle and an ideal Diesel cycle on $P - v$ diagram.

The efficiency of an ideal Otto cycle is given as

$$\eta_{Otto} = 1 - \frac{q_{out}}{q_{in}} = 1 - \frac{C_v(T_4 - T_1)}{C_v(T_3 - T_2)} = 1 - \frac{T_1}{T_2}\frac{(T_4/T_1 - 1)}{(T_3/T_2 - 1)} \quad (8.31)$$

Processes 2-3 and 4-1 are isochoric, thus,

$V_2 = V_3$ and $V_4 = V_1$, hence, $\frac{V_1}{V_2} = \frac{V_4}{V_3} = r$ where $r$ is known as **compression ratio**. For isentropic compression and expansion processes, we obtain

$\frac{T_2}{T_1} = \left(\frac{V_1}{V_2}\right)^{k-1} = r^{(k-1)}$ and $\frac{T_3}{T_4} = \left(\frac{V_4}{V_3}\right)^{k-1}$, where $k$ is the ratio of specific heats. Thus, we get $\frac{T_2}{T_1} = \frac{T_3}{T_4} \Rightarrow \frac{T_4}{T_1} = \frac{T_3}{T_2}$. Therefore, Eqn. (8.31) can be expressed as

$$\eta_{Otto} = 1 - \frac{1}{r^{(k-1)}} \tag{8.32}$$

The efficiency of an ideal Diesel cycle is given as

$$\eta_{Diesel} = 1 - \frac{q_{out}}{q_{in}} = 1 - \frac{C_v(T_4 - T_1)}{C_p(T_3 - T_2)} = 1 - \frac{1}{k}\frac{T_1}{T_2}\frac{(T_4/T_1 - 1)}{(T_3/T_2 - 1)} \tag{8.33}$$

Now by defining a new quantity called as **cut-off ratio** which is the ratio of cylinder volumes after and before the heat addition process, i.e. $r_c = \frac{V_{3'}}{V_2}$ and using isentropic relations of an ideal gas for processes 1-2 and 3'-4, we get

$$\eta_{Diesel} = 1 - \frac{1}{r^{(k-1)}}\left[\frac{r_c^k - 1}{k(r_c - 1)}\right] \tag{8.34}$$

Using the cold air standard assumptions, the thermal efficiency of an ideal Otto cycle is a function of compression ratio and ratio of specific heats. It is evident from Eqn. (8.32) that efficiency of an Otto cycle increases with increase in both compression ratio and ratio of specific heats. On the other hand, the efficiency of an ideal diesel cycle depends on its compression ratio, cut-off ratio and ratio of specific heats. A diesel cycle becomes identical to an Otto cycle for a cut-off ratio of one. It must be noted that a more advanced cycle that considers the combustion process of a reciprocating engine as the combination of part constant volume heat addition and part constant pressure heat addition is known as a **dual cycle**. The extent of contact volume and constant pressure fractions of the heat addition process can be decided as per the actual combustion process of the engine. One important parameter that is often used to compare the performance of same sized reciprocating engines is known as **mean effective pressure**. It is basically defined as the net work done by an engine per unit the volume swept by the piston during the power stroke. It is mathematical expressed as

$$P_{MEP} = \frac{W_{net}}{V_{max} - V_{min}} = \frac{w_{net}}{v_{max} - v_{min}} \tag{8.35}$$

## UNIT SUMMARY

In this chapter, we have introduced the basic power and refrigeration cycles that are used for thermodynamic analysis of actual cyclic energy conversion devices. An ideal cycle approximates an actual energy conversion device as the combination of internally reversible processes. It was found that the reversible Carnot cycle shows limitations in terms of its practical operations for operating a steam power plant. Hence, an ideal cycle that approximates a steam power plant is discussed and is known as an ideal Rankine cycle. In this cycle, water is used as a working fluid which operates in the two-phase regime. The effect of its operating parameters on the performance of the ideal Rankine cycle is also demonstrated. An ideal vapour compression refrigeration cycle that also operates in the two-phase regime is an ideal cycle for the operation of refrigerators, heat pumps and air conditioning units. It can have different working fluids depending upon the system requirements and environmental constraints.

Gas turbine engines and reciprocating engines are another class of power producing devices that uses the working fluid in gaseous phase throughout the cycle. In actual operation of these power cycles, air is mixed with fuel and energy is obtained from combustion of fuel to produce work and then products of combustion are exhausted. To facilitate the thermodynamic analysis of these devices, air standard assumptions are introduced which approximate their operation as the ideal closed cycle. The inclusion of air standard operations for a gas turbine engine results in an ideal cycle that is known as the ideal Brayton cycle. Similarly, the spark ignited reciprocating engines are approximated by an ideal Otto cycle. The operation of a compression ignition reciprocating engine is idealized by Diesel cycle. A slightly modified ideal cycle with two heat addition processes that more closely approximates the actual operation for both spark and compression ignited engines is called dual cycle.

It must be noted that the present book is mainly focussed on fundamental concepts of classical thermodynamics. This chapter is included to introduce the use of different concepts elucidated in the previous chapters to demonstrate their practical applications. Thus, we focused in this chapter only on the basic power and refrigeration cycles. In actual conditions, many modified versions of each of the basic ideal cycles are used. Therefore, the readers are suggested to refer to the text books of engineering thermodynamics and applied thermodynamics to study the concepts of applications of thermodynamics in detail.

## Solved Examples

**Example 8.1:** In a Rankine cycle based steam power plant, steam enters the turbine at 15 $MPa$ and 550°C. The condenser operates at a pressure of 20 $kPa$. Determine the cycle efficiency.

**Solution:** The thermal efficiency of the ideal Rankine cycle is to be determined.

**Assumptions:** All the processes are internally reversible. The cycle operates at steady flow conditions with negligible changes in kinetic and potential energies across each device.

**Analysis:** Based on the processes represented in Fig. 8.4, the given data is

$P_2 = P_3 = 15\ MPa, T_3 = 550°\text{C}, P_4 = P_1 = 20\ kPa.$

Using superheated steam table B.3, we get

$h_3 = 3450.4\ kJ/kg$ and $s_3 = 6.5230\ kJ/kg - K.$

Using saturation pressure table B.2, we get

$v_1 = v_{f@20\ kPa} = 0.001017\ m^3/kg,\ h_1 = h_{f@20\ kPa} = 251.42\ kJ/kg,$

$h_{fg@20\ kPa} = 2357.5\ kJ/kg,\ s_{f@20\ kPa} = 0.8320\ kJ/kg - K$ and

$$s_{fg@20\ kPa} = 7.0752\ kJ/kg - K$$

For isentropic expansion of steam in turbine, $s_3 = s_4 = 6.5230\ kJ/kg - K < s_{g@20\ kPa}$.

Thus, $s_4 = \left(s_f + x_4 s_{fg}\right)_{@20\ kPa} \Rightarrow x_4 = 0.804.$

Similarly, $h_4 = \left(h_f + x_4 h_{fg}\right)_{@20\ kPa} = 2146.85\ kJ/kg.$

For the isentropic process in pump, $h_2 - h_1 = \int_1^2 vdP = v_1(P_2 - P_1)$.

$\Rightarrow h_2 = 251.42 + 0.001017 \times (15 \times 1000 - 20) = 266.65\ kJ/kg$.

Specific heat added to working fluid in the boiler is $q_{in} = h_3 - h_2 = 3183.75\ kJ/kg$.

Specific heat removed from working fluid in the condenser is

$q_{out} = h_4 - h_1 = 1895.43\ kJ/kg$.

Therefore, the thermal efficiency of the cycle is $\eta_{Rankine} = 1 - \frac{q_{out}}{q_{in}} = 40.46\%$.

**Example 8.2:** R-134a is used as a refrigerant in an ideal refrigeration cycle that has evaporator temperature of $-15$°C and the condenser temperature of 45°C. The mass flow rate of refrigerant in the cycle is 0.025 $kg/s$. Determine the capacity of the refrigerating unit in terms of rate of refrigeration and its COP.

**Solution:** The rate of refrigeration and the COP of the given refrigerating plant are to be determined.

**Assumptions:** Consider that the refrigeration plant operates under the steady flow conditions and the changes in kinetic and potential energies of the refrigerant are negligible across each steady flow device.

**Analysis:** Based on the processes represented in Fig. 8.12, the given data is

$T_1 = T_4 = -15°C$, $T_3 = 45°C$, $\dot{m} = 0.025\ kg/s$.

Using saturation R-134a table, we get

$h_1 = h_{g@-15°C} = 389.20\ kJ/kg$, $s_1 = s_{g@-15°C} = 1.7354\ kJ/kg - K$,

$h_3 = h_{f@45°C} = 264.11\ kJ/kg$, $P_3 = P_{sat.@45°C} = 1160.2\ kPa = P_2$.

For isentropic compression process: $s_1 = s_2 = 1.7354\ kJ/kg - K$.

Using superheated R-134a table at $P_2 = 1160.2\ kPa$ and $s_2 = 1.7354\ kJ/kg - K$, we get

| $P = 1000\ kPa$ | | | $P = 1200\ kPa$ | | |
|---|---|---|---|---|---|
| $s\ (kJ/kg\ K)$ | $T$ (°C) | $h\ (kJ/kg)$ | $s\ (kJ/kg\ K)$ | $T$ (°C) | $h\ (kJ/kg)$ |
| 1.7148 | 40 | 420.25 | 1.7237 | 50 | 426.84 |
| **1.7354** ($\square_2$) | **45.95** (using interpolation) | **426.79** (using interpolation) | **1.7354** ($\square_2$) | **53.37** (using interpolation) | **430.67** (using interpolation) |
| 1.7494 | 50 | 431.24 | 1.7584 | 60 | 438.21 |

Using this, we obtain

| $s_2 = 1.7354\ kJ/kg\ K$ | | |
|---|---|---|
| $P\ (kPa)$ | $T$ (°C) | $h\ (kJ/kg)$ |
| 1000 | 45.95 | 426.79 |
| **1160.2** ($P_2$) | **51.89** (using interpolation) | **429.90** (using interpolation) |
| 1200 | 53.37 | 430.67 |

Thus, $T_2 = 51.89$°C and $h_2 = 429.90\ kJ/kg$.

For the expansion process, $h_4 \simeq h_3 = 264.11\ kJ/kg$.

Therefore, the rate of refrigeration is $\dot{Q}_L = \dot{m}(h_1 - h_4) = 3.13\ kW$.

The COP of the refrigeration plant is $\beta = \frac{q_L}{w_{in}} = \frac{h_1 - h_4}{h_2 - h_1} = 3.07$.

**Example 8.3:** Air enters the compressor of an ideal Brayton cycle at 100 $kPa$ and 25°C. It leaves the compressor at 1.2 $MPa$ and the maximum temperature in the cycle is 1150°C. Using the cold air standard assumptions, determine the compressor work, turbine work and cycle efficiency.

**Solution:** The compressor work, turbine work and efficiency of the Brayton cycle are to be determined.

**Assumptions:** Consider the cycle operates in steady flow conditions and the changes in kinetic and potential energies of the air are negligible across each steady flow device.

**Analysis:** By referring to Fig. 8.18 for knowing the state points across each steady flow device, given data is $P_1 = 100\ kPa$, $T_1 = 25°\text{C} = 298K$, $P_3 = 1.2\ MPa = 1200\ kPa$ and $T_3 = 1150°\text{C} = 1423K$. Also, $k = 1.4$ and $C_p = 1.005\ kJ/kg - K$ for air at room temperature. Since the heat addition process 2-3 is isobaric, thus $P_2 = P_3 = 1200\ kPa$. Similarly, heat rejection process 4-1 is also isobaric, therefore, $P_4 = P_1 = 100\ kPa$.

For isentropic compression process 1-2:

$$\frac{T_2}{T_1} = \left(\frac{P_2}{P_1}\right)^{\frac{k-1}{k}} \Rightarrow T_2 = 298 \times \left(\frac{1200}{100}\right)^{\frac{1.4-1}{1.4}} = 606.11K.$$

For isentropic expansion process 3-4:

$$\frac{T_3}{T_4} = \left(\frac{P_3}{P_4}\right)^{\frac{k-1}{k}} \Rightarrow T_4 = 1423 \times \left(\frac{1200}{100}\right)^{\frac{-(1.4-1)}{1.4}} = 699.63K.$$

The specific compressor work is $w_{comp} = h_2 - h_1 = C_p(T_2 - T_1) = 309.65\ kJ/kg$.

The specific turbine work is $w_{turbine} = h_3 - h_4 = C_p(T_3 - T_4) = 726.99\ kJ/kg$.

The net work output is $w_{net} = w_{turbine} - w_{comp} = 417.34\ kJ/kg$.

Heat supplied is $q_{in} = h_3 - h_2 = C_p(T_3 - T_2) = 820.97\ kJ/kg$.

Thus, cycle efficiency is $\eta_{Brayton} = \frac{w_{net}}{q_{in}} = \frac{417.34}{820.97} = 50.83\%$.

**Alternate approach:** The cycle efficiency is also expressed as $\eta_{Brayton} = \frac{1}{r_p^{(k-1)/k}}$ where $r_p = \frac{P_2}{P_1} = 12 \Rightarrow \eta_{Brayton} = 1 - \frac{1}{12^{(1.4-1)/1.4}} = 50.83\%$.

**Example 8.4:** A gasoline engine that operates based on an ideal Otto cycle has compression ratio of 12. The air has 100 $kPa$ pressure and 25°C temperature at the start of the compression process and heat of 1500 $kJ/kg$ is added after the compression process. Determine the pressure and temperature at each point of the cycle, mean effective pressure and its thermal efficiency.

**Solution:** The pressure and temperature at each point of the cycle, mean effective pressure and thermal efficiency are to be determined.

**Assumptions:** The kinetic and potential energies of working fluid is negligible and the cycle operates in a closed system.

**Analysis:** By referring to Fig. 8.22 for the ideal Otto cycle, the given data at different state points is $P_1 = 100\,kPa$, $T_1 = 25°C = 298K$, $q_{in} = 1500\,kJ/kg$ and $r = \frac{V_1}{V_2} = 12$. Also, $k = 1.4$, $C_p = 1.005\,kJ/kg - K$ and $C_v = 0.718\,kJ/kg - K$ for air at room temperature. Since the heat addition process 2-3 is isochoric, thus $V_2 = V_3$ and the heat rejection process 4-1 is isochoric, therefore, $V_4 = V_1 \Rightarrow r = \frac{V_4}{V_3}$.

Using ideal gas equation at state 1,

$$P_1 v_1 = RT_1 \Rightarrow v_1 = \frac{RT_1}{P_1} = \frac{0.287 \times 298}{100} = 0.855\,m^3/kg.$$

$$\Rightarrow v_2 = \frac{v_1}{12} = 0.07125\,m^3/kg.$$

For isentropic compression process 1-2:

$$\frac{T_2}{T_1} = \left(\frac{V_1}{V_2}\right)^{k-1} \Rightarrow T_2 = 298 \times (12)^{1.4-1} = 805.17K.$$

$$\frac{P_2}{P_1} = \left(\frac{V_1}{V_2}\right)^{k} \Rightarrow P_2 = 100 \times 12^{1.4} = 3242.30\,kPa.$$

For isochoric heat addition process 2-3:

$$q_{in} = u_3 - u_2 = C_v(T_3 - T_2) \Rightarrow T_3 = \frac{1500}{0.718} + 805.17 = 2894.31K.$$

$$\frac{P_2}{P_3} = \frac{T_2}{T_3} \Rightarrow P_3 = P_2\frac{T_3}{T_2} = 3242.30 \times \frac{2894.31}{805.17} = 11654.96\,kPa.$$

For isentropic expansion process 3-4:

$$\frac{T_3}{T_4} = \left(\frac{V_4}{V_3}\right)^{k-1} \Rightarrow T_4 = 2894.31 \times (12)^{-(1.4-1)} = 1071.20K.$$

$$\frac{P_3}{P_4} = \left(\frac{V_4}{V_3}\right)^{k} \Rightarrow P_4 = 11654.96 \times 12^{-1.4} = 359.46\,kPa.$$

The specific heat rejected is $q_{out} = u_4 - u_1 = C_v(T_4 - T_1) = 555.16\,kJ/kg$.

The net work output is $w_{net} = q_{net} = q_{in} - q_{out} = 944.84\,kJ/kg$.

Mean effective pressure is $P_{MEP} = \frac{w_{net}}{v_1 - v_2} = \frac{944.84}{0.855-0.07125} = 1205.54\ kPa.$

Thus, cycle efficiency is $\eta_{Otto} = \frac{w_{net}}{q_{in}} = \frac{944.84}{1500} = 62.99\%.$

**Alternate approach:** The cycle efficiency is also expressed as $\eta_{Otto} = \frac{1}{r^{(k-1)}}$.

$\Rightarrow \eta_{Otto} = 1 - \frac{1}{12^{(1.4-1)}} = 62.99\%.$

**Example 8.5:** A compression ignition engine operates on an ideal diesel cycle that has compression ratio of 20. The air has 100 $kPa$ pressure and 25°C temperature at the start of the compression process and heat of 1500 $kJ/kg$ is added after the compression process. Determine the pressure and temperature at each point of the cycle, mean effective pressure and its thermal efficiency.

**Solution:** The pressure and temperature at each point of the cycle, mean effective pressure and thermal efficiency are to be determined.

**Assumptions:** The kinetic and potential energies of working fluid is negligible and the cycle operates in a closed system.

**Analysis:** By referring to Fig. 8.22 for the ideal diesel cycle, the given data at different state points is $P_1 = 100\ kPa,\ T_1 = 25°C = 298K,\ q_{in} = 1500\ kJ/kg$ and $r = \frac{V_1}{V_2} = 20$. Also, $k = 1.4,\ C_p = 1.005\ kJ/kg-K$ and $C_v = 0.718\ kJ/kg-K$ for air at room temperature. Since the heat addition process 2-3 is isobaric, thus $P_2 = P_3$ and the heat rejection process 4-1 is isochoric, therefore, $V_4 = V_1$. Using ideal gas equation at state 1,

$P_1 v_1 = RT_1 \Rightarrow v_1 = \frac{RT_1}{P_1} = \frac{0.287\times298}{100} = 0.855\ m^3/kg.$

$\Rightarrow v_2 = \frac{v_1}{20} = 0.04275\ m^3/kg.$

For isentropic compression process 1-2:

$\frac{T_2}{T_1} = \left(\frac{V_1}{V_2}\right)^{k-1} \Rightarrow T_2 = 298 \times (20)^{1.4-1} = 987.71K.$

$\frac{P_2}{P_1} = \left(\frac{V_1}{V_2}\right)^{k} \Rightarrow P_2 = 100 \times 20^{1.4} = 6628.91\ kPa.$

For isobaric heat addition process 2-3:

$q_{in} = h_3 - h_2 = C_p(T_3 - T_2) \Rightarrow T_3 = \frac{1500}{1.005} + 987.71 = 2480.25K.$

$P_3 = P_2 = 2480.25\ kPa.$

$\frac{V_3}{V_2} = \frac{T_3}{T_2} \Rightarrow v_3 = 0.1074\ m^3/kg.$

For isentropic expansion process 3-4:

$\frac{T_3}{T_4} = \left(\frac{V_4}{V_3}\right)^{k-1} \Rightarrow T_4 = 2480.25 \times \left(\frac{0.855}{0.1074}\right)^{-(1.4-1)} = 1081.71K.$

$\frac{P_3}{P_4} = \left(\frac{V_4}{V_3}\right)^{k} \Rightarrow P_4 = 2480.25 \times \left(\frac{0.855}{0.1074}\right)^{-1.4} = 135.88\ kPa.$

The specific heat rejected is $q_{out} = u_4 - u_1 = C_v(T_4 - T_1) = 562.70\ kJ/kg$.

The net work output is $w_{net} = q_{net} = q_{in} - q_{out} = 937.30\ kJ/kg$.

Mean effective pressure is $P_{MEP} = \frac{w_{net}}{v_1 - v_2} = \frac{937.30}{0.855-0.04275} = 1153.96\ kPa$.

Thus, cycle efficiency is $\eta_{Diesel} = \frac{w_{net}}{q_{in}} = \frac{937.30}{1500} = 62.49\%$.

## EXERCISES

### Multiple Choice Questions

| | | | | |
|---|---|---|---|---|
| 8.1 | An ideal cycle for a power producing device consists of all | | | |
| | A. reversible processes | B. internally reversible processes | C. externally reversible processes | D. internally irreversible reversible processes |
| 8.2 | The operation of a steam power plant is most closely approximated by the | | | |
| | A. Carnot vapor cycle | B. Carnot gas cycle | C. Rankine cycle | D. Brayton cycle |
| 8.3 | The efficiency of an ideal Rankine cycle increases with increase of boiler pressure due to increase of | | | |
| | A. mean temperature of heat addition | B. highest temperature of the cycle | C. lowest temperature of the cycle | D. degree of superheat |
| 8.4 | The increase of boiler pressure results in | | | |
| | A. less net work output | B. more heat rejection | C. increase in steam quality at turbine exit | D. decrease in steam quality at turbine exit |
| 8.5 | The pump work in a steam power plant is very less in comparison to turbine work due to | | | |
| | A. low specific volume of liquid | B. low pressure ratio | C. low enthalpy of compressed liquid | D. high specific volume of liquid |

| 8.6 | The efficiency of an ideal Rankine cycle is________________than the efficiency of Carnot vapour cycle operating under similar conditions. | | | |
|---|---|---|---|---|
| | A. more | B. less | C. equal | D. cannot be compared |
| 8.7 | An ideal vapour compression refrigeration cycle is not an internally reversible cycle due to | | | |
| | A. isobaric heat addition | B. isentropic compression | C. isentropic expansion | D. isenthalpic expansion |
| 8.8 | The efficiency of an ideal Brayton cycle, that is based on cold air standard assumptions, depends upon<br>i. pressure ratio<br>ii. maximum cycle temperature<br>iii. specific heat ratio<br>iv. maximum cycle pressure | | | |
| | A. i and ii | B. i and iii | C. i and iv | D. iii and iv |
| 8.9 | A very high pressure ratio is not used in Brayton cycle because of | | | |
| | A. low net work output | B. low thermal efficiency | C. high net work output | D. high thermal efficiency |
| 8.10 | The efficiency of an Otto cycle increase with | | | |
| | A. decrease of compression ratio | B. decrease of specific heat ratio | C. increase of specific heat ratio | D. increase in specific heat |
| 8.11 | The efficiency of a Diesel cycle becomes identical to an Otto cycle at | | | |
| | A. cut-off ratio greater than one | B. cut-off ratio less than one | C. cut-off ratio equals to one | D. cut-off ratio equals to zero |
| 8.12 | The combustion process in a Diesel cycle is replaced with an | | | |
| | A. isochoric heat addition | B. isobaric heat addition | C. isothermal heat addition | D. isentropic heat addition |

**Answers of Multiple Choice Questions**

| **8.1 B, 8.2 C, 8.3 A, 8.4 D, 8.5 A, 8.6 B, 8.7 D, 8.8 B, 8.9 A, 8.10 C, 8.11 C, 8.12 B** |
|---|

**Short and Long Answer Type Questions**

8.1 What is the difference between an open cycle and a closed cycle?

8.2 Why is the Carnot cycle not a suitable ideal cycle for a steam power plant?

8.3 What are the differences between Carnot vapour cycle and ideal Rankine cycle? How is their efficiency compared?

8.4 How do boiler and condenser pressures influence the performance of an ideal Rankine cycle?

8.5 Why does the condenser of a steam power plant operate at less pressure than atmospheric pressure?

8.6 Explain air standard assumptions.

8.7 How do air standard assumptions replace the combustion and exhaust processes of an actual engine operation?

8.8 What do you mean by back work ratio? Why is the back work ratio high for gas turbine engines?

8.9 What happens to the back work ratio of a gas turbine engine upon decrease in isentropic efficiency of its compressor and turbine?

8.10 What is the relationship of revolutions per minute (rpm) of a four stroke spark ignition engine with the thermodynamic cycle?

8.11 How does the efficiency of an Otto cycle vary with compression ratio and specific heat ratio?

8.12 Why is high compression ratio not used in spark ignited gasoline engines?

8.13 What are the differences between a gasoline engine and a diesel engine?

8.14 Which among Otto and Diesel cycles is more efficient at the same compression ratio?

8.15 Among gasoline and diesel engines, which one is used at high compression ratio. Why?

**Numerical Problems**

8.1 A steam power plant develops 250 $MW$ of electricity. It uses coal having heating value of 30,000 $kJ/kg$ as the fuel. The plant is based on an ideal Rankine cycle in which steam enters the turbine at 4.5 $MPa$ and 440°C and leaves at a pressure of 20 $kPa$. Consider that combustion efficiency of coal is 80% and the efficiency of the electric generator is 92%. Determine (a) the rate of coal supply required to operate the plant, (b) required steam mass flow rate, (c) the overall efficiency of the plant and (d) the mass flow rate of cooling water required in the condenser if its temperature does not change by more than 20°C in the condenser.

8.2 Water leaves the boiler of a Rankine cycle at 3 $MPa$ and 480°C. The condenser of the cycle operates at 90°C and steam leaves the turbine as saturated vapour at 90°C. The isentropic efficiency of the pump is 85%. Show the cycle on $T - s$ diagram and determine its thermal efficiency.

8.3 An ideal vapour compression refrigeration cycle uses R-134a refrigerant and operates between pressures 0.1 $MPa$ and 0.65 $MPa$. Represent the cycle on $T - s$ diagram. The mass flow rate of refrigerant is 2.5 $kg/min$. Determine (a) the rate of refrigerating effect, (b) input power required by the compressor, (c) the rate of heat rejection to the surroundings and (d) COP of the refrigerator.

8.4 An air conditioning unit uses R-134a as the working fluid and has a minimum temperature of -8°C. The condenser pressure is 0.95 $MPa$ and the refrigerant leaves the adiabatic compressor at 48°C. The pressure loss in the evaporator and the condenser is negligible. Determine (a) the specific heat transfer from the controlled space, (b) specific heat transfer to the atmosphere, (c) isentropic efficiency of the compressor and (d) the COP of the device.

8.5 Air is used as working fluid in an ideal Brayton cycle based gas turbine engine. Atmospheric air enters the compressor at 96 $kPa$ and 20°C and it enters the turbine at 750 $kPa$ and 825°C. Heat is added to air at a rate of 30 $MJ/s$ in an isobaric heat exchanger. Determine (a) the mass flow rate of air, (b) the power developed by the gas turbine engine and (c) the efficiency of the cycle if cold air standard assumptions are considered. Also, find the changes in the calculated parameters if the variation of specific heat with temperature is considered.

8.6 A Brayton cycle based gas turbine power plant develops 30 $MW$ power using air as the working fluid. The pressure ratio used in the cycle is 8. The maximum and minimum temperatures in the cycle are $950K$ and $300K$, respectively. Consider the isentropic efficiency of both the compressor and the turbine is 85%. Represents the cycle on $T - s$ diagram and determine (a) the mass flow rate of air through the cycle and (b) the efficiency of the cycle. Assume the specific heat is estimated at room temperature.

8.7 Air volume of 580 $cm^3$ enters an ideal Otto cycle at 100 $kPa$ and 30°C. The temperature of air at the exit of the isentropic expansion is 520°C. The compression ratio of the cycle is 10. Consider specific heat values at room temperature. Determine (a) the highest pressure and temperature in the cycle, (b) the mean effective pressure, (c) the amount of heat added and (d) the thermal efficiency.

8.8 A four stroke, four cylinder 2.20 $L$ petrol engine works in an Otto cycle with a compression ratio of 11. The maximum pressure in the cycle is 8.5 $MPa$ and air at the starting of compression process is at 101 $kPa$ and 55°C. Consider the polytropic index of 1.3 for compression and expansion processes and use the specific heats of air at 550°C. Determine (a) the temperature at the completion of expansion process, (b) the net work output of the engine, (c) the mean effective pressure, (d ) thermal efficiency and (e) the engine speed required for power output of 60 $kW$.

8.9 Air is used as working fluid in an ideal diesel engine that operates at a compression ratio of 20. Air at the start of compression process is at 100 $kPa$ and 25°C. The maximum cycle temperature is limited by $2000K$ and assume the specific heats for air at room temperature. Determine (a) the mean effective pressure and (b) the thermal efficiency of the engine.

8.10 A four stroke, six cylinder diesel engine of 4.5 $L$ volume works on the ideal diesel cycle and has a compression ratio of 18 and engine speed of 1800 rpm. Air at the beginning of the compression process is at 100 $kPa$ and 50°C. The heating value of the used diesel fuel is 42 $MJ/kg$ and a combustion efficiency of 96%. The air-fuel ratio used in the engine operation is 24. Consider the specific heats of air obtained at room temperature. Determine (a) the cutoff ratio (b) maximum temperature of the cycle, (c) the net work output per cycle, (d) the mean effective pressure, (d) the thermal efficiency and (e) the net power output of the engine.

## PRACTICAL

**Activity 1:** Find the detailed specifications of an automobile engine of any vehicle of your choice. Approximate it to the closest ideal cycle possible and perform the thermodynamic analysis to find its thermal efficiency. Also, determine the second law efficiency of the same engine.

**Activity 2:** Approximate your home refrigerator as an ideal vapour compression refrigeration cycle and perform the thermodynamic analysis to find its COP. Also, determine the second law efficiency of the refrigerator.

## REFERENCES AND SUGGESTED READINGS

1. Cengel, Y.A., Boles, M.A. and Kanoğlu, M., 2011. *Thermodynamics: an engineering approach.* New York: McGraw-hill.
2. Moran, M.J., Shapiro, H.N., Boettner, D.D. and Bailey, M.B., 2010. *Fundamentals of engineering thermodynamics*. John Wiley & Sons.
3. Borgnakke, C. and Sonntag, R.E., 2022. *Fundamentals of thermodynamics*. John Wiley & Sons.
4. Nag, P.K., 2002. *Bàsic and applied thermodynamics*. Tata McGraw-Hill.

# APPENDICES

## APPENDIX-A: Properties of air

**Table A.1: Ideal gas properties of air, standard entropy is calculated at 1 bar pressure.**

| $T$ (K) | $C_v$ (kJ/kg-K) | $C_p$ (kJ/kg-K) | $u$ (kJ/kg) | $h$ (kJ/kg) | $s_T^0$ (kJ/kg-K) |
|---|---|---|---|---|---|
| 200 | 0.7153 | 1.002 | 142.7 | 200.1 | 6.488315 |
| 220 | 0.7155 | 1.003 | 157 | 220.2 | 6.575659 |
| 240 | 0.7158 | 1.003 | 171.3 | 240.2 | 6.656425 |
| 260 | 0.7162 | 1.003 | 185.6 | 260.3 | 6.731598 |
| 280 | 0.7168 | 1.004 | 200 | 280.3 | 6.801954 |
| 300 | 0.7177 | 1.005 | 214.3 | 300.4 | 6.868114 |
| 320 | 0.7188 | 1.006 | 228.7 | 320.5 | 6.930586 |
| 340 | 0.7202 | 1.007 | 243.1 | 340.7 | 6.989786 |
| 360 | 0.7219 | 1.009 | 257.5 | 360.8 | 7.046065 |
| 380 | 0.7239 | 1.011 | 272 | 381 | 7.099717 |
| 400 | 0.7262 | 1.013 | 286.5 | 401.3 | 7.150994 |
| 420 | 0.7289 | 1.016 | 301 | 421.6 | 7.200112 |
| 440 | 0.7318 | 1.019 | 315.6 | 441.9 | 7.247259 |
| 460 | 0.735 | 1.022 | 330.3 | 462.3 | 7.292599 |
| 480 | 0.7385 | 1.026 | 345 | 482.8 | 7.336274 |
| 500 | 0.7423 | 1.029 | 359.8 | 503.3 | 7.378412 |
| 520 | 0.7462 | 1.033 | 374.7 | 524 | 7.419125 |
| 540 | 0.7504 | 1.037 | 389.7 | 544.7 | 7.458513 |
| 560 | 0.7547 | 1.042 | 404.7 | 565.5 | 7.496666 |
| 580 | 0.7592 | 1.046 | 419.9 | 586.3 | 7.533666 |
| 600 | 0.7638 | 1.051 | 435.1 | 607.3 | 7.569584 |
| 620 | 0.7685 | 1.055 | 450.4 | 628.4 | 7.604488 |
| 640 | 0.7732 | 1.06 | 465.8 | 649.5 | 7.638436 |
| 660 | 0.778 | 1.065 | 481.3 | 670.8 | 7.671485 |
| 680 | 0.7828 | 1.07 | 497 | 692.1 | 7.703683 |
| 700 | 0.7876 | 1.075 | 512.7 | 713.6 | 7.735077 |
| 720 | 0.7925 | 1.079 | 528.5 | 735.1 | 7.76571 |
| 740 | 0.7973 | 1.084 | 544.4 | 756.8 | 7.795619 |

| $T$ (K) | $C_v$ (kJ/kg-K) | $C_p$ (kJ/kg-K) | $u$ (kJ/kg) | $h$ (kJ/kg) | $s_T^0$ (kJ/kg-K) |
|---|---|---|---|---|---|
| 760 | 0.802 | 1.089 | 560.3 | 778.5 | 7.824841 |
| 780 | 0.8068 | 1.094 | 576.4 | 800.3 | 7.85341 |
| 800 | 0.8114 | 1.098 | 592.6 | 822.2 | 7.881356 |
| 820 | 0.816 | 1.103 | 608.9 | 844.3 | 7.908707 |
| 840 | 0.8206 | 1.108 | 625.3 | 866.4 | 7.935491 |
| 860 | 0.825 | 1.112 | 641.7 | 888.6 | 7.961732 |
| 880 | 0.8294 | 1.116 | 658.3 | 910.8 | 7.987453 |
| 900 | 0.8337 | 1.121 | 674.9 | 933.2 | 8.012677 |
| 920 | 0.8379 | 1.125 | 691.6 | 955.7 | 8.037423 |
| 940 | 0.842 | 1.129 | 708.4 | 978.2 | 8.061711 |
| 960 | 0.846 | 1.133 | 725.3 | 1001 | 8.085559 |
| 980 | 0.85 | 1.137 | 742.2 | 1024 | 8.108984 |
| 1000 | 0.8538 | 1.141 | 759.3 | 1046 | 8.132001 |
| 1020 | 0.8575 | 1.145 | 776.4 | 1069 | 8.154625 |
| 1040 | 0.8612 | 1.148 | 793.6 | 1092 | 8.176872 |
| 1060 | 0.8648 | 1.152 | 810.8 | 1115 | 8.198753 |
| 1080 | 0.8682 | 1.155 | 828.2 | 1138 | 8.220283 |
| 1100 | 0.8716 | 1.159 | 845.6 | 1161 | 8.241472 |
| 1120 | 0.8749 | 1.162 | 863 | 1185 | 8.262333 |
| 1140 | 0.8782 | 1.165 | 880.6 | 1208 | 8.282876 |
| 1160 | 0.8813 | 1.168 | 898.2 | 1231 | 8.303111 |
| 1180 | 0.8843 | 1.171 | 915.8 | 1255 | 8.323049 |
| 1200 | 0.8873 | 1.174 | 933.5 | 1278 | 8.342698 |
| 1220 | 0.8902 | 1.177 | 951.3 | 1301 | 8.362068 |
| 1240 | 0.893 | 1.18 | 969.1 | 1325 | 8.381166 |
| 1260 | 0.8958 | 1.183 | 987 | 1349 | 8.400002 |
| 1280 | 0.8985 | 1.185 | 1005 | 1372 | 8.418582 |
| 1300 | 0.9011 | 1.188 | 1023 | 1396 | 8.436915 |
| 1320 | 0.9036 | 1.191 | 1041 | 1420 | 8.455006 |
| 1340 | 0.9061 | 1.193 | 1059 | 1444 | 8.472863 |
| 1360 | 0.9085 | 1.196 | 1077 | 1468 | 8.490493 |
| 1380 | 0.9109 | 1.198 | 1095 | 1492 | 8.507901 |
| 1400 | 0.9132 | 1.2 | 1114 | 1516 | 8.525094 |
| 1420 | 0.9154 | 1.202 | 1132 | 1540 | 8.542076 |

| $T$ (K) | $C_v$ (kJ/kg-K) | $C_p$ (kJ/kg-K) | $u$ (kJ/kg) | $h$ (kJ/kg) | $s_T^0$ (kJ/kg-K) |
|---|---|---|---|---|---|
| 1440 | 0.9176 | 1.205 | 1150 | 1564 | 8.558855 |
| 1460 | 0.9197 | 1.207 | 1169 | 1588 | 8.575434 |
| 1480 | 0.9218 | 1.209 | 1187 | 1612 | 8.591819 |
| 1500 | 0.9239 | 1.211 | 1206 | 1636 | 8.608015 |
| 1520 | 0.9259 | 1.213 | 1224 | 1660 | 8.624026 |
| 1540 | 0.9278 | 1.215 | 1243 | 1685 | 8.639857 |
| 1560 | 0.9297 | 1.217 | 1261 | 1709 | 8.655512 |
| 1580 | 0.9316 | 1.219 | 1280 | 1733 | 8.670996 |
| 1600 | 0.9334 | 1.22 | 1298 | 1758 | 8.686312 |
| 1620 | 0.9352 | 1.222 | 1317 | 1782 | 8.701465 |
| 1640 | 0.9369 | 1.224 | 1336 | 1807 | 8.716457 |
| 1660 | 0.9386 | 1.226 | 1355 | 1831 | 8.731293 |
| 1680 | 0.9403 | 1.227 | 1373 | 1856 | 8.745976 |
| 1700 | 0.9419 | 1.229 | 1392 | 1880 | 8.76051 |
| 1720 | 0.9435 | 1.231 | 1411 | 1905 | 8.774897 |
| 1740 | 0.9451 | 1.232 | 1430 | 1929 | 8.789142 |
| 1760 | 0.9466 | 1.234 | 1449 | 1954 | 8.803246 |
| 1780 | 0.9481 | 1.235 | 1468 | 1979 | 8.817213 |
| 1800 | 0.9496 | 1.237 | 1487 | 2003 | 8.831046 |
| 1820 | 0.9511 | 1.238 | 1506 | 2028 | 8.844747 |
| 1840 | 0.9525 | 1.24 | 1525 | 2053 | 8.858319 |
| 1860 | 0.9539 | 1.241 | 1544 | 2078 | 8.871766 |
| 1880 | 0.9553 | 1.242 | 1563 | 2103 | 8.885088 |
| 1900 | 0.9566 | 1.244 | 1582 | 2127 | 8.898289 |
| 1920 | 0.9579 | 1.245 | 1601 | 2152 | 8.911372 |
| 1940 | 0.9592 | 1.246 | 1620 | 2177 | 8.924337 |
| 1960 | 0.9605 | 1.248 | 1640 | 2202 | 8.937189 |
| 1980 | 0.9618 | 1.249 | 1659 | 2227 | 8.949928 |
| 2000 | 0.963 | 1.25 | 1678 | 2252 | 8.962557 |
| 2020 | 0.9633 | 1.25 | 1698 | 2277 | 8.975077 |
| 2040 | 0.9645 | 1.252 | 1717 | 2303 | 8.987492 |
| 2060 | 0.9656 | 1.253 | 1736 | 2328 | 8.999802 |
| 2080 | 0.9668 | 1.254 | 1756 | 2353 | 9.01201 |
| 2100 | 0.9679 | 1.255 | 1775 | 2378 | 9.024118 |

| $T$ (K) | $C_v$ (kJ/kg-K) | $C_p$ (kJ/kg-K) | $u$ (kJ/kg) | $h$ (kJ/kg) | $s_T^0$ (kJ/kg-K) |
|---|---|---|---|---|---|
| 2120 | 0.9689 | 1.256 | 1794 | 2403 | 9.036127 |
| 2140 | 0.97 | 1.257 | 1814 | 2428 | 9.048039 |
| 2160 | 0.9711 | 1.258 | 1833 | 2453 | 9.059855 |
| 2180 | 0.9721 | 1.259 | 1853 | 2478 | 9.071578 |
| 2200 | 0.9731 | 1.26 | 1872 | 2503 | 9.083209 |
| 2220 | 0.9741 | 1.261 | 1891 | 2529 | 9.094749 |
| 2240 | 0.9751 | 1.262 | 1911 | 2554 | 9.1062 |
| 2260 | 0.9761 | 1.263 | 1930 | 2579 | 9.117564 |
| 2280 | 0.977 | 1.264 | 1950 | 2604 | 9.128841 |
| 2300 | 0.9779 | 1.265 | 1970 | 2630 | 9.140034 |
| 2320 | 0.9789 | 1.266 | 1989 | 2655 | 9.151143 |
| 2340 | 0.9798 | 1.267 | 2009 | 2680 | 9.162171 |
| 2360 | 0.9807 | 1.268 | 2028 | 2706 | 9.173117 |
| 2380 | 0.9815 | 1.269 | 2048 | 2731 | 9.183985 |
| 2400 | 0.9824 | 1.269 | 2068 | 2756 | 9.194774 |
| 2420 | 0.9833 | 1.27 | 2087 | 2782 | 9.205485 |
| 2440 | 0.9841 | 1.271 | 2107 | 2807 | 9.216122 |
| 2460 | 0.985 | 1.272 | 2127 | 2833 | 9.226683 |
| 2480 | 0.9858 | 1.273 | 2146 | 2858 | 9.237171 |
| 2500 | 0.9866 | 1.274 | 2166 | 2884 | 9.247586 |
| 2520 | 0.9874 | 1.274 | 2186 | 2909 | 9.25793 |
| 2540 | 0.9882 | 1.275 | 2206 | 2935 | 9.268204 |
| 2560 | 0.989 | 1.276 | 2225 | 2960 | 9.278408 |
| 2580 | 0.9897 | 1.277 | 2245 | 2986 | 9.288544 |
| 2600 | 0.9905 | 1.278 | 2265 | 3011 | 9.298613 |
| 2650 | 0.9924 | 1.279 | 2314 | 3075 | 9.323496 |
| 2700 | 0.9942 | 1.281 | 2364 | 3139 | 9.347979 |
| 2750 | 0.996 | 1.283 | 2414 | 3203 | 9.372075 |
| 2800 | 0.9977 | 1.285 | 2464 | 3267 | 9.395798 |
| 2850 | 0.9994 | 1.286 | 2514 | 3332 | 9.419159 |
| 2900 | 1.001 | 1.288 | 2564 | 3396 | 9.44217 |
| 2950 | 1.003 | 1.29 | 2614 | 3460 | 9.464843 |
| 3000 | 1.004 | 1.291 | 2664 | 3525 | 9.487188 |

# APPENDIX-B: Properties of water

## Table B.1: Saturated water temperature table.

| Temp. (°C) | Sat. Press. (kPa) | Specific volume (m³/kg) | | Specific internal energy (kJ/kg) | | | Specific enthalpy (kJ/kg) | | | Specific entropy (kJ/kg-K) | | |
|---|---|---|---|---|---|---|---|---|---|---|---|---|
| $T$ | $P$ | $v_f \times 10^3$ | $v_g$ | $u_f$ | $u_{fg}$ | $u_g$ | $h_f$ | $h_{fg}$ | $h_g$ | $s_f$ | $s_{fg}$ | $s_g$ |
| 0.01 | 0.6113 | 1 | 206.132 | 0 | 2375.33 | 2375.33 | 0 | 2501.35 | 2501.35 | 0 | 9.1562 | 9.1562 |
| 5 | 0.8721 | 1 | 147.118 | 20.97 | 2361.27 | 2382.24 | 20.98 | 2489.57 | 2510.54 | 0.0761 | 8.9496 | 9.0257 |
| 10 | 1.2276 | 1 | 106.377 | 41.99 | 2347.16 | 2389.15 | 41.99 | 2477.75 | 2519.74 | 0.151 | 8.7498 | 8.9007 |
| 15 | 1.705 | 1.001 | 77.925 | 62.98 | 2333.06 | 2396.04 | 62.98 | 2465.93 | 2528.91 | 0.2245 | 8.5569 | 8.7813 |
| 20 | 2.339 | 1.002 | 57.7897 | 83.94 | 2318.98 | 2402.91 | 83.94 | 2454.12 | 2538.06 | 0.2966 | 8.3706 | 8.6671 |
| 25 | 3.169 | 1.003 | 43.3593 | 104.86 | 2304.9 | 2409.76 | 104.87 | 2442.3 | 2547.17 | 0.3673 | 8.1905 | 8.5579 |
| 30 | 4.246 | 1.004 | 32.8932 | 125.77 | 2290.81 | 2416.58 | 125.77 | 2430.48 | 2556.25 | 0.4369 | 8.0164 | 8.4533 |
| 35 | 5.628 | 1.006 | 25.2158 | 146.65 | 2276.71 | 2423.36 | 146.66 | 2418.62 | 2565.28 | 0.5052 | 7.8478 | 8.353 |
| 40 | 7.384 | 1.008 | 19.5229 | 167.53 | 2262.57 | 2430.11 | 167.54 | 2406.72 | 2574.26 | 0.5724 | 7.6845 | 8.2569 |
| 45 | 9.593 | 1.01 | 15.2581 | 188.41 | 2248.4 | 2436.81 | 188.42 | 2394.77 | 2583.19 | 0.6386 | 7.5261 | 8.1647 |
| 50 | 12.35 | 1.012 | 12.0318 | 209.3 | 2234.17 | 2443.47 | 209.31 | 2382.75 | 2592.06 | 0.7037 | 7.3725 | 8.0762 |
| 55 | 15.758 | 1.015 | 9.56835 | 230.19 | 2219.89 | 2450.08 | 230.2 | 2370.66 | 2600.86 | 0.7679 | 7.2234 | 7.9912 |
| 60 | 19.941 | 1.017 | 7.67071 | 251.09 | 2205.54 | 2456.63 | 251.11 | 2358.48 | 2609.59 | 0.8311 | 7.0784 | 7.9095 |
| 65 | 25.03 | 1.02 | 6.19656 | 272 | 2191.12 | 2463.12 | 272.03 | 2346.21 | 2618.24 | 0.8934 | 6.9375 | 7.8309 |
| 70 | 31.19 | 1.023 | 5.04217 | 292.93 | 2176.62 | 2469.55 | 292.96 | 2333.85 | 2626.8 | 0.9548 | 6.8004 | 7.7552 |
| 75 | 38.58 | 1.026 | 4.13123 | 313.87 | 2162.03 | 2475.91 | 313.91 | 2321.37 | 2635.28 | 1.0154 | 6.667 | 7.6824 |
| 80 | 47.39 | 1.029 | 3.40715 | 334.84 | 2147.36 | 2482.19 | 334.88 | 2308.77 | 2643.66 | 1.0752 | 6.5369 | 7.6121 |
| 85 | 57.83 | 1.032 | 2.82757 | 355.82 | 2132.58 | 2488.4 | 355.88 | 2296.05 | 2651.93 | 1.1342 | 6.4102 | 7.5444 |
| 90 | 70.14 | 1.036 | 2.36056 | 376.82 | 2117.7 | 2494.52 | 376.9 | 2283.19 | 2660.09 | 1.1924 | 6.2866 | 7.479 |
| 95 | 84.55 | 1.04 | 1.98186 | 397.86 | 2102.7 | 2500.56 | 397.94 | 2270.19 | 2668.13 | 1.25 | 6.1659 | 7.4158 |
| 100 | 101.3 | 1.044 | 1.6729 | 418.91 | 2087.58 | 2506.5 | 419.02 | 2257.03 | 2676.05 | 1.3068 | 6.048 | 7.3548 |
| 105 | 120.8 | 1.047 | 1.41936 | 440 | 2072.34 | 2512.34 | 440.13 | 2243.7 | 2683.83 | 1.3629 | 5.9328 | 7.2958 |
| 110 | 143.3 | 1.052 | 1.21014 | 461.12 | 2056.96 | 2518.09 | 461.27 | 2230.2 | 2691.47 | 1.4184 | 5.8202 | 7.2386 |
| 115 | 169.1 | 1.056 | 1.03658 | 482.28 | 2041.44 | 2523.72 | 482.46 | 2216.5 | 2698.96 | 1.4733 | 5.71 | 7.1832 |
| 120 | 198.5 | 1.06 | 0.89186 | 503.48 | 2025.76 | 2529.24 | 503.69 | 2202.61 | 2706.3 | 1.5275 | 5.602 | 7.1295 |
| 125 | 232.1 | 1.065 | 0.77059 | 524.72 | 2009.91 | 2534.63 | 524.96 | 2188.5 | 2713.46 | 1.5812 | 5.4962 | 7.0774 |
| 130 | 270.1 | 1.07 | 0.6685 | 546 | 1993.9 | 2539.9 | 546.29 | 2174.16 | 2720.46 | 1.6343 | 5.3925 | 7.0269 |
| 135 | 313 | 1.075 | 0.58217 | 567.34 | 1977.69 | 2545.03 | 567.67 | 2159.59 | 2727.26 | 1.6869 | 5.2907 | 6.9777 |
| 140 | 361.3 | 1.08 | 0.50885 | 588.72 | 1961.3 | 2550.02 | 589.11 | 2144.75 | 2733.87 | 1.739 | 5.1908 | 6.9298 |
| 145 | 415.4 | 1.085 | 0.44632 | 610.16 | 1944.69 | 2554.86 | 610.61 | 2129.65 | 2740.26 | 1.7906 | 5.0926 | 6.8832 |
| 150 | 475.9 | 1.09 | 0.39278 | 631.66 | 1927.87 | 2559.54 | 632.18 | 2114.26 | 2746.44 | 1.8417 | 4.996 | 6.8378 |
| 155 | 543.1 | 1.096 | 0.34676 | 653.23 | 1910.82 | 2564.04 | 653.82 | 2098.56 | 2752.39 | 1.8924 | 4.901 | 6.7934 |
| 160 | 617.8 | 1.102 | 0.30706 | 674.85 | 1893.52 | 2568.37 | 675.53 | 2082.55 | 2758.09 | 1.9426 | 4.8075 | 6.7501 |
| 165 | 700.5 | 1.108 | 0.27269 | 696.55 | 1875.97 | 2572.51 | 697.32 | 2066.2 | 2763.53 | 1.9924 | 4.7153 | 6.7078 |
| 170 | 791.7 | 1.114 | 0.24283 | 718.31 | 1858.14 | 2576.46 | 719.2 | 2049.5 | 2768.7 | 2.0418 | 4.6244 | 6.6663 |
| 175 | 892 | 1.121 | 0.2168 | 740.16 | 1840.03 | 2580.19 | 741.16 | 2032.42 | 2773.58 | 2.0909 | 4.5347 | 6.6256 |
| 180 | 1002.2 | 1.127 | 0.19405 | 762.08 | 1821.62 | 2583.7 | 763.21 | 2014.96 | 2778.16 | 2.1395 | 4.4461 | 6.5857 |
| 185 | 1122.7 | 1.134 | 0.17409 | 784.08 | 1802.9 | 2586.98 | 785.36 | 1997.07 | 2782.43 | 2.1878 | 4.3586 | 6.5464 |
| 190 | 1254.4 | 1.141 | 0.15654 | 806.17 | 1783.84 | 2590.01 | 807.61 | 1978.76 | 2786.37 | 2.2358 | 4.272 | 6.5078 |

| Temp. (°C) | Sat. Press. (kPa) | Specific volume ($m^3$/kg) | | Specific internal energy (kJ/kg) | | | Specific enthalpy (kJ/kg) | | | Specific entropy (kJ/kg-K) | | |
|---|---|---|---|---|---|---|---|---|---|---|---|---|
| $T$ | $P$ | $v_f \times 10^3$ | $v_g$ | $u_f$ | $u_{fg}$ | $u_g$ | $h_f$ | $h_{fg}$ | $h_g$ | $s_f$ | $s_{fg}$ | $s_g$ |
| 195 | 1397.8 | 1.149 | 0.14105 | 828.36 | 1764.43 | 2592.79 | 829.96 | 1959.99 | 2789.96 | 2.2835 | 4.1863 | 6.4697 |
| 200 | 1553.8 | 1.156 | 0.12736 | 850.64 | 1744.66 | 2595.29 | 852.43 | 1940.75 | 2793.18 | 2.3308 | 4.1014 | 6.4322 |
| 205 | 1723 | 1.164 | 0.11521 | 873.02 | 1724.49 | 2597.52 | 875.03 | 1921 | 2796.03 | 2.3779 | 4.0172 | 6.3951 |
| 210 | 1906.3 | 1.173 | 0.10441 | 895.51 | 1703.93 | 2599.44 | 897.75 | 1900.73 | 2798.48 | 2.4247 | 3.9337 | 6.3584 |
| 215 | 2104.2 | 1.181 | 0.09479 | 918.12 | 1682.94 | 2601.06 | 920.61 | 1879.91 | 2800.51 | 2.4713 | 3.8507 | 6.3221 |
| 220 | 2317.8 | 1.19 | 0.08619 | 940.85 | 1661.49 | 2602.35 | 943.61 | 1858.51 | 2802.12 | 2.5177 | 3.7683 | 6.286 |
| 225 | 2547.7 | 1.199 | 0.07849 | 963.72 | 1639.58 | 2603.3 | 966.77 | 1836.5 | 2803.27 | 2.5639 | 3.6863 | 6.2502 |
| 230 | 2794.9 | 1.209 | 0.07158 | 986.72 | 1617.17 | 2603.89 | 990.1 | 1813.85 | 2803.95 | 2.6099 | 3.6047 | 6.2146 |
| 235 | 3060.1 | 1.219 | 0.06536 | 1009.9 | 1594.24 | 2604.11 | 1013.6 | 1790.53 | 2804.13 | 2.6557 | 3.5233 | 6.1791 |
| 240 | 3344.2 | 1.229 | 0.05976 | 1033.2 | 1570.75 | 2603.95 | 1037.3 | 1766.5 | 2803.81 | 2.7015 | 3.4422 | 6.1436 |
| 245 | 3648.2 | 1.24 | 0.0547 | 1056.7 | 1546.68 | 2603.37 | 1061.2 | 1741.73 | 2802.95 | 2.7471 | 3.3612 | 6.1083 |
| 250 | 3973 | 1.251 | 0.05013 | 1080.4 | 1522 | 2602.37 | 1085.3 | 1716.18 | 2801.52 | 2.7927 | 3.2802 | 6.0729 |
| 255 | 4319.5 | 1.263 | 0.04598 | 1104.3 | 1496.66 | 2600.93 | 1109.7 | 1689.8 | 2799.51 | 2.8382 | 3.1992 | 6.0374 |
| 260 | 4688.6 | 1.276 | 0.0422 | 1128.4 | 1470.64 | 2599.01 | 1134.4 | 1662.54 | 2796.89 | 2.8837 | 3.1181 | 6.0018 |
| 265 | 5081.3 | 1.289 | 0.03877 | 1152.7 | 1443.87 | 2596.6 | 1159.3 | 1634.34 | 2793.61 | 2.9293 | 3.0368 | 5.9661 |
| 270 | 5498.7 | 1.302 | 0.03564 | 1177.3 | 1416.33 | 2593.66 | 1184.5 | 1605.16 | 2789.65 | 2.975 | 2.9551 | 5.9301 |
| 275 | 5941.8 | 1.317 | 0.03279 | 1202.2 | 1387.94 | 2590.17 | 1210.1 | 1574.92 | 2784.97 | 3.0208 | 2.873 | 5.8937 |
| 280 | 6411.7 | 1.332 | 0.03017 | 1227.4 | 1358.66 | 2586.09 | 1236 | 1543.55 | 2779.53 | 3.0667 | 2.7903 | 5.857 |
| 285 | 6909.4 | 1.348 | 0.02777 | 1253 | 1328.41 | 2581.38 | 1262.3 | 1510.97 | 2773.27 | 3.1129 | 2.7069 | 5.8198 |
| 290 | 7436 | 1.366 | 0.02557 | 1278.9 | 1297.11 | 2575.99 | 1289 | 1477.08 | 2766.13 | 3.1593 | 2.6227 | 5.7821 |
| 295 | 7992.8 | 1.384 | 0.02354 | 1305.2 | 1264.67 | 2569.87 | 1316.3 | 1441.78 | 2758.05 | 3.2061 | 2.5375 | 5.7436 |
| 300 | 8581 | 1.404 | 0.02167 | 1332 | 1230.99 | 2562.96 | 1344 | 1404.93 | 2748.94 | 3.2533 | 2.4511 | 5.7044 |
| 305 | 9201.8 | 1.425 | 0.01995 | 1359.2 | 1195.94 | 2555.16 | 1372.3 | 1366.38 | 2738.72 | 3.3009 | 2.3633 | 5.6642 |
| 310 | 9856.6 | 1.447 | 0.01835 | 1387 | 1159.37 | 2546.4 | 1401.3 | 1325.97 | 2727.27 | 3.3492 | 2.2737 | 5.6229 |
| 315 | 10547 | 1.472 | 0.01687 | 1415.4 | 1121.11 | 2536.55 | 1431 | 1283.48 | 2714.44 | 3.3981 | 2.1821 | 5.5803 |
| 320 | 11274 | 1.499 | 0.01549 | 1444.6 | 1080.93 | 2525.48 | 1461.5 | 1238.64 | 2700.08 | 3.4479 | 2.0882 | 5.5361 |
| 325 | 12040 | 1.528 | 0.0142 | 1474.4 | 1038.57 | 2513.01 | 1492.8 | 1191.13 | 2683.97 | 3.4987 | 1.9913 | 5.49 |
| 330 | 12845 | 1.561 | 0.013 | 1505.2 | 993.66 | 2498.91 | 1525.3 | 1140.56 | 2665.85 | 3.5506 | 1.8909 | 5.4416 |
| 335 | 13694 | 1.597 | 0.01186 | 1537.1 | 945.77 | 2482.88 | 1559 | 1086.37 | 2645.35 | 3.604 | 1.7863 | 5.3903 |
| 340 | 14586 | 1.638 | 0.0108 | 1570.3 | 894.26 | 2464.53 | 1594.2 | 1027.86 | 2622.01 | 3.6593 | 1.6763 | 5.3356 |
| 345 | 15525 | 1.685 | 0.00978 | 1605 | 838.29 | 2443.3 | 1631.2 | 964.02 | 2595.19 | 3.7169 | 1.5594 | 5.2763 |
| 350 | 16514 | 1.74 | 0.00881 | 1641.8 | 776.58 | 2418.39 | 1670.5 | 893.38 | 2563.92 | 3.7776 | 1.4336 | 5.2111 |
| 355 | 17554 | 1.807 | 0.00787 | 1681.4 | 707.11 | 2388.52 | 1713.1 | 813.59 | 2526.72 | 3.8427 | 1.2951 | 5.1378 |
| 360 | 18651 | 1.892 | 0.00694 | 1725.2 | 626.29 | 2351.47 | 1760.5 | 720.52 | 2481 | 3.9146 | 1.1379 | 5.0525 |
| 365 | 19807 | 2.011 | 0.00599 | 1776.1 | 526.54 | 2302.67 | 1816 | 605.44 | 2421.4 | 3.9983 | 0.9487 | 4.947 |
| 370 | 21028 | 2.213 | 0.00493 | 1843.8 | 384.69 | 2228.53 | 1890.4 | 441.75 | 2332.12 | 4.1104 | 0.6868 | 4.7972 |
| 374.1 | 22089 | 3.155 | 0.00315 | 2029.6 | 0 | 2029.58 | 2099.3 | 0 | 2099.26 | 4.4297 | 0 | 4.4297 |

**Table B.2: Saturated water pressure table.**

| Press. (kPa) | Sat. Temp. (°C) | Specific volume ($m^3$/kg) | | Specific internal energy (kJ/kg) | | | Specific enthalpy (kJ/kg) | | | Specific entropy (kJ/kg-K) | | |
|---|---|---|---|---|---|---|---|---|---|---|---|---|
| $P$ | $T$ | $v_f \times 10^3$ | $v_g$ | $u_f$ | $u_{fg}$ | $u_g$ | $h_f$ | $h_{fg}$ | $h_g$ | $s_f$ | $s_{fg}$ | $s_g$ |
| 0.6113 | 0.01 | 1 | 206.132 | 0 | 2375.3 | 2375.3 | 0 | 2501.3 | 2501.3 | 0 | 9.1562 | 9.1562 |
| 1 | 6.98 | 1 | 129.208 | 29.29 | 2355.69 | 2384.98 | 29.29 | 2484.89 | 2514.18 | 0.1059 | 8.8697 | 8.9756 |
| 1.5 | 13.03 | 1.001 | 87.98013 | 54.7 | 2338.63 | 2393.32 | 54.7 | 2470.59 | 2525.3 | 0.1956 | 8.6322 | 8.8278 |
| 2 | 17.5 | 1.001 | 67.00385 | 73.47 | 2326.02 | 2399.48 | 73.47 | 2460.02 | 2533.49 | 0.2607 | 8.4629 | 8.7236 |
| 2.5 | 21.08 | 1.002 | 54.25385 | 88.47 | 2315.93 | 2404.4 | 88.47 | 2451.56 | 2540.03 | 0.312 | 8.3311 | 8.6431 |
| 3 | 24.08 | 1.003 | 45.66502 | 101.03 | 2307.48 | 2408.51 | 101.03 | 2444.47 | 2545.5 | 0.3545 | 8.2231 | 8.5775 |
| 4 | 28.96 | 1.004 | 34.80015 | 121.44 | 2293.73 | 2415.17 | 121.44 | 2432.93 | 2554.37 | 0.4226 | 8.052 | 8.4746 |
| 5 | 32.88 | 1.005 | 28.19251 | 137.79 | 2282.7 | 2420.49 | 137.79 | 2423.66 | 2561.45 | 0.4763 | 7.9187 | 8.395 |
| 7.5 | 40.29 | 1.008 | 19.23775 | 168.76 | 2261.74 | 2430.5 | 168.77 | 2406.02 | 2574.79 | 0.5763 | 7.6751 | 8.2514 |
| 10 | 45.81 | 1.01 | 14.67355 | 191.79 | 2246.1 | 2437.89 | 191.81 | 2392.82 | 2584.63 | 0.6492 | 7.501 | 8.1501 |
| 15 | 53.97 | 1.014 | 10.02218 | 225.9 | 2222.83 | 2448.73 | 225.91 | 2373.14 | 2599.06 | 0.7548 | 7.2536 | 8.0084 |
| 20 | 60.06 | 1.017 | 7.64937 | 251.35 | 2205.36 | 2456.71 | 251.38 | 2358.33 | 2609.7 | 0.8319 | 7.0766 | 7.9085 |
| 25 | 64.97 | 1.02 | 6.20424 | 271.88 | 2191.21 | 2463.08 | 271.9 | 2346.29 | 2618.19 | 0.893 | 6.9383 | 7.8313 |
| 30 | 69.1 | 1.022 | 5.22918 | 289.18 | 2179.22 | 2468.4 | 289.21 | 2336.07 | 2625.28 | 0.9439 | 6.8247 | 7.7686 |
| 40 | 75.87 | 1.026 | 3.99345 | 317.51 | 2159.49 | 2477 | 317.55 | 2319.19 | 2636.74 | 1.0258 | 6.6441 | 7.67 |
| 50 | 81.33 | 1.03 | 3.24034 | 340.42 | 2143.43 | 2483.85 | 340.47 | 2305.4 | 2645.87 | 1.091 | 6.5029 | 7.5939 |
| 75 | 91.77 | 1.037 | 2.21711 | 394.29 | 2112.39 | 2496.67 | 384.36 | 2278.59 | 2662.96 | 1.2129 | 6.2434 | 7.4563 |
| 100 | 99.62 | 1.043 | 1.694 | 417.33 | 2088.72 | 2506.06 | 417.44 | 2258.02 | 2675.46 | 1.3025 | 6.0568 | 7.3593 |
| 125 | 105.99 | 1.048 | 1.3749 | 444.16 | 2069.32 | 2513.48 | 444.3 | 2241.05 | 2685.35 | 1.3739 | 5.9104 | 7.2843 |
| 150 | 111.37 | 1.053 | 1.15933 | 466.92 | 2052.72 | 2519.64 | 467.08 | 2226.46 | 2693.54 | 1.4335 | 5.7897 | 7.2232 |
| 175 | 116.06 | 1.057 | 1.00363 | 486.78 | 2038.12 | 2524.9 | 486.97 | 2213.57 | 2700.53 | 1.4848 | 5.6868 | 7.1717 |
| 200 | 120.23 | 1.061 | 0.88573 | 504.47 | 2025.02 | 2529.49 | 504.68 | 2201.96 | 2706.63 | 1.53 | 5.597 | 7.1271 |
| 225 | 124 | 1.064 | 0.79325 | 520.45 | 2013.1 | 2533.56 | 520.69 | 2191.35 | 2712.04 | 1.5705 | 5.5173 | 7.0878 |
| 250 | 127.43 | 1.067 | 0.71871 | 535.08 | 2002.14 | 2537.21 | 535.34 | 2181.55 | 2716.89 | 1.6072 | 5.4455 | 7.0526 |
| 275 | 130.6 | 1.07 | 0.65731 | 548.57 | 1991.95 | 2540.53 | 548.87 | 2172.42 | 2721.29 | 1.6407 | 5.3801 | 7.0208 |
| 300 | 133.55 | 1.073 | 0.60582 | 561.13 | 1982.43 | 2543.55 | 561.45 | 2163.85 | 2725.3 | 1.6717 | 5.3201 | 6.9918 |
| 325 | 136.3 | 1.076 | 0.56201 | 572.88 | 1973.46 | 2546.34 | 573.23 | 2155.76 | 2728.99 | 1.7005 | 5.2646 | 6.9651 |
| 350 | 138.88 | 1.079 | 0.52425 | 583.93 | 1964.98 | 2548.92 | 584.31 | 2148.1 | 2732.4 | 1.7274 | 5.213 | 6.9404 |
| 375 | 141.32 | 1.081 | 0.49137 | 594.38 | 1956.93 | 2551.31 | 594.79 | 2140.79 | 2735.58 | 1.7527 | 5.1647 | 6.9174 |
| 400 | 143.63 | 1.084 | 0.46246 | 604.29 | 1949.26 | 2553.55 | 604.73 | 2133.81 | 2738.53 | 1.7766 | 5.1193 | 6.8958 |
| 450 | 147.93 | 1.088 | 0.41398 | 622.75 | 1934.87 | 2557.62 | 623.24 | 2120.67 | 2743.91 | 1.8206 | 5.0359 | 6.8565 |
| 500 | 151.86 | 1.093 | 0.37489 | 639.66 | 1921.57 | 2561.23 | 640.21 | 2108.47 | 2748.67 | 1.8606 | 4.9606 | 6.8212 |
| 550 | 155.48 | 1.097 | 0.34268 | 655.3 | 1909.17 | 2564.47 | 655.91 | 2097.04 | 2752.94 | 1.8972 | 4.892 | 6.7892 |
| 600 | 158.85 | 1.101 | 0.31567 | 669.88 | 1897.52 | 2567.4 | 670.54 | 2086.26 | 2756.8 | 1.9311 | 4.8289 | 6.76 |
| 650 | 162.01 | 1.104 | 0.29268 | 683.55 | 1886.51 | 2570.06 | 684.26 | 2076.04 | 2760.3 | 1.9627 | 4.7704 | 6.733 |
| 700 | 164.97 | 1.108 | 0.27286 | 696.43 | 1876.07 | 2572.49 | 697.2 | 2066.3 | 2763.5 | 1.9922 | 4.7158 | 6.708 |
| 750 | 167.77 | 1.111 | 0.2556 | 708.62 | 1866.11 | 2574.73 | 709.45 | 2056.98 | 2766.43 | 2.0199 | 4.6647 | 6.6846 |
| 800 | 170.43 | 1.115 | 0.24043 | 720.2 | 1856.58 | 2576.79 | 721.1 | 2048.04 | 2769.13 | 2.0461 | 4.6166 | 6.6627 |
| 850 | 172.96 | 1.118 | 0.22698 | 731.25 | 1847.45 | 2578.69 | 732.2 | 2039.43 | 2771.63 | 2.0709 | 4.5711 | 6.6421 |
| 900 | 175.38 | 1.121 | 0.21497 | 741.81 | 1838.65 | 2580.46 | 742.82 | 2031.12 | 2773.94 | 2.0946 | 4.528 | 6.6225 |

| Press. (kPa) | Sat. Temp. (°C) | Specific volume (m³/kg) | | Specific internal energy (kJ/kg) | | | Specific enthalpy (kJ/kg) | | | Specific entropy (kJ/kg-K) | | |
|---|---|---|---|---|---|---|---|---|---|---|---|---|
| $P$ | $T$ | $v_f \times 10^3$ | $v_g$ | $u_f$ | $u_{fg}$ | $u_g$ | $h_f$ | $h_{fg}$ | $h_g$ | $s_f$ | $s_{fg}$ | $s_g$ |
| 950 | 177.69 | 1.124 | 0.20419 | 751.94 | 1830.17 | 2582.11 | 753 | 2023.08 | 2776.08 | 2.1171 | 4.4869 | 6.604 |
| 1000 | 179.91 | 1.127 | 0.19444 | 761.67 | 1821.97 | 2583.64 | 762.79 | 2015.29 | 2778.08 | 2.1386 | 4.4478 | 6.5864 |
| 1100 | 184.09 | 1.133 | 0.17753 | 780.08 | 1806.32 | 2586.4 | 781.32 | 2000.36 | 2781.68 | 2.1791 | 4.3744 | 6.5535 |
| 1200 | 187.99 | 1.139 | 0.16333 | 797.27 | 1791.55 | 2588.82 | 798.64 | 1986.19 | 2784.82 | 2.2165 | 4.3067 | 6.5233 |
| 1300 | 191.64 | 1.144 | 0.15125 | 813.42 | 1777.53 | 2590.95 | 814.91 | 1972.67 | 2787.58 | 2.2514 | 4.2438 | 6.4953 |
| 1400 | 195.07 | 1.149 | 0.14084 | 828.68 | 1764.15 | 2592.83 | 830.29 | 1959.72 | 2790 | 2.2842 | 4.185 | 6.4692 |
| 1500 | 198.32 | 1.154 | 0.13177 | 843.14 | 1751.3 | 2594.5 | 844.87 | 1947.28 | 2792.15 | 2.315 | 4.1298 | 6.4448 |
| 1750 | 205.76 | 1.166 | 0.11349 | 876.44 | 1721.39 | 2597.83 | 878.48 | 1917.95 | 2796.43 | 2.3851 | 4.0044 | 6.3895 |
| 2000 | 212.42 | 1.177 | 0.09963 | 906.42 | 1693.84 | 2600.26 | 908.77 | 1890.74 | 2799.51 | 2.4473 | 3.8935 | 6.3408 |
| 2250 | 218.45 | 1.187 | 0.08875 | 933.81 | 1668.18 | 2601.98 | 936.48 | 1865.19 | 2801.67 | 2.5034 | 3.7938 | 6.2971 |
| 2500 | 223.99 | 1.197 | 0.07998 | 959.09 | 1644.04 | 2603.13 | 962.09 | 1840.98 | 2803.07 | 2.5546 | 3.7028 | 6.2574 |
| 2750 | 229.12 | 1.207 | 0.07275 | 982.65 | 1621.16 | 2603.81 | 985.97 | 1817.89 | 2803.86 | 2.6018 | 3.619 | 6.2208 |
| 3000 | 233.9 | 1.216 | 0.06668 | 1004.76 | 1599.34 | 2604.1 | 1008.41 | 1795.73 | 2804.14 | 2.6456 | 3.5412 | 6.1869 |
| 3250 | 238.38 | 1.226 | 0.06152 | 1025.62 | 1578.43 | 2604.04 | 1029.6 | 1774.37 | 2803.97 | 2.6866 | 3.4685 | 6.1551 |
| 3500 | 242.6 | 1.235 | 0.05707 | 1045.41 | 1558.29 | 2603.7 | 1049.73 | 1753.7 | 2803.43 | 2.7252 | 3.4 | 6.1252 |
| 4000 | 250.4 | 1.252 | 0.04978 | 1082.28 | 1519.99 | 2602.27 | 1087.29 | 1714.09 | 2801.38 | 2.7963 | 3.2737 | 6.07 |
| 5000 | 263.99 | 1.286 | 0.03944 | 1147.78 | 1449.34 | 2597.12 | 1154.21 | 1640.12 | 2794.33 | 2.9201 | 3.0532 | 5.9733 |
| 6000 | 275.64 | 1.319 | 0.03244 | 1205.41 | 1384.27 | 2589.69 | 1213.32 | 1571 | 2784.33 | 3.0266 | 2.8625 | 5.8891 |
| 7000 | 285.88 | 1.351 | 0.02737 | 1257.51 | 1322.97 | 2580.48 | 1266.97 | 1505.1 | 2772.07 | 3.121 | 2.6922 | 5.8132 |
| 8000 | 295.06 | 1.384 | 0.02352 | 1305.54 | 1264.25 | 2569.79 | 1316.61 | 1441.33 | 2757.94 | 3.2067 | 2.5365 | 5.7431 |
| 9000 | 303.4 | 1.418 | 0.02048 | 1350.47 | 1207.28 | 2557.75 | 1363.23 | 1378.88 | 2742.11 | 3.2857 | 2.3915 | 5.6771 |
| 10000 | 311.06 | 1.452 | 0.01803 | 1393 | 1151.4 | 2544.41 | 1407.53 | 1317.14 | 2724.67 | 3.3595 | 2.2545 | 5.614 |
| 11000 | 318.15 | 1.489 | 0.01599 | 1433.68 | 1096.06 | 2529.74 | 1450.05 | 1255.55 | 2705.6 | 3.4294 | 2.1233 | 5.5527 |
| 12000 | 324.75 | 1.527 | 0.01426 | 1472.92 | 1040.76 | 2513.67 | 1491.24 | 1193.59 | 2684.83 | 3.4961 | 1.9962 | 5.4923 |
| 13000 | 330.93 | 1.567 | 0.01278 | 1511.09 | 984.99 | 2496.08 | 1531.46 | 1130.76 | 2662.22 | 3.5604 | 1.8718 | 5.4323 |
| 14000 | 336.75 | 1.611 | 0.01149 | 1548.53 | 928.23 | 2476.76 | 1571.08 | 1066.47 | 2637.55 | 3.6231 | 1.7485 | 5.3716 |
| 15000 | 342.24 | 1.658 | 0.01034 | 1585.58 | 869.85 | 2455.43 | 1610.45 | 1000.04 | 2610.49 | 3.6847 | 1.625 | 5.3097 |
| 16000 | 347.43 | 1.711 | 0.00931 | 1622.63 | 809.07 | 2431.7 | 1650 | 930.59 | 2580.59 | 3.746 | 1.4995 | 5.2454 |
| 17000 | 352.37 | 1.77 | 0.00836 | 1660.16 | 744.8 | 2404.96 | 1690.25 | 856.9 | 2547.15 | 3.8078 | 1.3698 | 5.1776 |
| 18000 | 357.06 | 1.84 | 0.00749 | 1698.86 | 675.42 | 2374.28 | 1731.97 | 777.13 | 2509.09 | 3.8713 | 1.233 | 5.1044 |
| 19000 | 361.54 | 1.924 | 0.00666 | 1739.87 | 598.18 | 2338.05 | 1776.43 | 688.11 | 2464.54 | 3.9387 | 1.0841 | 5.0227 |
| 20000 | 365.81 | 2.035 | 0.00583 | 1785.47 | 507.58 | 2293.05 | 1826.18 | 583.56 | 2409.74 | 4.0137 | 0.9132 | 4.9269 |
| 21000 | 369.89 | 2.206 | 0.00495 | 1841.97 | 388.74 | 2230.71 | 1888.3 | 446.42 | 2334.72 | 4.1073 | 0.6942 | 4.8015 |
| 22000 | 373.8 | 2.808 | 0.00353 | 1973.16 | 108.24 | 2081.39 | 2034.92 | 124.04 | 2158.97 | 4.3307 | 0.1917 | 4.5224 |
| 22089 | 374.14 | 3.155 | 0.00315 | 2029.58 | 0 | 2029.58 | 2099.26 | 0 | 2099.26 | 4.4297 | 0 | 4.4297 |

**Table B.3: Superheated steam tables.**

| T (°C) | v (m³/kg) | u (kJ/kg) | h (kJ/kg) | s (kJ/kg-K) | v (m³/kg) | u (kJ/kg) | h (kJ/kg) | s (kJ/kg-K) |
|---|---|---|---|---|---|---|---|---|
| | P = 10 kPa (45.81°C) | | | | P = 50 kPa (81.33°C) | | | |
| Sat. | 14.67355 | 2437.89 | 2584.63 | 8.1501 | 3.24034 | 2483.85 | 2645.87 | 7.5939 |
| 50 | 14.8692 | 2443.87 | 2592.56 | 8.1749 | — | — | — | — |
| 100 | 17.19561 | 2515.5 | 2687.46 | 8.4479 | 3.41833 | 2511.61 | 2682.52 | 7.6947 |
| 150 | 19.51251 | 2587.86 | 2782.99 | 8.6881 | 3.88937 | 2585.61 | 2780.08 | 7.94 |
| 200 | 21.82507 | 2661.27 | 2879.52 | 8.9037 | 4.35595 | 2659.85 | 2877.64 | 8.1579 |
| 250 | 24.13559 | 2735.95 | 2977.31 | 9.1002 | 4.82045 | 2734.97 | 2975.99 | 8.3555 |
| 300 | 26.44508 | 2812.06 | 3076.51 | 9.2812 | 5.28391 | 2811.33 | 3075.52 | 8.5372 |
| 400 | 31.06252 | 2968.89 | 3279.51 | 9.6076 | 6.20929 | 2968.43 | 3278.89 | 8.8641 |
| 500 | 35.67896 | 3132.26 | 3489.05 | 9.8977 | 7.13364 | 3131.94 | 3488.62 | 9.1545 |
| 600 | 40.29488 | 3302.45 | 3705.4 | 10.1608 | 8.05748 | 3302.22 | 3705.1 | 9.4177 |
| 700 | 44.91052 | 3479.63 | 3928.73 | 10.4028 | 8.98104 | 3479.45 | 3928.51 | 9.6599 |
| 800 | 49.52599 | 3663.84 | 4159.1 | 10.6281 | 9.90444 | 3663.7 | 4158.92 | 9.8852 |
| 900 | 54.14137 | 3855.03 | 4396.44 | 10.8395 | 10.82773 | 3854.91 | 4396.3 | 10.0967 |
| 1000 | 58.75669 | 4053.01 | 4640.58 | 11.0392 | 11.75097 | 4052.91 | 4640.46 | 10.2964 |
| 1100 | 63.37198 | 4257.47 | 4891.19 | 11.2287 | 12.67418 | 4257.37 | 4891.08 | 10.4858 |
| 1200 | 67.98724 | 4467.91 | 5147.78 | 11.409 | 13.59737 | 4467.82 | 5147.69 | 10.6662 |
| 1300 | 72.6025 | 4683.68 | 5409.7 | 14.581 | 14.52054 | 4683.58 | 5409.61 | 10.8382 |
| | P = 100 kPa (99.62°C) | | | | P = 200 kPa (120.23°C) | | | |
| Sat. | 1.694 | 2506.06 | 2675.46 | 7.3593 | 0.88573 | 2529.49 | 2706.63 | 7.1271 |
| 150 | 1.93636 | 2582.75 | 2776.38 | 7.6133 | 0.95964 | 2576.87 | 2768.8 | 7.2795 |
| 200 | 2.17226 | 2658.05 | 2875.27 | 7.8342 | 1.08034 | 2654.39 | 2870.46 | 7.5066 |
| 250 | 2.40604 | 2733.73 | 2974.33 | 8.0332 | 1.1988 | 2731.22 | 2970.98 | 7.7085 |
| 300 | 2.63876 | 2810.41 | 3074.28 | 8.2157 | 1.31616 | 2808.55 | 3071.79 | 7.8926 |
| 400 | 3.10263 | 2967.85 | 3278.11 | 8.5434 | 1.5493 | 2966.69 | 3276.55 | 8.2217 |
| 500 | 3.56547 | 3131.54 | 3488.09 | 8.8341 | 1.78139 | 3130.75 | 3487.03 | 8.5132 |
| 600 | 4.02781 | 3301.94 | 3704.72 | 9.0975 | 2.01297 | 3301.36 | 3703.96 | 8.7769 |
| 700 | 4.48986 | 3479.24 | 3928.23 | 9.3398 | 2.24426 | 3478.81 | 3927.66 | 9.0194 |
| 800 | 4.95174 | 3663.53 | 4158.71 | 9.5652 | 2.47539 | 3663.19 | 4158.27 | 9.245 |
| 900 | 5.41353 | 3854.77 | 4396.12 | 9.7767 | 2.70643 | 3854.49 | 4395.77 | 9.4565 |
| 1000 | 5.87526 | 4052.78 | 4640.31 | 9.9764 | 2.9374 | 4052.53 | 4640.01 | 9.6563 |
| 1100 | 6.33696 | 4257.25 | 4890.95 | 10.1658 | 3.16834 | 4257.01 | 4890.68 | 9.8458 |
| 1200 | 6.79863 | 4467.7 | 5147.56 | 10.3462 | 3.39927 | 4467.46 | 5147.32 | 10.0262 |
| 1300 | 7.2603 | 4683.47 | 5409.49 | 10.5182 | 3.63018 | 4683.23 | 5409.26 | 10.1982 |
| | P = 300 kPa (133.55°C) | | | | P = 400 kPa (143.63°C) | | | |
| Sat. | 0.60582 | 2543.55 | 2725.3 | 6.9918 | 0.46246 | 2553.55 | 2738.53 | 6.8958 |
| 150 | 0.63388 | 2570.79 | 2760.95 | 7.0778 | 0.47084 | 2564.48 | 2752.82 | 6.9299 |
| 200 | 0.71629 | 2650.65 | 2865.54 | 7.3115 | 0.53422 | 2646.83 | 2860.51 | 7.1706 |
| 250 | 0.79636 | 2728.69 | 2967.59 | 7.5165 | 0.59512 | 2726.11 | 2964.16 | 7.3788 |
| 300 | 0.87529 | 2806.69 | 3069.28 | 7.7022 | 0.65484 | 2804.81 | 3066.75 | 7.5661 |
| 400 | 1.03151 | 2965.53 | 3274.98 | 8.0329 | 0.77262 | 2964.36 | 3273.41 | 7.8984 |

| $T$ (°C) | $v$ (m³/kg) | $u$ (kJ/kg) | $h$ (kJ/kg) | $s$ (kJ/kg-K) | $v$ (m³/kg) | $u$ (kJ/kg) | $h$ (kJ/kg) | $s$ (kJ/kg-K) |
|---|---|---|---|---|---|---|---|---|
| 500 | 1.18669 | 3129.95 | 3485.96 | 8.325 | 0.88934 | 3129.15 | 3484.89 | 8.1912 |
| 600 | 1.34136 | 3300.79 | 3703.2 | 8.5892 | 1.00555 | 3300.22 | 3702.44 | 8.4557 |
| 700 | 1.49573 | 3478.38 | 3927.1 | 8.8319 | 1.12147 | 3477.95 | 3926.53 | 8.6987 |
| 800 | 1.64994 | 3662.85 | 4157.83 | 9.0575 | 1.23722 | 3662.51 | 4157.4 | 8.9244 |
| 900 | 1.80406 | 3854.2 | 4395.42 | 9.2691 | 1.35288 | 3853.91 | 4395.06 | 9.1361 |
| 1000 | 1.95812 | 4052.27 | 4639.71 | 9.4689 | 1.46847 | 4052.02 | 4639.41 | 9.336 |
| 1100 | 2.11214 | 4256.77 | 4890.41 | 9.6585 | 1.58404 | 4256.53 | 4890.15 | 9.5255 |
| 1200 | 2.26614 | 4467.23 | 5147.07 | 9.8389 | 1.69958 | 4466.99 | 5146.83 | 9.7059 |
| 1300 | 2.42013 | 4682.99 | 5409.03 | 10.0109 | 1.81511 | 4682.75 | 5408.8 | 9.878 |
| | P = 500 kPa (151.86°C) | | | | P = 600 kPa (158.85°C) | | | |
| Sat. | 0.37489 | 2561.23 | 2748.67 | 6.8212 | 0.31567 | 2567.4 | 2756.8 | 6.76 |
| 200 | 0.42492 | 2642.91 | 2855.37 | 7.0592 | 0.35202 | 2638.91 | 2850.12 | 6.9665 |
| 250 | 0.47436 | 2723.5 | 2960.68 | 7.2708 | 0.39383 | 2720.86 | 2957.16 | 7.1816 |
| 300 | 0.52256 | 2802.91 | 3064.2 | 7.4598 | 0.43437 | 2801 | 3061.63 | 7.3723 |
| 350 | 0.57012 | 2882.59 | 3167.65 | 7.6328 | 0.47424 | 2881.12 | 3165.66 | 7.5463 |
| 400 | 0.61728 | 2963.19 | 3271.83 | 7.7937 | 0.51372 | 2962.02 | 3270.25 | 7.7078 |
| 500 | 0.71093 | 3128.35 | 3483.82 | 8.0872 | 0.59199 | 3127.55 | 3482.75 | 8.002 |
| 600 | 0.80406 | 3299.64 | 3701.67 | 8.3521 | 0.66974 | 3299.07 | 3700.91 | 8.2673 |
| 700 | 0.89691 | 3477.52 | 3925.97 | 8.5952 | 0.7472 | 3477.08 | 3925.41 | 8.5107 |
| 800 | 0.98959 | 3662.17 | 4156.96 | 8.8211 | 0.8245 | 3661.83 | 4156.52 | 8.7367 |
| 900 | 1.08217 | 3853.63 | 4394.71 | 9.0329 | 0.90169 | 3853.34 | 4394.36 | 8.9485 |
| 1000 | 1.17469 | 4051.76 | 4639.11 | 9.2328 | 0.97883 | 4051.51 | 4638.81 | 9.1484 |
| 1100 | 1.26718 | 4256.29 | 4889.88 | 9.4224 | 1.05594 | 4256.05 | 4889.61 | 9.3381 |
| 1200 | 1.35964 | 4466.76 | 5146.58 | 9.6028 | 1.13302 | 4466.52 | 5146.34 | 9.5185 |
| 1300 | 1.4521 | 4682.52 | 5408.57 | 9.7749 | 1.21009 | 4682.28 | 5408.34 | 9.6906 |
| | P = 800 kPa (170.43°C) | | | | P = 1000 kPa (179.91°C) | | | |
| Sat. | 0.24043 | 2576.79 | 2769.13 | 6.6627 | 0.19444 | 2583.64 | 2778.08 | 6.5864 |
| 200 | 0.2608 | 2630.61 | 2839.25 | 6.8158 | 0.20596 | 2621.9 | 2827.86 | 6.6939 |
| 250 | 0.29314 | 2715.46 | 2949.97 | 7.0384 | 0.23268 | 2709.91 | 2942.59 | 6.9246 |
| 300 | 0.32411 | 2797.14 | 3056.43 | 7.2327 | 0.25794 | 2793.21 | 3051.15 | 7.1228 |
| 350 | 0.35439 | 2878.16 | 3161.68 | 7.4088 | 0.28247 | 2875.18 | 3157.65 | 7.301 |
| 400 | 0.38426 | 2959.66 | 3267.07 | 7.5715 | 0.30659 | 2957.29 | 3263.88 | 7.465 |
| 500 | 0.44331 | 3125.95 | 3480.6 | 7.8672 | 0.35411 | 3124.34 | 3478.44 | 7.7621 |
| 600 | 0.50184 | 3297.91 | 3699.38 | 8.1332 | 0.40109 | 3296.76 | 3697.85 | 8.0289 |
| 700 | 0.56007 | 3476.22 | 3924.27 | 8.377 | 0.44779 | 3475.35 | 3923.14 | 8.2731 |
| 800 | 0.61813 | 3661.14 | 4155.65 | 8.6033 | 0.49432 | 3660.46 | 4154.78 | 8.4996 |
| 900 | 0.6761 | 3852.77 | 4393.65 | 8.8153 | 0.54075 | 3852.19 | 4392.94 | 8.7118 |
| 1000 | 0.73401 | 4051 | 4638.2 | 9.0153 | 0.58712 | 4050.49 | 4637.6 | 8.9119 |
| 1100 | 0.79188 | 4255.57 | 4889.08 | 9.2049 | 0.63345 | 4255.09 | 4888.55 | 9.1016 |
| 1200 | 0.84974 | 4466.05 | 5145.85 | 9.3854 | 0.67977 | 4465.58 | 5145.36 | 9.2821 |
| 1300 | 0.90758 | 4681.81 | 5407.87 | 9.5575 | 0.72608 | 4681.33 | 5407.41 | 9.4542 |

| *T* (°C) | *v* (m³/kg) | *u* (kJ/kg) | *h* (kJ/kg) | *s* (kJ/kg-K) | *v* (m³/kg) | *u* (kJ/kg) | *h* (kJ/kg) | *s* (kJ/kg-K) |
|---|---|---|---|---|---|---|---|---|
| | **P = 1200 kPa (187.99°C)** | | | | **P = 1400 kPa (195.07°C)** | | | |
| Sat. | 0.16333 | 2588.82 | 2784.82 | 6.5233 | 0.14084 | 2592.83 | 2790 | 6.4692 |
| 200 | 0.1693 | 2612.74 | 2815.9 | 6.5898 | 0.14302 | 2603.09 | 2803.32 | 6.4975 |
| 250 | 0.19235 | 2704.2 | 2935.01 | 6.8293 | 0.1635 | 2698.32 | 2927.22 | 6.7467 |
| 300 | 0.21382 | 2789.22 | 3045.8 | 7.0316 | 0.18228 | 2785.16 | 3040.35 | 6.9533 |
| 350 | 0.23452 | 2872.16 | 3153.59 | 7.212 | 0.20026 | 2869.12 | 3149.49 | 7.1359 |
| 400 | 0.2548 | 2954.9 | 3260.66 | 7.3773 | 0.2178 | 2952.5 | 3257.42 | 7.3025 |
| 500 | 0.29463 | 3122.72 | 3476.28 | 7.6758 | 0.25215 | 3121.1 | 3474.11 | 7.6026 |
| 600 | 0.33393 | 3295.6 | 3696.32 | 7.9434 | 0.28596 | 3294.44 | 3694.78 | 7.871 |
| 700 | 0.37294 | 3474.48 | 3922.01 | 8.1881 | 0.31947 | 3473.61 | 3920.87 | 8.116 |
| 800 | 0.41177 | 3659.77 | 4153.9 | 8.4149 | 0.35281 | 3659.09 | 4153.03 | 8.3431 |
| 900 | 0.45051 | 3851.62 | 4392.23 | 8.6272 | 0.38606 | 3851.05 | 4391.53 | 8.5555 |
| 1000 | 0.48919 | 4049.98 | 4637 | 8.8274 | 0.41924 | 4049.47 | 4636.41 | 8.7558 |
| 1100 | 0.52783 | 4254.61 | 4888.02 | 9.0171 | 0.45239 | 4254.14 | 4887.49 | 8.9456 |
| 1200 | 0.56646 | 4465.12 | 5144.87 | 9.1977 | 0.48552 | 4464.65 | 5144.38 | 9.1262 |
| 1300 | 0.60507 | 4680.86 | 5406.95 | 9.3698 | 0.51864 | 4680.39 | 5406.49 | 9.2983 |
| | **P = 1600 kPa (201.40°C)** | | | | **P = 1800 kPa (207.15°C)** | | | |
| Sat. | 0.1238 | 2595.95 | 2794.02 | 6.4217 | 0.11042 | 2598.38 | 2797.13 | 6.3793 |
| 250 | 0.14184 | 2692.26 | 2919.2 | 6.6732 | 0.12497 | 2686.02 | 2910.96 | 6.6066 |
| 300 | 0.15862 | 2781.03 | 3034.83 | 6.8844 | 0.14021 | 2776.83 | 3029.21 | 6.8226 |
| 350 | 0.17456 | 2866.05 | 3145.35 | 7.0693 | 0.15457 | 2862.95 | 3141.18 | 7.0099 |
| 400 | 0.19005 | 2950.09 | 3254.17 | 7.2373 | 0.16847 | 2947.66 | 3250.9 | 7.1793 |
| 500 | 0.22029 | 3119.47 | 3471.93 | 7.5389 | 0.1955 | 3117.84 | 3469.75 | 7.4824 |
| 600 | 0.24998 | 3293.27 | 3693.23 | 7.808 | 0.22199 | 3292.1 | 3691.69 | 7.7523 |
| 700 | 0.27937 | 3472.74 | 3919.73 | 8.0535 | 0.24818 | 3471.87 | 3918.59 | 7.9983 |
| 800 | 0.30859 | 3658.4 | 4152.15 | 8.2808 | 0.2742 | 3657.71 | 4151.27 | 8.2258 |
| 900 | 0.33772 | 3850.47 | 4390.82 | 8.4934 | 0.30012 | 3849.9 | 4390.11 | 8.4386 |
| 1000 | 0.36678 | 4048.96 | 4635.81 | 8.6938 | 0.32598 | 4048.45 | 4635.21 | 8.639 |
| 1100 | 0.39581 | 4253.66 | 4886.95 | 8.8837 | 0.3518 | 4253.18 | 4886.42 | 8.829 |
| 1200 | 0.42482 | 4464.18 | 5143.89 | 9.0642 | 0.37761 | 4463.71 | 5143.4 | 9.0096 |
| 1300 | 0.45382 | 4679.92 | 5406.02 | 9.2364 | 0.4034 | 4679.44 | 5405.56 | 9.1817 |
| | **P = 2000 kPa (212.42°C)** | | | | **P = 2500 kPa (223.99°C)** | | | |
| Sat. | 0.09963 | 2600.26 | 2799.51 | 6.3408 | 0.07998 | 2603.13 | 2803.07 | 6.2574 |
| 250 | 0.11144 | 2679.58 | 2902.46 | 6.5452 | 0.087 | 2662.55 | 2880.06 | 6.4084 |
| 300 | 0.12547 | 2772.56 | 3023.5 | 6.7663 | 0.0989 | 2761.56 | 3008.81 | 6.6437 |
| 350 | 0.13857 | 2859.81 | 3136.96 | 6.9562 | 0.10976 | 2851.84 | 3126.24 | 6.8402 |
| 400 | 0.1512 | 2945.21 | 3247.6 | 7.127 | 0.1201 | 2939.03 | 3239.28 | 7.0147 |
| 450 | 0.16353 | 3030.41 | 3357.48 | 7.2844 | 0.13014 | 3025.43 | 3350.77 | 7.1745 |
| 500 | 0.17568 | 3116.2 | 3467.55 | 7.4316 | 0.13998 | 3112.08 | 3462.04 | 7.3233 |
| 600 | 0.1996 | 3290.93 | 3690.14 | 7.7023 | 0.1593 | 3287.99 | 3686.25 | 7.596 |
| 700 | 0.22323 | 3470.99 | 3917.45 | 7.9487 | 0.17832 | 3468.8 | 3914.59 | 7.8435 |
| 800 | 0.24668 | 3657.03 | 4150.4 | 8.1766 | 0.19716 | 3655.3 | 4148.2 | 8.072 |

| *T* (°C) | *v* (m³/kg) | *u* (kJ/kg) | *h* (kJ/kg) | *s* (kJ/kg-K) | *v* (m³/kg) | *u* (kJ/kg) | *h* (kJ/kg) | *s* (kJ/kg-K) |
|---|---|---|---|---|---|---|---|---|
| 900 | 0.27004 | 3849.33 | 4389.4 | 8.3895 | 0.2159 | 3847.89 | 4387.64 | 8.2853 |
| 1000 | 0.29333 | 4047.94 | 4634.61 | 8.59 | 0.23458 | 4046.67 | 4633.12 | 8.486 |
| 1100 | 0.31659 | 4252.71 | 4885.89 | 8.78 | 0.25322 | 4251.52 | 4884.57 | 8.6761 |
| 1200 | 0.33984 | 4463.25 | 5142.92 | 8.9606 | 0.27185 | 4462.08 | 5141.7 | 8.8569 |
| 1300 | 0.36306 | 4678.97 | 5405.1 | 9.1328 | 0.29046 | 4677.8 | 5403.95 | 9.0291 |
| | **P = 3000 kPa (233.90°C)** | | | | **P = 4000 kPa (250.40°C)** | | | |
| Sat. | 0.06668 | 2604.1 | 2804.14 | 6.1869 | 0.04978 | 2602.27 | 2801.38 | 6.07 |
| 250 | 0.07058 | 2644 | 2855.75 | 6.2871 | — | — | — | — |
| 300 | 0.08114 | 2750.05 | 2993.48 | 6.5389 | 0.05884 | 2725.33 | 2960.68 | 6.3614 |
| 350 | 0.09053 | 2843.66 | 3115.25 | 6.7427 | 0.06645 | 2826.65 | 3092.43 | 6.582 |
| 400 | 0.09936 | 2932.75 | 3230.82 | 6.9211 | 0.07341 | 2919.88 | 3213.51 | 6.7689 |
| 450 | 0.10787 | 3020.38 | 3344 | 7.0833 | 0.08003 | 3010.13 | 3330.23 | 6.9362 |
| 500 | 0.11619 | 3107.92 | 3456.48 | 7.2337 | 0.08643 | 3099.49 | 3445.21 | 7.09 |
| 600 | 0.13243 | 3285.03 | 3682.34 | 7.5084 | 0.09885 | 3279.06 | 3674.44 | 7.3688 |
| 700 | 0.14838 | 3466.59 | 3911.72 | 7.7571 | 0.11095 | 3462.15 | 3905.94 | 7.6198 |
| 800 | 0.16414 | 3653.58 | 4146 | 7.9862 | 0.12287 | 3650.11 | 4141.59 | 7.8502 |
| 900 | 0.1798 | 3846.46 | 4385.87 | 8.1999 | 0.13469 | 3843.59 | 4382.34 | 8.0647 |
| 1000 | 0.19541 | 4045.4 | 4631.63 | 8.4009 | 0.14645 | 4042.87 | 4628.65 | 8.2661 |
| 1100 | 0.21098 | 4250.33 | 4883.26 | 8.5911 | 0.15817 | 4247.96 | 4880.63 | 8.4566 |
| 1200 | 0.22652 | 4460.92 | 5140.49 | 8.7719 | 0.16987 | 4458.6 | 5138.07 | 8.6376 |
| 1300 | 0.24206 | 4676.63 | 5402.81 | 8.9442 | 0.18156 | 4674.29 | 5400.52 | 8.8099 |
| | **P = 5000 kPa (263.99°C)** | | | | **P = 6000 kPa (275.64°C)** | | | |
| Sat. | 0.03944 | 2597.12 | 2794.33 | 5.9733 | 0.03244 | 2589.69 | 2784.33 | 5.8891 |
| 300 | 0.04532 | 2697.94 | 2924.53 | 6.2083 | 0.03616 | 2667.22 | 2884.19 | 6.0673 |
| 350 | 0.05194 | 2808.67 | 3068.39 | 6.4492 | 0.04223 | 2789.61 | 3042.97 | 6.3334 |
| 400 | 0.05781 | 2906.58 | 3195.64 | 6.6458 | 0.04739 | 2892.81 | 3177.17 | 6.5407 |
| 450 | 0.0633 | 2999.64 | 3316.15 | 6.8185 | 0.05214 | 2988.9 | 3301.76 | 6.7192 |
| 500 | 0.06857 | 3090.92 | 3433.76 | 6.9758 | 0.05665 | 3082.2 | 3422.12 | 6.8802 |
| 550 | 0.07368 | 3181.82 | 3550.23 | 7.1217 | 0.06101 | 3174.57 | 3540.62 | 7.0287 |
| 600 | 0.07869 | 3273.01 | 3666.47 | 7.2588 | 0.06525 | 3266.89 | 3658.4 | 7.1676 |
| 700 | 0.08849 | 3457.67 | 3900.13 | 7.5122 | 0.07352 | 3453.15 | 3894.28 | 7.4234 |
| 800 | 0.09811 | 3646.62 | 4137.17 | 7.744 | 0.0816 | 3643.12 | 4132.74 | 7.6566 |
| 900 | 0.10762 | 3840.71 | 4378.82 | 7.9593 | 0.08958 | 3837.84 | 4375.29 | 7.8727 |
| 1000 | 0.11707 | 4040.35 | 4625.69 | 8.1612 | 0.09749 | 4037.83 | 4622.74 | 8.0751 |
| 1100 | 0.12648 | 4245.61 | 4878.02 | 8.3519 | 0.10536 | 4243.26 | 4875.42 | 8.2661 |
| 1200 | 0.13587 | 4456.3 | 5135.67 | 8.533 | 0.11321 | 4454 | 5133.28 | 8.4473 |
| 1300 | 0.14526 | 4671.96 | 5398.24 | 8.7055 | 0.12106 | 4669.64 | 5395.97 | 8.6199 |
| | **P = 8000 kPa (295.06°C)** | | | | **P = 10000 kPa (311.06°C)** | | | |
| Sat. | 0.02352 | 2569.79 | 2757.94 | 5.7431 | 0.01803 | 2544.41 | 2724.67 | 5.614 |
| 300 | 0.02426 | 2590.93 | 2784.98 | 5.7905 | — | — | — | — |
| 350 | 0.02995 | 2747.67 | 2987.3 | 6.13 | 0.02242 | 2699.16 | 2923.39 | 5.9442 |
| 400 | 0.03432 | 2863.75 | 3138.28 | 6.3633 | 0.02641 | 2832.38 | 3096.46 | 6.2119 |

| *T* (°C) | *v* (m³/kg) | *u* (kJ/kg) | *h* (kJ/kg) | *s* (kJ/kg-K) | *v* (m³/kg) | *u* (kJ/kg) | *h* (kJ/kg) | *s* (kJ/kg-K) |
|---|---|---|---|---|---|---|---|---|
| 450 | 0.03817 | 2966.66 | 3271.99 | 6.555 | 0.02975 | 2943.32 | 3240.83 | 6.4189 |
| 500 | 0.04175 | 3064.3 | 3398.27 | 6.7239 | 0.03279 | 3045.77 | 3373.63 | 6.5965 |
| 550 | 0.04516 | 3159.76 | 3521.01 | 6.8778 | 0.03564 | 3144.54 | 3500.92 | 6.7561 |
| 600 | 0.04845 | 3254.43 | 3642.03 | 7.0205 | 0.03837 | 3241.68 | 3625.34 | 6.9028 |
| 700 | 0.05481 | 3444 | 3882.47 | 7.2812 | 0.04358 | 3434.72 | 3870.52 | 7.1687 |
| 800 | 0.06097 | 3636.08 | 4123.84 | 7.5173 | 0.04859 | 3628.97 | 4114.91 | 7.4077 |
| 900 | 0.06702 | 3832.08 | 4368.26 | 7.735 | 0.05349 | 3826.32 | 4361.24 | 7.6272 |
| 1000 | 0.07301 | 4032.81 | 4616.87 | 7.9384 | 0.05832 | 4027.81 | 4611.04 | 7.8315 |
| 1100 | 0.07896 | 4238.6 | 4870.25 | 8.1299 | 0.06312 | 4233.97 | 4865.14 | 8.0236 |
| 1200 | 0.08489 | 4449.45 | 5128.54 | 8.3115 | 0.06789 | 4444.93 | 5123.84 | 8.2054 |
| 1300 | 0.0908 | 4665.02 | 5391.46 | 8.4842 | 0.07265 | 4660.44 | 5386.99 | 8.3783 |
| | P = 15000 kPa (342.24°C) | | | | P = 20000 kPa (365.81°C) | | | |
| Sat. | 0.01034 | 2455.43 | 2610.49 | 5.3097 | 0.00583 | 2293.05 | 2409.74 | 4.9269 |
| 350 | 0.01147 | 2520.36 | 2692.41 | 5.442 | — | — | — | — |
| 400 | 0.01565 | 2740.7 | 2975.44 | 5.881 | 0.00994 | 2619.22 | 2818.07 | 5.5539 |
| 450 | 0.01845 | 2879.47 | 3156.15 | 6.1403 | 0.0127 | 2806.16 | 3060.06 | 5.9016 |
| 500 | 0.0208 | 2996.52 | 3308.53 | 6.3442 | 0.01477 | 2942.82 | 3238.18 | 6.14 |
| 550 | 0.02293 | 3104.71 | 3448.61 | 6.5198 | 0.01656 | 3062.34 | 3393.45 | 6.3347 |
| 600 | 0.02491 | 3208.64 | 3582.3 | 6.6775 | 0.01818 | 3174 | 3537.57 | 6.5048 |
| 650 | 0.0268 | 3310.37 | 3712.32 | 6.8223 | 0.01969 | 3281.46 | 3675.32 | 6.6582 |
| 700 | 0.02861 | 3410.94 | 3840.12 | 6.9572 | 0.02113 | 3386.46 | 3809.09 | 6.7993 |
| 800 | 0.0321 | 3610.99 | 4092.43 | 7.204 | 0.02385 | 3592.73 | 4069.8 | 7.0544 |
| 900 | 0.03546 | 3811.89 | 4343.75 | 7.4279 | 0.02645 | 3797.44 | 4326.37 | 7.283 |
| 1000 | 0.03875 | 4015.41 | 4596.63 | 7.6347 | 0.02897 | 4003.12 | 4582.45 | 7.4925 |
| 1100 | 0.042 | 4222.55 | 4852.56 | 7.8282 | 0.03145 | 4211.3 | 4840.24 | 7.6874 |
| 1200 | 0.04523 | 4433.78 | 5112.27 | 8.0108 | 0.03391 | 4422.81 | 5100.96 | 7.8706 |
| 1300 | 0.04845 | 4649.12 | 5375.94 | 8.1839 | 0.03636 | 4637.95 | 5365.1 | 8.0441 |
| | P = 30000 kPa | | | | P = 40000 kPa | | | |
| 375 | 0.001789 | 1737.75 | 1791.43 | 3.9303 | 0.001641 | 1677.09 | 1742.71 | 3.8289 |
| 400 | 0.00279 | 2067.34 | 2151.04 | 4.4728 | 0.001908 | 1854.52 | 1930.83 | 4.1134 |
| 425 | 0.005304 | 2455.06 | 2614.17 | 5.1503 | 0.002532 | 2096.83 | 2198.11 | 4.5028 |
| 450 | 0.006735 | 2619.3 | 2821.35 | 5.4423 | 0.003693 | 2365.07 | 2512.79 | 4.9459 |
| 500 | 0.008679 | 2820.67 | 3081.03 | 5.7904 | 0.005623 | 2678.36 | 2903.26 | 5.4699 |
| 550 | 0.010168 | 2970.31 | 3275.36 | 6.0342 | 0.006984 | 2869.69 | 3149.05 | 5.7784 |
| 600 | 0.011446 | 3100.53 | 3443.91 | 6.233 | 0.008094 | 3022.61 | 3346.38 | 6.0113 |
| 650 | 0.012596 | 3221.04 | 3598.93 | 6.4057 | 0.009064 | 3158.04 | 3520.58 | 6.2054 |
| 700 | 0.013661 | 3335.84 | 3745.67 | 6.5606 | 0.009942 | 3283.63 | 3681.29 | 6.375 |
| 800 | 0.015623 | 3555.6 | 4024.31 | 6.8332 | 0.011523 | 3517.89 | 3978.8 | 6.6662 |
| 900 | 0.017448 | 3768.48 | 4291.93 | 7.0717 | 0.012963 | 3739.42 | 4257.93 | 6.915 |
| 1000 | 0.019196 | 3978.79 | 4554.68 | 7.2867 | 0.014324 | 3954.64 | 4527.59 | 7.1356 |
| 1100 | 0.020903 | 4189.18 | 4816.28 | 7.4845 | 0.015643 | 4167.38 | 4793.08 | 7.3364 |
| 1200 | 0.022589 | 4401.29 | 5078.97 | 7.6691 | 0.01694 | 4380.11 | 5057.72 | 7.5224 |

| T (°C) | v (m³/kg) | u (kJ/kg) | h (kJ/kg) | s (kJ/kg-K) | v (m³/kg) | u (kJ/kg) | h (kJ/kg) | s (kJ/kg-K) |
|---|---|---|---|---|---|---|---|---|
| 1300 | 0.024266 | 4615.96 | 5343.95 | 7.8432 | 0.018229 | 4594.28 | 5323.45 | 7.6969 |

## Table B.4: Compressed steam tables.

| T (°C) | v (m³/kg) | u (kJ/kg) | h (kJ/kg) | s (kJ/kg-K) | v (m³/kg) | u (kJ/kg) | h (kJ/kg) | s (kJ/kg-K) |
|---|---|---|---|---|---|---|---|---|
| | P = 500 kPa (151.86°C) | | | | P = 2000 kPa (212.42°C) | | | |
| Sat. | 0.001093 | 639.66 | 640.21 | 1.8606 | 0.001177 | 906.42 | 908.77 | 2.4473 |
| 0.01 | 0.000999 | 0.01 | 0.51 | 0 | 0.000999 | 0.03 | 2.03 | 0.0001 |
| 20 | 0.001002 | 83.91 | 84.41 | 0.2965 | 0.001001 | 83.82 | 85.82 | 0.2962 |
| 40 | 0.001008 | 167.47 | 167.98 | 0.5722 | 0.001007 | 167.29 | 169.3 | 0.5716 |
| 60 | 0.001017 | 251 | 251.51 | 0.8308 | 0.001016 | 250.73 | 252.77 | 0.83 |
| 80 | 0.001029 | 334.73 | 335.24 | 1.0749 | 0.001028 | 334.38 | 336.44 | 1.0739 |
| 100 | 0.001043 | 418.8 | 419.32 | 1.3065 | 0.001043 | 418.36 | 420.45 | 1.3053 |
| 120 | 0.00106 | 503.37 | 503.9 | 1.5273 | 0.001059 | 502.84 | 504.96 | 1.5259 |
| 140 | 0.00108 | 588.66 | 589.2 | 1.7389 | 0.001079 | 588.02 | 590.18 | 1.7373 |
| 160 | — | — | — | — | 0.001101 | 674.14 | 676.34 | 1.941 |
| 180 | — | — | — | — | 0.001127 | 761.46 | 763.71 | 2.1382 |
| 200 | — | — | — | — | 0.001156 | 850.3 | 852.61 | 2.3301 |
| | P = 5000 kPa (263.99°C) | | | | P = 10000 kPa (311.06°C) | | | |
| Sat | 0.001286 | 1147.78 | 1154.21 | 2.9201 | 0.001452 | 1393 | 1407.53 | 3.3595 |
| 0 | 0.000998 | 0.03 | 5.02 | 0.0001 | 0.000995 | 0.1 | 10.05 | 0.0003 |
| 20 | 0.001 | 83.64 | 88.64 | 0.2955 | 0.000997 | 83.35 | 93.32 | 0.2945 |
| 40 | 0.001006 | 166.93 | 171.95 | 0.5705 | 0.001003 | 166.33 | 176.36 | 0.5685 |
| 60 | 0.001015 | 250.21 | 255.28 | 0.8284 | 0.001013 | 249.34 | 259.47 | 0.8258 |
| 80 | 0.001027 | 333.69 | 338.83 | 1.0719 | 0.001025 | 332.56 | 342.81 | 1.0687 |
| 100 | 0.001041 | 417.5 | 422.71 | 1.303 | 0.001039 | 416.09 | 426.48 | 1.2992 |
| 120 | 0.001058 | 501.79 | 507.07 | 1.5232 | 0.001055 | 500.07 | 510.61 | 1.5188 |
| 140 | 0.001077 | 586.74 | 592.13 | 1.7342 | 0.001074 | 584.67 | 595.4 | 1.7291 |
| 160 | 0.001099 | 672.61 | 678.1 | 1.9374 | 0.001195 | 670.11 | 681.07 | 1.9316 |
| 180 | 0.001124 | 759.62 | 765.24 | 2.1341 | 0.00112 | 756.63 | 767.83 | 2.1274 |
| 200 | 0.001153 | 848.08 | 853.85 | 2.3254 | 0.001148 | 844.49 | 855.97 | 2.3178 |
| 220 | 0.001187 | 938.43 | 944.36 | 2.5128 | 0.001181 | 934.07 | 945.88 | 2.5038 |
| 240 | 0.001226 | 1031.34 | 1037.47 | 2.6978 | 0.001219 | 1025.94 | 1038.13 | 2.6872 |
| 260 | 0.001275 | 1127.92 | 1134.3 | 2.8829 | 0.001265 | 1121.03 | 1133.68 | 2.8698 |
| 280 | 0.001322 | 1220.9 | 1234.11 | 3.0547 | | | | |
| 300 | 0.001397 | 1328.34 | 1342.31 | 3.2468 | | | | |
| | P = 15000 kPa (342.24°C) | | | | P = 20000 kPa (365.81°C) | | | |
| Sat. | 0.001658 | 1585.58 | 1610.45 | 3.6847 | 0.002035 | 1785.47 | 1826.18 | 4.0137 |
| 0 | 0.000993 | 0.15 | 15.04 | 0.0004 | 0.00099 | 0.2 | 20 | 0.0004 |
| 20 | 0.000995 | 83.05 | 97.97 | 0.2934 | 0.000993 | 82.75 | 102.61 | 0.2922 |
| 40 | 0.001001 | 165.73 | 180.75 | 0.5665 | 0.000999 | 165.15 | 185.14 | 0.5646 |

| *T* (°C) | *v* (m³/kg) | *u* (kJ/kg) | *h* (kJ/kg) | *s* (kJ/kg-K) | *v* (m³/kg) | *u* (kJ/kg) | *h* (kJ/kg) | *s* (kJ/kg-K) |
|---|---|---|---|---|---|---|---|---|
| 60 | 0.001011 | 248.49 | 263.65 | 0.8231 | 0.001008 | 247.66 | 267.82 | 0.8205 |
| 80 | 0.001022 | 331.46 | 346.79 | 1.0655 | 0.00102 | 330.38 | 350.78 | 1.0623 |
| 100 | 0.001036 | 414.72 | 430.26 | 1.2954 | 0.001034 | 413.37 | 434.04 | 1.2917 |
| 120 | 0.001052 | 498.39 | 514.17 | 1.5144 | 0.00105 | 496.75 | 517.74 | 1.5101 |
| 140 | 0.001071 | 582.64 | 598.7 | 1.7241 | 0.001068 | 580.67 | 602.03 | 1.7192 |
| 160 | 0.001092 | 667.69 | 684.07 | 1.9259 | 0.001089 | 665.34 | 687.11 | 1.9203 |
| 180 | 0.001116 | 753.74 | 770.48 | 2.1209 | 0.001112 | 750.94 | 773.18 | 2.1146 |
| 200 | 0.001143 | 841.04 | 858.18 | 2.3103 | 0.001139 | 837.7 | 860.47 | 2.3031 |
| 220 | 0.001175 | 929.89 | 947.52 | 2.4952 | 0.001169 | 925.89 | 949.27 | 2.4869 |
| 240 | 0.001211 | 1020.82 | 1038.99 | 2.677 | 0.001205 | 1015.94 | 1040.04 | 2.6673 |
| 260 | 0.001255 | 1114.59 | 1133.41 | 2.8575 | 0.001246 | 1108.53 | 1133.45 | 2.8459 |
| 280 | 0.001308 | 1212.47 | 1232.09 | 3.0392 | 0.001297 | 1204.69 | 1230.62 | 3.0248 |
| 300 | 0.001377 | 1316.58 | 1337.23 | 3.2259 | 0.00136 | 1306.1 | 1333.29 | 3.2071 |
| 320 | 0.001472 | 1431.05 | 1453.13 | 3.4246 | 0.001444 | 1415.66 | 1444.53 | 3.3978 |
| 340 | 0.001631 | 1567.42 | 1591.88 | 3.6545 | 0.001568 | 1539.64 | 1571.01 | 3.6074 |
| 360 | 0.001823 | 1702.78 | 1739.23 | 3.877 | | | | |
| | P = 30000 kPa | | | | P = 50000 kPa | | | |
| 0 | 0.000986 | 0.25 | 29.82 | 0.0001 | 0.000977 | 0.2 | 49.03 | −0.0014 |
| 20 | 0.000989 | 82.16 | 111.82 | 0.2898 | 0.00098 | 80.98 | 130 | 0.2847 |
| 40 | 0.000995 | 164.01 | 193.87 | 0.5606 | 0.000987 | 161.84 | 211.2 | 0.5526 |
| 60 | 0.001004 | 246.03 | 276.16 | 0.8153 | 0.000996 | 242.96 | 292.77 | 0.8051 |
| 80 | 0.001016 | 328.28 | 358.75 | 1.0561 | 0.001007 | 324.32 | 374.68 | 1.0439 |
| 100 | 0.001029 | 410.76 | 441.63 | 1.2844 | 0.00102 | 405.86 | 456.87 | 1.2703 |
| 120 | 0.001044 | 493.58 | 524.91 | 1.5017 | 0.001035 | 487.63 | 539.37 | 1.4857 |
| 140 | 0.001062 | 576.86 | 608.73 | 1.7097 | 0.001052 | 569.76 | 622.33 | 1.6915 |
| 160 | 0.001082 | 660.81 | 693.27 | 1.9095 | 0.00107 | 652.39 | 705.91 | 1.889 |
| 180 | 0.001105 | 745.57 | 778.71 | 2.1024 | 0.001091 | 735.68 | 790.24 | 2.0793 |
| 200 | 0.00113 | 831.34 | 865.24 | 2.2892 | 0.001115 | 819.73 | 875.46 | 2.2634 |
| 220 | 0.001159 | 918.32 | 953.09 | 2.471 | 0.001141 | 904.67 | 961.71 | 2.4419 |
| 240 | 0.001192 | 1006.84 | 1042.6 | 2.6489 | 0.00117 | 990.69 | 1049.2 | 2.6158 |
| 260 | 0.00123 | 1097.38 | 1134.29 | 2.8242 | 0.001203 | 1078.06 | 1138.23 | 2.786 |
| 280 | 0.001275 | 1190.69 | 1228.96 | 2.9985 | 0.001242 | 1167.19 | 1229.26 | 2.9536 |
| 300 | 0.00133 | 1287.89 | 1327.8 | 3.174 | 0.001286 | 1258.66 | 1322.95 | 3.12 |
| 320 | 0.0014 | 1390.64 | 1432.63 | 3.3538 | 0.001339 | 1353.23 | 1420.17 | 3.2867 |
| 340 | 0.001492 | 1501.71 | 1546.47 | 3.5425 | 0.001403 | 1451.91 | 1522.07 | 3.4556 |
| 360 | 0.001627 | 1626.57 | 1675.36 | 3.7492 | 0.001484 | 1555.97 | 1630.16 | 3.629 |
| 380 | 0.001869 | 1781.35 | 1837.43 | 4.001 | 0.001588 | 1667.13 | 1746.54 | 3.81 |

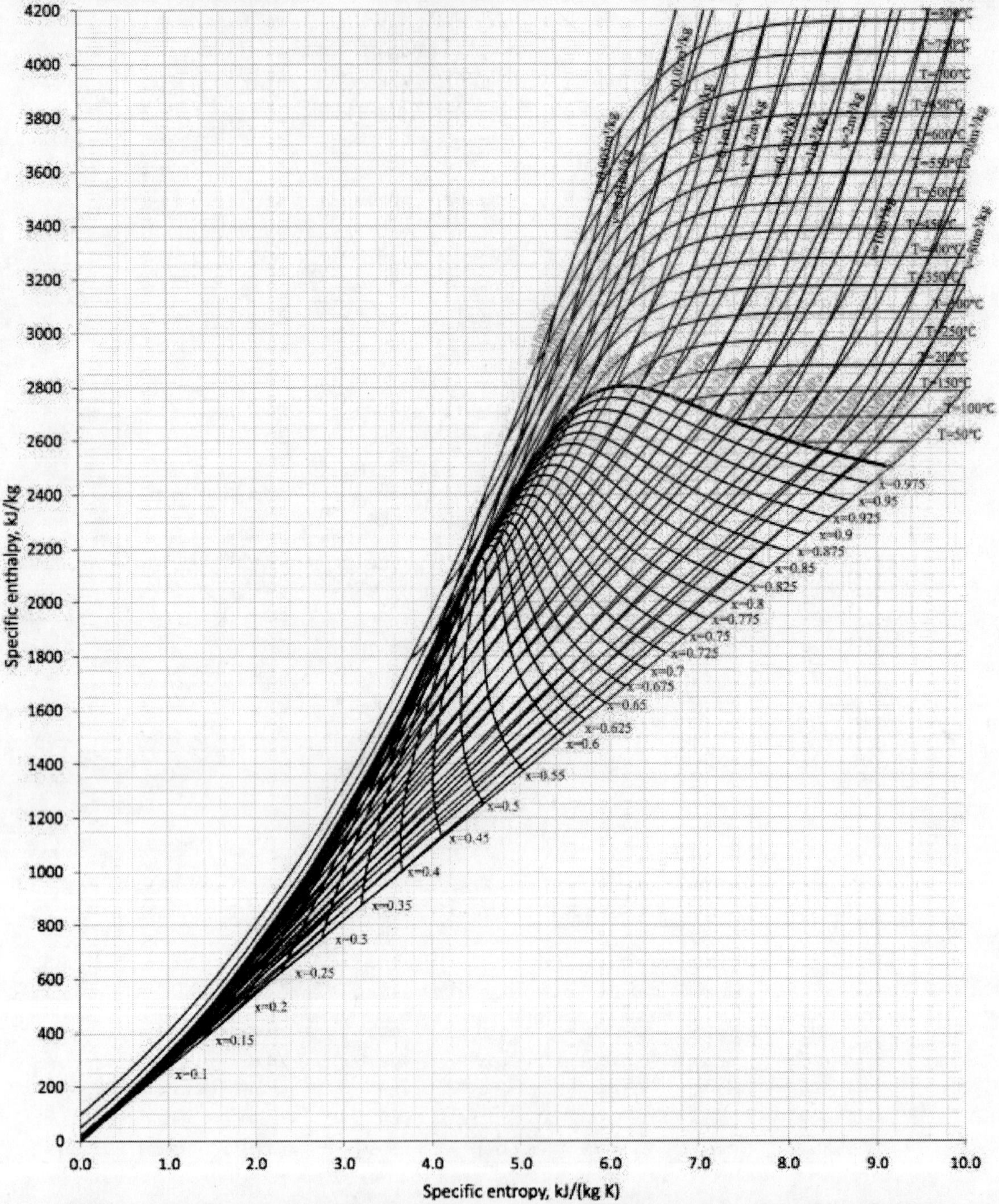


**Fig. B.1:** Mollier diagram for water (Source: Kim, C., Kim, H. and Mun, K., 2020. Use of the international association for the properties of water and steam (IAPWS) formulations, IAPWS-95 & IAPWS-IF97: Making of Mollier diagram and Ts diagram of water and steam. *Thermal Science and Engineering Progress*, *20*, p.100691).

# APPENDIX-C: Properties of refrigerant R-134a

**Table C.1: Saturated R-134a table.**

| Temp. (°C) | Sat. Press. (kPa) | Specific volume ($m^3$/kg) | | Specific internal energy (kJ/kg) | | | Specific enthalpy (kJ/kg) | | | Specific entropy (kJ/kg-K) | | |
|---|---|---|---|---|---|---|---|---|---|---|---|---|
| $T$ | $P$ | $v_f \times 10^3$ | $v_g$ | $u_f$ | $u_{fg}$ | $u_g$ | $h_f$ | $h_{fg}$ | $h_g$ | $s_f$ | $s_{fg}$ | $s_g$ |
| −70 | 8.3 | 0.675 | 1.97274 | 119.46 | 218.74 | 338.2 | 119.47 | 235.15 | 354.62 | 0.6645 | 1.1575 | 1.822 |
| −65 | 11.7 | 0.679 | 1.42983 | 123.18 | 217.76 | 340.94 | 123.18 | 234.55 | 357.73 | 0.6825 | 1.1268 | 1.8094 |
| −60 | 16.3 | 0.684 | 1.05268 | 127.52 | 216.19 | 343.71 | 127.53 | 233.33 | 360.86 | 0.7031 | 1.0947 | 1.7978 |
| −55 | 22.2 | 0.689 | 0.78678 | 132.36 | 214.14 | 346.5 | 132.37 | 231.63 | 364 | 0.7256 | 1.0618 | 1.7874 |
| −50 | 29.9 | 0.695 | 0.59657 | 137.6 | 211.71 | 349.31 | 137.62 | 229.54 | 367.16 | 0.7493 | 1.0286 | 1.778 |
| −45 | 39.6 | 0.701 | 0.45853 | 143.15 | 208.99 | 352.15 | 143.18 | 227.14 | 370.32 | 0.774 | 0.9956 | 1.7695 |
| −40 | 51.8 | 0.708 | 0.35696 | 148.95 | 206.05 | 355 | 148.98 | 224.5 | 373.48 | 0.7991 | 0.9629 | 1.762 |
| −35 | 66.8 | 0.715 | 0.28122 | 154.93 | 202.93 | 357.86 | 154.98 | 221.67 | 376.64 | 0.8245 | 0.9308 | 1.7553 |
| −30 | 85.1 | 0.722 | 0.22402 | 161.06 | 199.67 | 360.73 | 161.12 | 218.68 | 379.8 | 0.8499 | 0.8994 | 1.7493 |
| −26.3 | 101.3 | 0.728 | 0.1902 | 165.73 | 197.16 | 362.89 | 165.8 | 216.36 | 382.16 | 0.869 | 0.8763 | 1.7453 |
| −25 | 107.2 | 0.73 | 0.1803 | 167.3 | 196.31 | 363.61 | 167.38 | 215.57 | 382.95 | 0.8754 | 0.8687 | 1.7441 |
| −20 | 133.7 | 0.738 | 0.14649 | 173.65 | 192.85 | 366.5 | 173.74 | 212.34 | 386.08 | 0.9007 | 0.8388 | 1.7395 |
| −15 | 165 | 0.746 | 0.12007 | 180.07 | 189.32 | 369.39 | 180.19 | 209 | 389.2 | 0.9258 | 0.8096 | 1.7354 |
| −10 | 201.7 | 0.755 | 0.09921 | 186.57 | 185.7 | 372.27 | 186.72 | 205.56 | 392.28 | 0.9507 | 0.7812 | 1.7319 |
| −5 | 244.5 | 0.764 | 0.08257 | 193.14 | 182.01 | 375.15 | 193.32 | 202.02 | 395.34 | 0.9755 | 0.7534 | 1.7288 |
| 0 | 294 | 0.773 | 0.06919 | 199.77 | 178.24 | 378.01 | 200 | 198.36 | 398.36 | 1 | 0.7262 | 1.7262 |
| 5 | 350.9 | 0.783 | 0.05833 | 206.48 | 174.38 | 380.85 | 206.75 | 194.57 | 401.32 | 1.0243 | 0.6995 | 1.7239 |
| 10 | 415.8 | 0.794 | 0.04945 | 213.25 | 170.42 | 383.67 | 213.58 | 190.65 | 404.23 | 1.0485 | 0.6733 | 1.7218 |
| 15 | 489.5 | 0.805 | 0.04213 | 220.1 | 166.35 | 386.45 | 220.49 | 186.58 | 407.07 | 1.0725 | 0.6475 | 1.72 |
| 20 | 572.8 | 0.817 | 0.03606 | 227.03 | 162.16 | 389.19 | 227.49 | 182.35 | 409.84 | 1.0963 | 0.622 | 1.7183 |
| 25 | 666.3 | 0.829 | 0.03098 | 234.04 | 157.83 | 391.87 | 234.59 | 177.92 | 412.51 | 1.1201 | 0.5967 | 1.7168 |
| 30 | 771 | 0.843 | 0.02671 | 241.14 | 153.34 | 394.48 | 241.79 | 173.29 | 415.08 | 1.1437 | 0.5716 | 1.7153 |
| 35 | 887.6 | 0.857 | 0.0231 | 248.34 | 148.68 | 397.02 | 249.1 | 168.42 | 417.52 | 1.1673 | 0.5465 | 1.7139 |
| 40 | 1017 | 0.873 | 0.02002 | 255.65 | 143.81 | 399.46 | 256.54 | 163.28 | 419.82 | 1.1909 | 0.5214 | 1.7123 |
| 45 | 1160.2 | 0.89 | 0.01739 | 263.08 | 138.71 | 401.79 | 264.11 | 157.85 | 421.96 | 1.2145 | 0.4962 | 1.7106 |
| 50 | 1318.1 | 0.908 | 0.01512 | 270.63 | 133.35 | 403.98 | 271.83 | 152.08 | 423.91 | 1.2381 | 0.4706 | 1.7088 |
| 55 | 1491.6 | 0.928 | 0.01316 | 278.33 | 127.68 | 406.01 | 279.72 | 145.93 | 425.65 | 1.2619 | 0.4447 | 1.7066 |
| 60 | 1681.8 | 0.951 | 0.01146 | 286.19 | 121.66 | 407.85 | 287.79 | 139.33 | 427.13 | 1.2857 | 0.4182 | 1.704 |
| 65 | 1889.9 | 0.976 | 0.00997 | 294.24 | 115.22 | 409.46 | 296.09 | 132.21 | 428.3 | 1.3099 | 0.391 | 1.7008 |
| 70 | 2117 | 1.005 | 0.00866 | 302.51 | 108.27 | 410.78 | 304.64 | 124.47 | 429.11 | 1.3343 | 0.3627 | 1.697 |
| 75 | 2364.4 | 1.038 | 0.00749 | 311.06 | 100.68 | 411.74 | 313.51 | 115.94 | 429.45 | 1.3592 | 0.333 | 1.6923 |
| 80 | 2633.6 | 1.078 | 0.00645 | 319.96 | 92.26 | 412.22 | 322.79 | 106.4 | 429.19 | 1.3849 | 0.3013 | 1.6862 |
| 85 | 2926.2 | 1.128 | 0.0055 | 329.35 | 82.67 | 412.01 | 332.65 | 95.45 | 428.1 | 1.4117 | 0.2665 | 1.6782 |
| 90 | 3244.5 | 1.195 | 0.00461 | 339.51 | 71.24 | 410.75 | 343.38 | 82.31 | 425.7 | 1.4404 | 0.2267 | 1.6671 |
| 95 | 3591.5 | 1.297 | 0.00373 | 351.17 | 56.25 | 407.42 | 355.83 | 64.98 | 420.81 | 1.4733 | 0.1765 | 1.6498 |
| 100 | 3973.2 | 1.557 | 0.00264 | 368.55 | 28.19 | 396.74 | 374.74 | 32.47 | 407.21 | 1.5228 | 0.087 | 1.6098 |
| 101.2 | 4064 | 1.969 | 0.00197 | 382.97 | 0 | 382.97 | 390.98 | 0 | 390.98 | 1.5658 | 0 | 1.5658 |

**Table C.2: Superheated R-134a tables.**

| T (°C) | v (m3/kg) | u (kJ/kg) | h (kJ/kg) | s (kJ/kg-K) | v (m3/kg) | u (kJ/kg) | h (kJ/kg) | s (kJ/kg-K) |
|---|---|---|---|---|---|---|---|---|
| | P = 50 kPa (−40.67°C) | | | | P = 100 kPa (−26.54°C) | | | |
| Sat. | 0.36889 | 354.61 | 373.06 | 1.7629 | 0.19257 | 362.73 | 381.98 | 1.7456 |
| −20 | 0.40507 | 368.57 | 388.82 | 1.8279 | 0.1986 | 367.36 | 387.22 | 1.7665 |
| −10 | 0.42222 | 375.53 | 396.64 | 1.8582 | 0.20765 | 374.51 | 395.27 | 1.7978 |
| 0 | 0.43921 | 382.63 | 404.59 | 1.8878 | 0.21652 | 381.76 | 403.41 | 1.8281 |
| 10 | 0.45608 | 389.9 | 412.7 | 1.917 | 0.22527 | 389.14 | 411.67 | 1.8578 |
| 20 | 0.47287 | 397.32 | 420.96 | 1.9456 | 0.23392 | 396.66 | 420.05 | 1.8869 |
| 30 | 0.48958 | 404.9 | 429.38 | 1.9739 | 0.2425 | 404.31 | 428.56 | 1.9155 |
| 40 | 0.50623 | 412.64 | 437.96 | 2.0017 | 0.25101 | 412.12 | 437.22 | 1.9436 |
| 50 | 0.52284 | 420.55 | 446.7 | 2.0292 | 0.25948 | 420.08 | 446.03 | 1.9712 |
| 60 | 0.53941 | 428.63 | 455.6 | 2.0563 | 0.26791 | 428.2 | 454.99 | 1.9985 |
| 70 | 0.55595 | 436.86 | 464.66 | 2.0831 | 0.27631 | 436.47 | 464.1 | 2.0255 |
| 80 | 0.57247 | 445.26 | 473.88 | 2.1096 | 0.28468 | 444.89 | 473.36 | 2.0521 |
| 90 | 0.58896 | 453.82 | 483.26 | 2.1358 | 0.29302 | 453.47 | 482.78 | 2.0784 |
| 100 | 0.60544 | 462.53 | 492.81 | 2.1617 | 0.30135 | 462.21 | 492.35 | 2.1044 |
| 110 | 0.6219 | 471.41 | 502.5 | 2.1874 | 0.30967 | 471.11 | 502.07 | 2.1301 |
| 120 | 0.63835 | 480.44 | 512.36 | 2.2128 | 0.31797 | 480.16 | 511.95 | 2.1555 |
| 130 | 0.65479 | 489.63 | 522.37 | 2.2379 | 0.32626 | 489.36 | 521.98 | 2.1807 |
| | P = 150 kPa (−17.29°C) | | | | P = 200 kPa (−10.22°C) | | | |
| Sat. | 0.13139 | 368.06 | 387.77 | 1.7372 | 0.10002 | 372.15 | 392.15 | 1.732 |
| −10 | 0.13602 | 373.44 | 393.84 | 1.7606 | 0.10013 | 372.31 | 392.34 | 1.7328 |
| 0 | 0.14222 | 380.85 | 402.19 | 1.7917 | 0.10501 | 379.91 | 400.91 | 1.7647 |
| 10 | 0.14828 | 388.36 | 410.6 | 1.822 | 0.10974 | 387.55 | 409.5 | 1.7956 |
| 20 | 0.15424 | 395.98 | 419.11 | 1.8515 | 0.11436 | 395.27 | 418.15 | 1.8256 |
| 30 | 0.16011 | 403.71 | 427.73 | 1.8804 | 0.11889 | 403.1 | 426.87 | 1.8549 |
| 40 | 0.16592 | 411.59 | 436.47 | 1.9088 | 0.12335 | 411.04 | 435.71 | 1.8836 |
| 50 | 0.17168 | 419.6 | 445.35 | 1.9367 | 0.12776 | 419.11 | 444.66 | 1.9117 |
| 60 | 0.1774 | 427.76 | 454.37 | 1.9642 | 0.13213 | 427.31 | 453.74 | 1.9394 |
| 70 | 0.18308 | 436.06 | 463.53 | 1.9913 | 0.13646 | 435.65 | 462.95 | 1.9666 |
| 80 | 0.18874 | 444.52 | 472.83 | 2.018 | 0.14076 | 444.14 | 472.3 | 1.9935 |
| 90 | 0.19437 | 453.13 | 482.28 | 2.0444 | 0.14504 | 452.78 | 481.79 | 2.02 |
| 100 | 0.19999 | 461.89 | 491.89 | 2.0705 | 0.1493 | 461.56 | 491.42 | 2.0461 |
| 110 | 0.20559 | 470.8 | 501.64 | 2.0963 | 0.15355 | 470.5 | 501.21 | 2.072 |
| 120 | 0.21117 | 479.87 | 511.54 | 2.1218 | 0.15777 | 479.58 | 511.13 | 2.0976 |
| 130 | 0.21675 | 489.08 | 521.6 | 2.147 | 0.16199 | 488.81 | 521.21 | 2.1229 |
| 140 | 0.22231 | 498.45 | 531.8 | 2.172 | 0.1662 | 498.19 | 531.43 | 2.1479 |
| | P = 300 kPa (0.56°C) | | | | P = 400 kPa (8.84°C) | | | |
| Sat. | 0.06787 | 378.33 | 398.69 | 1.7259 | 0.05136 | 383.02 | 403.56 | 1.7223 |
| 10 | 0.07111 | 385.84 | 407.17 | 1.7564 | 0.05168 | 383.98 | 404.65 | 1.7261 |
| 20 | 0.07441 | 393.8 | 416.12 | 1.7874 | 0.05436 | 392.22 | 413.97 | 1.7584 |

| T (°C) | v (m3/kg) | u (kJ/kg) | h (kJ/kg) | s (kJ/kg-K) | v (m3/kg) | u (kJ/kg) | h (kJ/kg) | s (kJ/kg-K) |
|---|---|---|---|---|---|---|---|---|
| 30 | 0.07762 | 401.81 | 425.1 | 1.8175 | 0.05693 | 400.45 | 423.22 | 1.7895 |
| 40 | 0.08075 | 409.9 | 434.12 | 1.8468 | 0.0594 | 408.7 | 432.46 | 1.8195 |
| 50 | 0.08382 | 418.09 | 443.23 | 1.8755 | 0.06181 | 417.03 | 441.75 | 1.8487 |
| 60 | 0.08684 | 426.39 | 452.44 | 1.9035 | 0.06417 | 425.44 | 451.1 | 1.8772 |
| 70 | 0.08982 | 434.82 | 461.76 | 1.9311 | 0.06648 | 433.95 | 460.55 | 1.9051 |
| 80 | 0.09277 | 443.37 | 471.21 | 1.9582 | 0.06877 | 442.58 | 470.09 | 1.9325 |
| 90 | 0.0957 | 452.07 | 480.78 | 1.985 | 0.07102 | 451.34 | 479.75 | 1.9595 |
| 100 | 0.09861 | 460.9 | 490.48 | 2.0113 | 0.07325 | 460.22 | 489.52 | 1.986 |
| 110 | 0.1015 | 469.87 | 500.32 | 2.0373 | 0.07547 | 469.24 | 499.43 | 2.0122 |
| 120 | 0.10437 | 478.99 | 510.3 | 2.0631 | 0.07767 | 478.4 | 509.46 | 2.0381 |
| 130 | 0.10723 | 488.26 | 520.43 | 2.0885 | 0.07985 | 487.69 | 519.63 | 2.0636 |
| 140 | 0.11008 | 497.66 | 530.69 | 2.1136 | 0.08202 | 497.13 | 529.94 | 2.0889 |
| 150 | 0.11292 | 507.22 | 541.09 | 2.1385 | 0.08418 | 506.71 | 540.38 | 2.1139 |
| 160 | 0.11575 | 516.91 | 551.64 | 2.1631 | 0.08634 | 516.43 | 550.97 | 2.1386 |
| | P = 500 kPa (15.66°C) | | | | P = 600 kPa (21.52°C) | | | |
| Sat. | 0.04126 | 386.82 | 407.45 | 1.7198 | 0.03442 | 390.01 | 410.66 | 1.7179 |
| 20 | 0.04226 | 390.52 | 411.65 | 1.7342 | — | — | — | — |
| 30 | 0.04446 | 398.99 | 421.22 | 1.7663 | 0.03609 | 397.44 | 419.09 | 1.7461 |
| 40 | 0.04656 | 407.44 | 430.72 | 1.7971 | 0.03796 | 406.11 | 428.88 | 1.7779 |
| 50 | 0.04858 | 415.91 | 440.2 | 1.827 | 0.03974 | 414.75 | 438.59 | 1.8084 |
| 60 | 0.05055 | 424.44 | 449.72 | 1.856 | 0.04145 | 423.41 | 448.28 | 1.8379 |
| 70 | 0.05247 | 433.06 | 459.29 | 1.8843 | 0.04311 | 432.13 | 457.99 | 1.8666 |
| 80 | 0.05435 | 441.77 | 468.94 | 1.912 | 0.04473 | 440.93 | 467.76 | 1.8947 |
| 90 | 0.0562 | 450.59 | 478.69 | 1.9392 | 0.04632 | 449.82 | 477.61 | 1.9222 |
| 100 | 0.05804 | 459.53 | 488.55 | 1.966 | 0.04788 | 458.82 | 487.55 | 1.9492 |
| 110 | 0.05985 | 468.6 | 498.52 | 1.9924 | 0.04943 | 467.94 | 497.59 | 1.9758 |
| 120 | 0.06164 | 477.79 | 508.61 | 2.0184 | 0.05095 | 477.18 | 507.75 | 2.0019 |
| 130 | 0.06342 | 487.13 | 518.83 | 2.044 | 0.05246 | 486.55 | 518.03 | 2.0277 |
| 140 | 0.06518 | 496.59 | 529.19 | 2.0694 | 0.05396 | 496.05 | 528.43 | 2.0532 |
| 150 | 0.06694 | 506.2 | 539.67 | 2.0945 | 0.05544 | 505.69 | 538.95 | 2.0784 |
| 160 | 0.06869 | 515.95 | 550.29 | 2.1193 | 0.05692 | 515.46 | 549.61 | 2.1033 |
| 170 | 0.07043 | 525.83 | 561.04 | 2.1438 | 0.05839 | 525.36 | 560.4 | 2.1279 |
| | P = 800 kPa (31.30°C) | | | | P = 1000 kPa (39.37°C) | | | |
| Sat. | 0.02571 | 395.15 | 415.72 | 1.715 | 0.02038 | 399.16 | 419.54 | 1.7125 |
| 40 | 0.02711 | 403.17 | 424.86 | 1.7446 | 0.02047 | 399.78 | 420.25 | 1.7148 |
| 50 | 0.02861 | 412.23 | 435.11 | 1.7768 | 0.02185 | 409.39 | 431.24 | 1.7494 |
| 60 | 0.03002 | 421.2 | 445.22 | 1.8076 | 0.02311 | 418.78 | 441.89 | 1.7818 |
| 70 | 0.03137 | 430.17 | 455.27 | 1.8373 | 0.02429 | 428.05 | 452.34 | 1.8127 |
| 80 | 0.03268 | 439.17 | 465.31 | 1.8662 | 0.02542 | 437.29 | 462.7 | 1.8425 |
| 90 | 0.03394 | 448.22 | 475.38 | 1.8943 | 0.0265 | 446.53 | 473.03 | 1.8713 |
| 100 | 0.03518 | 457.35 | 485.5 | 1.9218 | 0.02754 | 455.82 | 483.36 | 1.8994 |
| 110 | 0.03639 | 466.58 | 495.7 | 1.9487 | 0.02856 | 465.18 | 493.74 | 1.9268 |

| *T* (°C) | *v* (m3/kg) | *u* (kJ/kg) | *h* (kJ/kg) | *s* (kJ/kg-K) | *v* (m3/kg) | *u* (kJ/kg) | *h* (kJ/kg) | *s* (kJ/kg-K) |
|---|---|---|---|---|---|---|---|---|
| 120 | 0.03758 | 475.92 | 505.99 | 1.9753 | 0.02956 | 474.62 | 504.17 | 1.9537 |
| 130 | 0.03876 | 485.37 | 516.38 | 2.0014 | 0.03053 | 484.16 | 514.69 | 1.9801 |
| 140 | 0.03992 | 494.94 | 526.88 | 2.0271 | 0.0315 | 493.81 | 525.3 | 2.0061 |
| 150 | 0.04107 | 504.64 | 537.5 | 2.0525 | 0.03244 | 503.57 | 536.02 | 2.0318 |
| 160 | 0.04221 | 514.46 | 548.23 | 2.0775 | 0.03338 | 513.46 | 546.84 | 2.057 |
| 170 | 0.04334 | 524.42 | 559.09 | 2.1023 | 0.03431 | 523.46 | 557.77 | 2.082 |
| 180 | 0.04446 | 534.51 | 570.08 | 2.1268 | 0.03523 | 533.6 | 568.83 | 2.1067 |
| | **P = 1200 kPa (46.31°C)** | | | | **P = 1400 kPa (52.42°C)** | | | |
| Sat. | 0.01676 | 402.37 | 422.49 | 1.7102 | 0.01414 | 404.98 | 424.78 | 1.7077 |
| 50 | 0.01724 | 406.15 | 426.84 | 1.7237 | — | — | — | — |
| 60 | 0.01844 | 416.08 | 438.21 | 1.7584 | 0.01503 | 413.03 | 434.08 | 1.736 |
| 70 | 0.01953 | 425.74 | 449.18 | 1.7908 | 0.01608 | 423.2 | 445.72 | 1.7704 |
| 80 | 0.02055 | 435.27 | 459.92 | 1.8217 | 0.01704 | 433.09 | 456.94 | 1.8026 |
| 90 | 0.02151 | 444.74 | 470.55 | 1.8514 | 0.01793 | 442.83 | 467.93 | 1.8333 |
| 100 | 0.02244 | 454.2 | 481.13 | 1.8801 | 0.01878 | 452.5 | 478.79 | 1.8628 |
| 110 | 0.02333 | 463.71 | 491.7 | 1.9081 | 0.01958 | 462.17 | 489.59 | 1.8914 |
| 120 | 0.0242 | 473.27 | 502.31 | 1.9354 | 0.02036 | 471.87 | 500.38 | 1.9192 |
| 130 | 0.02504 | 482.91 | 512.97 | 1.9621 | 0.02112 | 481.63 | 511.19 | 1.9463 |
| 140 | 0.02587 | 492.65 | 523.7 | 1.9884 | 0.02186 | 491.46 | 522.05 | 1.973 |
| 150 | 0.02669 | 502.48 | 534.51 | 2.0143 | 0.02258 | 501.37 | 532.98 | 1.9991 |
| 160 | 0.0275 | 512.43 | 545.43 | 2.0398 | 0.02329 | 511.39 | 543.99 | 2.0248 |
| 170 | 0.02829 | 522.5 | 556.44 | 2.0649 | 0.02399 | 521.51 | 555.1 | 2.0502 |
| 180 | 0.02907 | 532.68 | 567.57 | 2.0898 | 0.02468 | 531.75 | 566.3 | 2.0752 |
| | **P = 1600 kPa (57.90°C)** | | | | **P = 2000 kPa (67.48°C)** | | | |
| Sat. | 0.01215 | 407.11 | 426.54 | 1.7051 | 0.0093 | 410.15 | 428.75 | 1.6991 |
| 60 | 0.01239 | 409.49 | 429.32 | 1.7135 | — | — | — | — |
| 70 | 0.01345 | 420.37 | 441.89 | 1.7507 | 0.00958 | 413.37 | 432.53 | 1.7101 |
| 80 | 0.01438 | 430.72 | 453.72 | 1.7847 | 0.01055 | 425.2 | 446.3 | 1.7497 |
| 90 | 0.01522 | 440.79 | 465.15 | 1.8166 | 0.01137 | 436.2 | 458.95 | 1.785 |
| 100 | 0.01601 | 450.71 | 476.33 | 1.8469 | 0.01211 | 446.78 | 471 | 1.8177 |
| 110 | 0.01676 | 460.57 | 487.39 | 1.8762 | 0.01279 | 457.12 | 482.69 | 1.8487 |
| 120 | 0.01748 | 470.42 | 498.39 | 1.9045 | 0.01342 | 467.34 | 494.19 | 1.8783 |
| 130 | 0.01817 | 480.3 | 509.37 | 1.9321 | 0.01403 | 477.51 | 505.57 | 1.9069 |
| 140 | 0.01884 | 490.23 | 520.38 | 1.9591 | 0.01461 | 487.68 | 516.9 | 1.9346 |
| 150 | 0.01949 | 500.24 | 531.43 | 1.9855 | 0.01517 | 497.89 | 528.22 | 1.9617 |
| 160 | 0.02013 | 510.33 | 542.54 | 2.0115 | 0.01571 | 508.15 | 539.57 | 1.9882 |
| 170 | 0.02076 | 520.52 | 553.73 | 2.037 | 0.01624 | 518.48 | 550.96 | 2.0142 |
| 180 | 0.02138 | 530.81 | 565.02 | 2.0622 | 0.01676 | 528.89 | 562.42 | 2.0398 |
| | **P = 3000 kPa (86.20°C)** | | | | **P = 4000 kPa (100.33°C)** | | | |
| Sat. | 0.00528 | 411.83 | 427.67 | 1.6759 | 0.00252 | 394.86 | 404.94 | 1.6036 |
| 90 | 0.00575 | 418.93 | 436.19 | 1.6995 | — | — | — | — |
| 100 | 0.00665 | 433.77 | 453.73 | 1.7472 | — | — | — | — |

| *T* (°C) | *v* (m3/kg) | *u* (kJ/kg) | *h* (kJ/kg) | *s* (kJ/kg-K) | *v* (m3/kg) | *u* (kJ/kg) | *h* (kJ/kg) | *s* (kJ/kg-K) |
|---|---|---|---|---|---|---|---|---|
| 110 | 0.00734 | 446.48 | 468.5 | 1.7862 | 0.00428 | 429.74 | 446.84 | 1.7148 |
| 120 | 0.00792 | 458.27 | 482.04 | 1.8211 | 0.005 | 445.97 | 465.99 | 1.7642 |
| 130 | 0.00845 | 469.58 | 494.91 | 1.8535 | 0.00556 | 459.63 | 481.87 | 1.804 |
| 140 | 0.00893 | 480.61 | 507.39 | 1.884 | 0.00603 | 472.19 | 496.29 | 1.8394 |
| 150 | 0.00937 | 491.49 | 519.62 | 1.9133 | 0.00644 | 484.15 | 509.92 | 1.872 |
| 160 | 0.0098 | 502.3 | 531.7 | 1.9415 | 0.00683 | 495.77 | 523.07 | 1.9027 |
| 170 | 0.01021 | 513.09 | 543.71 | 1.9689 | 0.00718 | 507.19 | 535.92 | 1.932 |
| 180 | 0.0106 | 523.89 | 555.69 | 1.9956 | 0.00752 | 518.51 | 548.57 | 1.9603 |
| | **P = 6000 kPa** | | | | **P = 10000 kPa** | | | |
| 90 | 0.001059 | 328.34 | 334.7 | 1.4081 | 0.000991 | 320.72 | 330.62 | 1.3856 |
| 100 | 0.00115 | 346.71 | 353.61 | 1.4595 | 0.00104 | 336.45 | 346.85 | 1.4297 |
| 110 | 0.001307 | 368.06 | 375.9 | 1.5184 | 0.0011 | 352.74 | 363.73 | 1.4744 |
| 120 | 0.001698 | 396.59 | 406.78 | 1.5979 | 0.001175 | 369.69 | 381.44 | 1.52 |
| 130 | 0.002396 | 426.81 | 441.18 | 1.6843 | 0.001272 | 387.44 | 400.16 | 1.567 |
| 140 | 0.002985 | 448.34 | 466.25 | 1.7458 | 0.0014 | 405.97 | 419.98 | 1.6155 |
| 150 | 0.003439 | 465.19 | 485.82 | 1.7926 | 0.001564 | 424.99 | 440.63 | 1.6649 |
| 160 | 0.003814 | 479.89 | 502.77 | 1.8322 | 0.001758 | 443.77 | 461.34 | 1.7133 |
| 170 | 0.004141 | 493.45 | 518.3 | 1.8676 | 0.001965 | 461.65 | 481.3 | 1.7589 |
| 180 | 0.004435 | 506.35 | 532.96 | 1.9004 | 0.002172 | 478.4 | 500.12 | 1.8009 |

## CO AND PO ATTAINMENT TABLE

Course outcomes (COs) for this course can be mapped with the programme outcomes (POs) after the completion of the course and a correlation can be made for the attainment of POs to analyze the gap. After proper analysis of the gap in the attainment of POs necessary measures can be taken to overcome the gaps.

Table for CO and PO attainment

| Course Outcomes | Expected Mapping with Programme Outcomes (1- Weak Correlation; 2- Medium correlation; 3- Strong Correlation) | | | | | | | | | | | |
|---|---|---|---|---|---|---|---|---|---|---|---|---|
| | PO-1 | PO-2 | PO-3 | PO-4 | PO-5 | PO-6 | PO-7 | PO-8 | PO-9 | PO-10 | PO-11 | PO-12 |
| CO-1 | | | | | | | | | | | | |
| CO-2 | | | | | | | | | | | | |
| CO-3 | | | | | | | | | | | | |
| CO-4 | | | | | | | | | | | | |
| CO-5 | | | | | | | | | | | | |
| CO-6 | | | | | | | | | | | | |

The data filled in the above table can be used for gap analysis.

# INDEX